Biochemistry of
Chemical Carcinogenesis

Biochemistry of Chemical Carcinogenesis

Edited by
R. Colin Garner
University of York
York, United Kingdom

and

Jan Hradec
Research Institute for Tuberculosis and Respiratory Diseases
Prague, Czechoslovakia

PLENUM PRESS • NEW YORK AND LONDON

Library of Congress Cataloging-in-Publication Data

Biochemistry of chemical carcinogenesis / edited by R. Colin Garner
and Jan Hradec.
        p.   cm.
      Proceedings of a satellite symposium of the Fourteenth
International Congress of Biochemistry, on biochemistry of chemical
carcinogenesis, held July 7-9, 1988, in Prague, Czechoslovakia"-
-T.p. verso.
      Includes bibliographical references.
      ISBN-13:978-1-4612-7856-6   e-ISBN-13:978-1-4613-0539-2
      DOI: 10.1007/978-1-4613-0539-2
      1. Carcinogenesis--Congresses.   2. Carcinogens--Metabolism-
-Congresses.   I. Garner, R. C., 1944-   . II. Hradec, Jan.
III. International Congress of Biochemistry (14th : 1988 : Prague,
Czechoslovakia)
      [DNLM: 1. Carcinogens--congresses.   QZ 202 B6147]
RC268.5.B556   1989
616.99'4071--dc20
DNLM/DLC
for Library of Congress   89-23090
                                                             CIP

Proceedings of a Satellite Symposium of the Fourteenth International
Congress of Biochemistry, on Biochemistry of Chemical Carcinogenesis,
held July 7-9, 1988, in Prague, Czechoslovakia

© 1989 Plenum Press, New York
Softcover reprint of the hardcover 1st edition 1989

A Division of Plenum Publishing Corporation
233 Spring Street, New York, N.Y. 10013

PREFACE

Cancer is a disease which knows no boundaries. It strikes at the old, the young, the rich, the poor, the East, the West and the Third World. In the West cancer accounts for up to thirty percent of all deaths. The major cause of these is associated with the smoking of cigarettes, thus establishing an important link between chemical exposure and human cancer. With the advent of suitable human biomonitoring techniques there is little doubt that much of what has been discovered from experimental studies applies to man. In recent years major advances have been made in our understanding of the molecular mechanisms of cancer which give hope that we may be able to either prevent or interfere with the process of the disease.

In the summer of 1988 the Czechoslovak Research Institute for Tuberclu- osis and Respiratory Diseases, Prague, organized a meeting on the Biochem- istry of Chemical Carcinogenesis as a satellite symposium of the 4th Inter- national Congress of Biochemistry. The satellite meeting consisted of invited lectures and submitted oral presentations and posters. This book contains the proceedings of this meeting. The subjects covered are wide- ranging and include the relevance of animal carcinogenicity testing to human risk assessment, the involvement of the drug metabolising enzymes in carcinogen activation ad detoxification, studies on a variety of experimental and human carcinogens, human biomonitoring and the involvement of oncogene activation in carcinogenesis.

The meeting enabled scientists from all parts of the world to discuss their latest research findings as well as allowing students to hear from internationally respected scientists. The meeting was sucessful at both an academic and social level, helping to demonstrate the variety of research taking place in chemical carcinogenesis.

R.C. Garner
J. Hradec

CONTENTS

CHEMICAL CARCINOGENESIS AND THE PRIMARY PREVENTION OF HUMAN CANCER

Lorenzo Tomatis

International Agency for Research on Cancer
150 Cours Albert Thomas
69372 Lyon Cedex 08
France

GOALS FOR CANCER PREVENTION

A tentative estimate of the reduction in the incidence of cancer that can be achieved in a given population or region of the world could be made by calculating the differences between the highest and the lowest observed rates for a given site in populations which are not too dissimilar genetically. One can assume that, at least in theory, the lowest rates observed in populations where a reliable cancer registry exists, would be suitable target levels for primary prevention programme. Within Europe for instance this would imply that in males a reduction in incidence of 72% could be achieved for large bowel cancer (from 44.0 to 12.1 cases per 10^5), of 78% for lung cancer (from 110.4 to 24.5 per 10^5), of 69% for bladder cancer (from 24.7 to 7.7 per 10^5). If we push this theoretical estimate a bit further and extend it to the incidence of cancers at all sites, the total reduction within Europe could be from the present maximum rate of between 330 and 240 cases per 10^5 per annum, to an incidence of between 164 and 105 cases per 10^5 per annum, an incidence calculated by adding together the lowest incidences observed in European populations for each site (1).

This theoretical exercise has the merit of giving an idea of the order of magnitude of the variability in risks in different regions and in different populations of the world, and therefore, indirectly of the important role environmental factors may play in determining cancer risks. It certainly does not provide any indication as to the actual feasibility and efficacy of specific preventive measures, nor more importantly, does it take into account our still limited knowledge of the causes of human cancer, nor, for the causes we know, the extent to which they would be amenable to primary prevention. Although many more etiological agents of human cancer are known today than 50 years ago, they explain only an ill-defined proportion of all human cancers, and not the majority. The list of recognized carcinogenic agents for humans seems to reflect the male domination of our society, as it is to a large extent composed of agents related to cancers occurring predominantly in males. There are at least three reasons for such sex-related unevenness:

1 not much is known, beyond a number of credible but as yet still unproven hypotheses, on the etiology of the two most important female cancers, namely of the breast and of the cervix;

Biochemistry of Chemical Carcinogenesis
Edited by R. Colin Garner and Jan Hradec
Plenum Press, New York, 1990

2 many of the recognized human carcinogens are related to occupational
exposures, which are of greater concern to males than to females;

3 females have taken up the habit of cigarette smoking much later than males

This latter fact has made the lung an important target organ for cancer in
females in Western societies only in quite recent years, an importance that
still appears to be growing.

The IARC Monograph Programme

     The IARC has contributed considerably (Tables 1, 2, 3, 4) to the prepa-
ration of this list with a programme initiated in 1969, and centered on the
preparation of monographs on individual chemical compounds, groups of
compounds or complex exposures. In these assessments, all available data
relevant to the carcinogenicity of the exposure in question are critically
analysed before an actual evaluation of carcinogenic risk for humans is made
(2). A first tentative list of chemical agents definitely carcinogenic to
humans was made in 1978, summarizing the data analysed in the first 20 Monographs volumes (3). This list was updated in 1982 using the data contained
in Supplement no. 4 to the IARC Monographs which summarized and updated the
information contained in the 29 volumes of monographs published until then
(4). A further updating, carried out in 1987, resulted in the publication of
Supplement 7, covering all the data contained in the 42 volumes of the
Monographs published until then (5).

     According to the criteria for carcinogenicity used within the IARC pro-
gramme, elaborated in collaboration with expert advisors, chemicals, groups
of chemicals or complex exposures are assigned to four groups, according to
the evidence for carcinogenicity that exist for each of them: Group 1 -
carcinogenic to humans; Group 2A - probably carcinogenic to humans; Group
2B - possibly carcinogenic to humans; Group 3 - not classifiable as to
carcinogenicity to humans; and Group 4 - probably not carcinogenic to humans.
The assignment of an agent to a group is made according to a scientific
judgement that reflects the strength of the evidence derived from studies in
humans and in experimental animals, and from other relevant data (5).

     The assigment of the chemicals, groups of chemicals or complex exposures
to Group 1 is based, however, solely on evidence for carcinogenicity in
humans provided by epidemiological studies, and does not take into account
any experimental data.

Experimental versus epidemiological data

     There could be a variety of reasons why a chemical, for which there is
sufficient experimental evidence of carcinogenicity and to which humans are
exposed, has not been the subject of an epidemiological investigation (6).
Some reflect the difficulties intrinsic to the epidemiological approach: the
number of individuals exposed may be too small to provide statistically meaningful results, the duration of exposure may be too short to permit any conclusion to be drawn from a retrospective study, the chemical of interest may
be just one among the many to which the individuals are exposed. Other
reasons are of a different nature, among which is prominent the considerable
socio-economic importance of certain substances. The changeable and, to a
certain extent, unpredictable way people perceive risk may also play a role.
Most people, in fact, support and urge rapid decisions on risks that are
sharply peaked in time, and tend to disregard risks which span a long period
of time - an attitude which is, for instance, at the root of the persistence
of smokers to maintain their habit, in spite of the many warnings of its
long-term adverse effects.

Table 1. Industrial Processes Causally Related to Human Cancer (from ref 5)

Exposure	Target Organs
Aluminium production	Lung, bladder (lymphoma, oesophagus, stomach)
Auramine, manufacture of	Bladder
Boot and shoe manufacture and repair (certain occupations)	Leukemia, nasal sinus (bladder, digestive tract)
Coal gasification (older processes)	Skin, lung, bladder
Coke production	Skin, lung, kidney
Furniture and cabinet making	Nasal sinus
Haematite mining, underground, with exposure to radon	Lung
Iron, steel and steel founding	Lung (digestive tract, genito-urinary tract, leukemia)
Isopropyl alcohol manufacture (strong acid process)	Nasal sinus (larynx)
Magenta, manufacture of	Bladder
Rubber industry (certain occupations)	Bladder, leukemia (lymphoma, lung, renal tract, digestive tract, skin, liver, larynx, brain, stomach)

(suspected target organs in parentheses)

It is worth noting the considerable change in emphasis concerning the relative importance attributed to epidemiological and experimental results in providing evidence for carcinogenicity in humans. The results of the experiment of Yamagiwa and Ichikawa (7) on tars in 1915 were taken as the final confirmation that soot and tars were indeed carcinogenic, 130 years after Pott's report on chimney sweeps (8). In the WHO document of 1964 (9), it was still recommended that tobacco smoke be further investigated experimentally, in spite of the overwhelming evidence of carcinogenicity in humans; the opinion that experimental results can provide definitive confirmation to epidemiological observations still prevailed. Today the prevailing attitude is that only epidemiological data provide sufficient evidence for carcinogenicity to humans, as is reflected in the criteria for assignment of an agent to Group 1 (carcinogenic to humans) within the IARC Monographs programme.

The IARC recommends, however, that in the absence of adequate human data, chemicals for which there is sufficient experimental evidence of carcinogenicity should be regarded, for practical purposes, as if they were carcinogenic to humans. Thus the IARC makes an attempt to strike a balance between a rigidly objective scientific view of the data, and a public health-orientated interpretation of the experimental evidence of carcinogenicity.

Table 2.   Chemicals and groups of chemicals causally related to human cancer for which exposure has been mostly occupational (from Ref 5)

Exposure	Target organs
4-Aminobiphenyl	Bladder
*Arsenic and arsenic compounds	Skin, lung (liver, haematopoietic system, gastrointestinal tract, kydney)
Asbestos	Lung, pleura, peritoneum (gastrointestinal tract, larynx)
Benzene	Leukaemia
Benzidine	Bladder
Bis(chloromethyl)ether and chloromethyl methyl ether (technical grade)	Lung
*Chromium compounds, hexavalent	Lung (gastrontestinal tract)
Coal-tars (and iatrogenic exposures)	Skin, lung (bladder)
Coal-tar pitches	Skin, lung, bladder
Mineral oils (untreated and mildly treated)	Skin (respiratory tract, bladder, gastrointestinal tract)
Mustard gas	Lung, larynx, pharynx
2-Naphthylamine	Bladder (liver)
*Nickel and nickel compounds	Nasal sinus, lung (larynx)
Shale-oils	Skin (colon)
Soots	Skin, lung
Talc containing asbestos fibres	Lung (pleura)
Vinyl chloride	Liver, lung, brain, lymphatic and haematopoietic system (gastrointestinal)

*The evaluation of carcinogenicity to humans applies to the group of agents as a whole and not necessarily to all individual agents within the group. (Suspected organs in parentheses)

Regulatory aspects

    It is not always easy to understand the criteria of the regulatory authorities and follow the reasoning that triggers them to take action against a substance, to delay such action, or to decide to ignore it. Regulatory agencies have been reported to show inconsistencies in their actions towards

4

Table 3. Drugs causally associated with cancer in humans   (from ref 5)

Exposure	Target organs
Analgesic mixtures containing penacetin	Renal pelvis (ureter, bladder)
Azathioprine	Lymphoma (skin, hepato-biliary system, mesenchymal tumors)
Clorambucil	Leukemia
Chlornaphazine	Bladder, leukemia
1-(2-Chloroethyl)-3-(4-methyl-cyclohexyl)-1-nitrosourea (Methyl-CCNU)	Leukemia
Diethylstilboestrol	Cervix, vagina, breast, endometrium (testis)
Melphalan	Leukemia
8-Methoxyproralen plus ultraviolet radiation	Skin
MOPP and other combined chemotherapy including alkylating agents	Leukemia
Myleran	Leukemia
Oestrogen replacement therapy	Endometrium, breast
*Oestrogens non-steroidal	Cervix, vagina, breast (testis)
*Oestrogens steroidal	Endometrium (breast)
**Oral contraceptives, combined	Liver
Oral contraceptives, sequential	Liver
Treosulfan	Leukemia

*The evaluation of carcinogenicity to humans applies to the group.of agents as a whole and not necessarily to all individual agents within the group.
**There is also conclusive evidence that these agents protect against cancer of the ovary and endometrium. (Suspected target organs in parentheses)

environmental chemicals, as Ames et al.(10) have pointed out in the case of ethylene dibromide. Among the factors that may influence a regulatory deci-sion are the degree of the risk and the number of individuals at risk, as well as the magnitude of the socio-economic importance of the substance, bal-anced against the estimated cost of the lives that a regulation might save.

   In the USA, it appears that the primary determinant of regulation is cost-effectiveness, if the risk is not of obvious and immediate concern (called in the USA a "de-manifestis" risk) but is above an acceptable level

Table 4. Environmental and cultural risk factors causally associated with human cancer.

Exposure	Target organs
* Erionite	Pleura
** Ionizing radiations	Leukemia, skin, various internal organs
** UV light	Skin
* Aflatoxins	Liver
* Alcoholic beverages	Oral cavity (pharynx, larynx, oesophagus, liver (breast)
* Betel-quid chewing with tobacco	Oral cavity (pharynx, larynx, oesophagus)
* Smokeless tobacco use (chewing and oral snuff)	Oral cavity (pharynx, oesophagus)
* Tobacco smoke	Lung, bladder, oral cavity, larynx, pharynx, oesophagus, paancreas, renal pelvis (stomach, liver, cervix)
** Hepatitus B virus infection	Liver
** Human T-cell leukaemia virus	T-cell leukemia

*   From refs 3 and 5
** Not evaluated in the IARC Monographs (Suspected organs in parentheses)

(called "de minimis") and therefore below regulatory concern (11). Similar considerations are probably applied elsewhere, if less explicitly.

Old and recent chemical carcinogens

The historical development of our knowledge concerning carcinogenic agents to some extent justifies the view of cancer as largely a "chemical" disease (Table 5) (12-34). While radiation is probably the oldest carcino-genic agent on this planet, chemicals were the first to be recognized as etiological factors for human cancer, and have remained the predominant factors, at least in numerical terms.

It does however seem justified to ask whether human cancer really is a disease predominantly related to environmental chemicals. Two subordinate questions concern the relative importance of those chemical agents that have been identified, and the experimental approaches adopted for their identifi-cation. Are the chemical agents so far identified the most important ones and are the tests used for their identification appropriate to identify the important chemical agents as the origin of human cancer?

The group of chemicals found to be carcinogenic to humans in the last few decades appears to be clearly related to the industrial development, in particular of the chemical industry, which began to take off in the second half of the last century and has not ceased to grow since, with a spectacular

Table 5. Identification of cancer causing agents (an abridged chronological sequence)

1761	Tobacco snuff and nasal cancer, J. Hill (12)
1775/1921/1954	Soot and scrotal cancer, P. Pott (8); Yamagiwa and Ichikawa (7); Kennaway (13); Cook et al. (14)
1895/1921/1954	"Aniline" and bladder cancer, L. Rehn (15); I.L.O. document (16); the exhaustive epidemiological study of R.A.M. Case et al. (17)
1909/1911	X-rays, Wolbach (18); Hesse (19)
1933/1955/1960	Asbestos and lung cancer, mesotheliomas. K.M. Lynch (20); R. Doll (21); C. Wagner et al.(22)
1939/1950	Tobacco smoke and lung cancer, F.H. Muller (23); R. Doll and B. Hill (24); E.L. Wynder and E.A. Graham (25)
1932/1971	Oestrogens and mammary and genito-urinary tumours; A. Lacassagne (26); A Herbst et al. (27)
1973	BCME and lung cancer, W.G. Figueroa et al. (28); A. Thiess et al. (29)
1970/1974	Vinyl chloride and hemagioendothemiomas of the liver, P.L. Viola et al.(30); C Maltoni et al. (31); J.L. Creech and M.N. Johnson (32)
1911/1981	Viruses and human cancer, P. Rous (33); B.J. Poiesz and R. Gallo (34)

leap after World War II. This industrial development was carried out without much, if any, concern for possible health effects, partly because of genuine ignorance, especially of the long-term adverse effects such as cancer, and later by dismissing them as being either unavoidable evils, or of such minimal importance compared to the benefits (and profits) as not to justify expensive modifications of the production procedures. It is also interesting that the industrial production of cigarettes began at the same time as the expansion of the chemical industry. The first cigarette factories were in fact built in 1853 in Havana, Cuba, in 1856 in London and in 1860 in Virginia (35).

The etiological agents of human cancer that have been firmly identified up to now (with the exception of ionizing radiations, UV light, asbestos, certain metals, combustion products, mycotoxins and viruses) have been therefore with us a relatively short time, since the human species has been confronted with the massive presence of man-made chemicals in the environment and with the expansion of the most hazardous cultural habit only since the middle of the last century.

Furthermore, the distinction between natural and man-made carcinogens is often difficult to make. Tobacco is obviously a natural product, but the cigarette could hardly be seen as such. Similarly asbestos is a natural product, but it is not so natural to mine millions of tons of it, so helping to disseminate it into the environment with which we have direct contact. An example of how an old carcinogen may become a new hazard comes from Japan,

Table 6. The most frequent target organs for the agents or exposures recognized as causally associated with human cancer.

Target Organ	No of Agents or Exposures	Agents or Exposures
Lung	18	Aluminium production, coal gasification, coke production, underground hematite mining (radon), iron/steel founding
		Arsenic and arsenic compounds, asbestos, BCME chromium compounds (hexavlent), coal-tars, coal-tar pitches, mustard gas, nickel compounds, soots, talc containing asbestiform fibres, vinyl chloride, tobacco smoke, ionizing radiation
Bladder	12	Aluminium production, auramine production, coal gasification, magenta production, rubber industry
		4-aminobiphenyl, benzidine, 2-naphthyla mine, chlornaphizine, cylophosphamide, tobacco smoke.
Leukemia	12	Boots and shoes manufacture and repair (certain occupations), rubber industry
		Benzene, chlorambucil, methyl-CCNU, cyclo-phosphamide, melphalen, MOPP, myleran, treosulphan, ionizing radiations, HTLV-1
Skin	11	Coal gasification, coke production
		Arsenic and arsenic compounds, coal tars, coal-tar pitches, mineral oils, shale oils, soots, 8-methoxypsoralene, UV light, ionizing radiations.
Liver	5	Vinyl chloride, oral contraceptives, aflatoxin, hepatitis B virus, alcholic beverages
		Alcoholic beverages, betel quid, chewing with tobacco, smokeless tobacco use, tobacco smoke
Oral cavity	4	Alcoholic beverages, betal quid, chewing with tobacco, smokeless tobacco use, tobacco smoke
Nasal sinuses	4	Furniture and cabinet making, Boot and shoe manufacture and repair, isopropyl alcohol manufacturing (strong acid process)
Pharynx	3	Tobacco smoke, alcoholic beverages, mustard gas
Larynx	3	Tobacco smoke, alcoholic beverages, mustard gas
Oesophagus	2	Alcoholic beverages, tobacco smoke

which imports most of the asbestos it uses. The increase in its use there paralleled its importation, which therefore provides a relatively accurate surrogate measure of exposure. The importations have risen from less than 5000 tons a year in the 1950s, to 100,000 tons in the 1960s and to about 300,000 tons in 1970s. An analysis of lung specimens of individuals autopsied or operated on in the area of Tokyo in five year periods between 1937 and 1981 shows a dramatic increase of ferruginous bodies from the late 1950s (36).

Very rarely can trends in exposure be documented as well as in this instance, although this could at least in theory be attempted for other natural carcinogens, such as metals. In other cases, even if it were important to learn if and to what extent the widespread use of a chemical or a chemical mixture had an impact on health, it cannot be done. A typical case is that of bitumens, the quantity of which used worldwide annually is over 60 million tons, with the largest share taken by road construction and repair. In the USA alone the production of bitumens rose from 18,000 tons to 1902 to 5 million tons in 1938, and up to the present level of about 20 million tons (37). A not insignificant fraction of the most inhabited regions of our planet has been covered with bitumens, a mixture containing chemicals for which there is sufficient experimental evidence of carcinogenicity. Yet we are in no position to say whether or not this represents a hazard for human health. One might assume that such a hazard is not significant for the general population, but it would certainly be better if we were able to confirm it, since it is known that there has been, and possibly still is, a cancer risk for a few individuals occupationally exposed to bitumens.

The common cancers

The etiological agents for cancer that have been firmly identified to date are predominantly associated with tumors occuring at certain sites and are as a whole, as said above, of more importance for cancers occurring in males than for those occurring in females. The organs involved are lung, bladder, leukemia and skin. Of the 55 factors casually associated with cancer in humans, 18 induce lung cancer, 12 bladder cancer, 12 leukemia, 11 skin tumors and tumors of the oral cavity (Table 6).

The preference for lung and skin can be largely explained by the fact that they are two of the most important routes of exposure, and are of particular importance in the occupational setting. In the case of bladder cancer and leukemia, the preference is possibly due to clustering of two particular types of compound, namely of aromatic amines in the case of the bladder, and of active alkylating agents in the case of leukemia. Tobacco smoke appears as the most important agent for the lung and bladder, while ionizing radiation is associated with cancer of three of the four sites, namely lung, skin and leukemia.

If we now look at the most frequent target sites for tumors worldwide (38), we cannot fail to note that some of them are conspiciouly absent from the list of the main target organs for recognized human carcinogens. Missing in fact are tumors of the breast, cervix, colon-rectum, stomach and prostrate Although this absence can be slightly mitigated by the fact that several studies have indicated a possible causal association between certain exposures and cancer at those sites, nevertheless we can hardly ignore this wide gap in our understanding of the etiology of human cancer.

CONCLUSIONS

1. There is clear evidence that environmental chemicals are important factors in the origin of human cancers, playing a role in the various stages

of the process of carcinogenesis and of the development of the clinical manifestation of the disease.

2. Some of the chemicals and chemical mixtures identified as carcinogenic to humans (such as tobacco smoke, asbestos, certain industrial chemicals and some medicinal drugs) are definitely very significant but there is no compelling evidence that we have already identified all the most important chemical carcinogens. It is important that we should recognize the limitations of our approaches and of our actual knowledge of the causes of cancer and of the mechanisims of carcinogenesis. This, however, does not mean that we have to discuss the approaches followed until now, nor that we should denigrate what we have achieved already. Indeed, there is considerable evidence that animal tests can be valid predictors of similiar effects in humans (39, 40). Rather than looking for radical alternatives, we should carefully consider what we can do in addition to what we are already doing.

3. The tests so far used for the identification of carcinogenic agents are mainly designed to discover which play a predominant role in the early initiation phases of the carcinogenesis process, and much less in the subsequent phases. The development of new tests that would be capable of identifying factors that contribute to the inactivation of the so-called suppressor genes, or that cause gene amplification (41) or gene recombinations, will probably permit the identification of agents that play an important role in the genesis of human cancer.

REFERENCES

1 Muir, C.S., Waterhouse, J., Mack, T., Powell, J., and Whelan, S., eds., Cancer Incidence in Five Continents, Volume 5, IARC Scientific Publications No. 88, International Agency for Research on Cancer, Lyon, 1987
2 IARC Monographs on the Evaluation of the Carcinogenic Risk of Chemicals to Humans, Volumes 1-44, International Agency for Research on Cancer, Lyon, 1972-1988
3 IARC Monographs on the Evaluation of the Carcinogenic Risk of Chemicals to Humans, Supplement No 1, International Agency for Research on Cancer, Lyon, 1979
4 IARC Monographs on the Evaluation of the Carcinogenic Risk of Chemicals to Humans, Supplement No 4, International Agency for Research on Cancer Lyon, 1982
5 IARC Monographs on the Evaluation of Carcinogenic Risks to Humans, Supplement No 7, International Agency for Research on Cancer, Lyon, 1987
6 Karstadt, M., Bobal, R., Selikoff, I.J. A Survey of availability of epidemiologic data on humans exposed to animal carcinogens. In: Peto, R. and Schneiderman, M., eds., Quantification of Occupational Cancer, New York, Cold Spring Harbor Laboratory, Banbury Report No., pp. 223-245, 1981
7 Yamagiwa, K. and Ichikawa, K. Experimentelle Studie uber die Pathologenese der Epithelialgeschwulste. Mitteilungen Med. Facult Kaiserl. Univ. Tokyo, 15 (2): 295-344, 1915
8 Percivall Pott. Chirurgical observations relative to the cataract, the polypus of the nose, the cancer of the scrotum, the different kinds of ruptures, and the mortifiction of the toes and feet. London: Hawes, Clarke and Collins, 1775
9 World Health Organization: Prevention of Cancer, Tech. Rep. Series No 226, Geneva, Switzerland, 1964
10 Ames, B., Magaw, R., Gold, L.W., Ranking possible carcinogenic hazards. Science, 236: 271-280, 1987
11 Travis, C.C., Richter, S.A., Crouch, E.A.C., Wilson, R., Klema, E.D. Cancer risk management - a review of 132 federal regulatory decisions.

Environmental Science & Technology, 21: 415-420, 1987

12 Hill, J. Cautions against the immoderate use of snuff. In: Redmond, D.E. Tobacco and cancer: the first clinical report, 1761. New Engl. J. Med., 282: 18-23, 1970

13 Kennaway, E.L. Carcinogenic substances and their fluorescence spectra. Biochem. J., 24, 497-504, 1930

14 Cook, J.W., Hewett, C.L. and Hieger, I. The isolation of a cancer producing hydrocarbon from coal tar, J. Chem. Soc., 395-405, 1933

15 Rehn, L. Blasengeschwulste bein Fuchsin-Arbeitern. Arch. Klin. Chir., 50: 588-600, 1895

16 International Labour Office (ILO). Cancer of the bladder among workers in aniline factors. In: Studies and Reports, series F, no 1. ILO, Geneva, Switzerland, 1921

17 Case, R.A.M., M.E. Hosker, D.B. McDonald and J.D. Pearson. Tumors of the urinary bladder in workmen engaged in the manufacture and use of certain dyestuff intermediates in the British chemical industry. Br. J. Med., 11: 75-104, 1954

18 Wolbach, S.B. The pathological histology of chronic x-ray dermatitis and early x-ray cancer. J. Med. Res., 1909, 21: 415. In: Hayward, O.S. The History of Oncology. Surgery, 58: 745-757, 1965

19 Hesse, O. Syptomatologie, Pathogenesse und Therapie des Rontenkarzinoma. Heftio, Zwanglose Abhandlungen aus dem Geniete der Medizinischen elektrologic und rontenkunde, Leipzig 1911, J.A. Barth. In: Hayward, O.S. The History of Oncology. Surgery, 58: 745-757, 1965

20 Lynch, K.M. and Smith, W.A. Pulmonary asbestos. III. Carcinoma of the lung in asbestos-silicosis. Amer. J. Cancer, 24: 56-65, 1935

21 Doll, R. Mortality from lung cancer in asbestos workers. Brit. J. Industr. Med., 12: 81-86, 1955

22 Wagner, J.C., Sleggs, C.A. and Marchand, P. Diffuse pleural mesothelioma and asbestos exposure in the North Western Cape Province. Brit. J. Industr. Med., 17: 260-271, 1960

23 Muller, F.h. Tabakmisbrauch und Lingenkarzinom. Z. Krebsforsch, 49: 57-85, 1940

24 Doll, R., Hill, A.b. A study of the aetiology of carcinoma of the lung. Brit. Med. J., 2: 1271-1286, 1952

25 Wynder, E.L. and Graham, E.A. Tobacco smoking as a possible etiologic factor in bronchogenic carcinoma. A study of six hundred and eighty-four proved cases. JAMA, 143: 329-336, 1950

26 Lacassagne, A. Apparition de cancers de la mamelle chez la souris male, soumise a des injections de folliculine. C.R. Acad. Sci. (Paris), 195: 630-632, 1932

27 Herbst, A.L., Ulfelder, H., Poskanzer, D.C. Adenocarcinoma of the vagina. Association of maternal stilbestrol therapy with tumor appearance in young women. New Engl. J. Med., 284: 878-881, 1971

28 Figueroa, W.G., Raskowski, R., Weiss, W. Lung Cancer in chloromethyl methyl ether workers. New Engl. J. Med., 288: 1096-1097, 1973

29 Theiss, A.M., Hey, W., Zeller, H. Zur tokikologie von Dichlorodimethylather-Verdacht auf kanzerogene Wirkung auch beim Menschen. Zbl. Arbeitsmed., 23: 97-102, 1973

30 Viola, P.L., Bigotti, A. and Caputo, A. Oncogenic response of rat skin, lungs and bones to vinyl chloride. Cancer Res., 31: 516-522

31 Maltoni, C. and Lefemine, G. Carcinogenicity bioassays of vinyl chloride. Environ. Res., 7: 387-405, 1974

32 Creech, J.L. and Johnson, M.N. Angiosarcoma of liver in the manufacture of polyvinyl chloride. J. Occup. Med., 16: 150-151, 174

33 Rous, P. A sarcoma of the fowl transmissible by an agent separable from the tumor cells. J. Exp. Med., 13: 397-411, 1911

34 Poiesz, B.J., Ruscetti, F.W., Reitz, M.S., Kalyanaraman, V.S. and Gallo, R.C. Isolation of a new type-C retovirus (HTLV) in primary uncultured cells of a patient with Sezary T-cell leukemia. Nature, 294: 268-271, 1981

35 IARC Monographs on the Evaluation of the Carcinogenic Risk of Chemicals to Humans, Vol. 38: Tobacco Smoking. Internatinal Agency for Research on Cancer, Lyon, 1986

36 Shishido, S., Iwai, K. and Tukagoshi, K. Incidence of ferruginous bodies in the lungs during a 45-year period and mineralogical analysis of the core fibres and uncoated fibres. In: Bignon, J., Peto, J. and Saracci, R., eds., Non-Occupational EXposure to Mineral Fibres, IARC Scientific Publications No. 90, International Agency for Research on Cancer, Lyon, 1988, pp 229-238

37 IARC Monographs on the Evaluation of the Carcinogenic Risk of Chemicals to Humans, Vol. 35: Polynuclear Aromatic Compounds, Part 4. Bitumens, Coal-tars and Derived Products, Shale-oils and soots. International Agency for Research on Cancer, Lyon, 1985

38 Parkin, D.M., Stjernsward, J., Muir, C.S. Estimates of the worldwide frequency of twelve major cancers. Bull. WHO, 62: 163-182, 1984

39 Tomatis , L., Breslow, N.E and Bartsch. Experimental Studies in the Assessment of Hman Risk. Chapter Four. In: Schottenfeld, D. and Fraumeni, J.F. Jr., Cancer Epidemiology and Prevention, W.B. Saunders Company, pp 44-73, 1982

40 Kaldor, J., Day, N.E. and Hemminki, K. Quantifying the carcinogenicity of antineoplastic drugs. European J. Cancer Clin. Onc., 24: 703-711, 1988

41 Te-Chang Lee, Tanaka, N., Lamb, P.W., Gilmer, T.M., Barrett, J.C. Induction of gene amplification by arsenic. Science, 241: 79-81, 1988

METABOLISM OF CHEMICAL CARCINOGENS

Franz Oesch and Helmut Thomas

Institute of Toxicology, University of Mainz
Obere Zahlbacher Strasse 67, D-6500 Mainz
Federal Republic of Germany

ABSTRACT

Most chemical carcinogens are chemically unreactive per se and need metabolic activation to the ultimate carcinogenic species. The enzyme pattern responsible for the generation and disposition of reactive metabolites constitutes one important early contribution to the control of chemical carcinogenesis. Especially well studied is the group of enzymes responsible for the control of reactive epoxides. Many natural as well as manmade foreign compounds, including pharmaceuticals, possess olefinic or aromatic double bonds. Such compounds can be transformed to epoxides by microsomal monooxygenases present in many mammalian organs. By virtue of their electrophilic reactivity such epoxides may spontaneously react with nucleophilic centers in the cell and thus covalently bind to DNA, RNA and protein. Such alterations of critical cellular macromolecules may disturb the normal biochemistry of the cell and lead to cytotoxic, mutagenic and/or carcinogenic effects. Enzymes controlling the concentration of such epoxides are an important contributing factor in the control of chemical carcinogenesis. Several microsomal monooxygenases exist differing in activity and substrate specificity. With respect to large substrates, some monooxygenases preferentially attack on one specific site different from that attacked by others. Some of these pathways lead to reactive products, others are detoxification pathways. Enzymes metabolizing such epoxides represent a further determining factor. These enzymes include epoxide hydrolases, glutathione transferases and dihydrodiol dehydrogenases. These enzymes do not play a pure inactivating role but can in some cases also act as co-activating enzymes. Enzymes involved in biosynthesis and further metabolism of epoxides differ in quantity and also in substrate specificity between organs, developmental stages, sexes and animal species. Differences in susceptibilities between species and individuals are often causally linked to these metabolic differences.

INTRODUCTION

Many foreign compounds which have been identified as chemical carcinogens were found to elicit their biological effects only after metabolic conversion to reactive intermediates (1, 2, 3). Frequently occurring structural elements in a wide variety of drugs, industrial chemicals, pesticides and pollutants are aromatic moieties and olefinic double bonds, which give rise

Biochemistry of Chemical Carcinogenesis
Edited by R. Colin Garner and Jan Hradec
Plenum Press, New York, 1990

to the metabolic formation of epoxides by the microsomal cytochrome P-450
dependent monooxygenases. By virtue of their electrophilicity, these epoxides may covalently bind to essential nucleophilic constituents of the cell
including DNA, RNA and protein. Such alterations of critical cellular macromolecules may finally lead to the well known expression of cytotoxity and
carcinogenicity.

Therefore major interest in the investigation of mechanisms of chemical
carcinogenicity has focussed recently on enzyme systems responsible for the
control of reactive epoxides. Three enzyme families, epoxide hydrolases (4),
glutathione S-transferases (5) and dihydrodioldehydrogenases (6), each comprising a number of isoenzymes with significantly different substrate specificities and catalytic properties, have been shown to play a crucial role in
the detoxification of these metabolites. However, depending on their subcellular localization, organ distribution and expression in various species
including man, different protection may be provided for different organs and
organisms. Information on the isoenzyme pattern, tissue distribution and
function in different species is therefore a prerequisite for any risk
assessment concerning chemical carcinogens and the extra polation of results
obtained from animal systems to man.

Enzymes responsible for the formation of reactive epoxides

A major pathway in the metabolism of aromatic and olefinic compounds
consists of the transformation to epoxides by microsomal cytochrome P-450
dependent monoxygenases (7, 8, 9). However, among the enzymes reponsible for
further metabolism of these epoxides are forms which play a dual role of
either detoxification or (10) by providing substrates to certain monooxygenases leading to the formation of even more hazardous secondary oxidation
products. Thus various sets of reactive metabolites are generated, some of
which may be detoxified by phase II enzymes while others are converted by the
same phase II or phase III enzyme(s) to even more reactive tertiary metabolites. Benzo(a)pyrene (BP) is metabolized by monooxygenase to reactive
epoxides including the 7, 8-oxide which is inactivated by microsomal epoxide
hydrolase to the the corresponding dihydrodiol. This is, however, a substrate for a second monooxygenation step, which introduces a further epoxide
moiety yielding a dihydrodiol bay region epoxide.

Chemical quantum calculations by Lehr and Jerina (11) revealed a particular chemical reactivity of this compound, and it has in fact, been shown to
be an ultimate carcinogen (8, 12, 13, 14) and responsible for the formation
of the major DNA-adducts of this class of carcinogens (15).

Similarly, benz(a)anthracene (BA) itself is a weakly carcinogenic polycyclic aromatic hydrocarbon. It is predominantly metabolized via the corresponding epoxides to the 5, 6- and 8, 9- dihydrodiols by rat liver homogenates and microsomal preparations (16, 17, 18). The dihydrodiol at the
3, 4-position derived from the respective epoxide by an epoxide hydrolase
catalyzed reaction was a minor metabolite in the cited studies, but displayed
extraordinary high carcinogenic activity. In both the newborn mouse assay as
well as the mouse skin assay, BA-3, 4-dihydrodiol expressed more than tenfold
the activity of the parent hydrocarbon and of all other dihydrodiols of BA
(19, 20), presumably due to the formation of the ultimate carcinogen BA-3, 4-
dihydrodiol-1, 2-epoxide.

Thus a single enzyme, in this case microsomal epoxide hydrolase can play
multiple roles in inactivating some metabolites but producing precursors for
reactive species from other intermediates. In order to perform a rational
risk assessment information is therefore not only needed on the relative
efficiencies of individual enzymes for a certain substrate but also on their

impact on the further metabolism of the compound in question.

To study this problem we have successfully used _in vitro_ systems, which are capable of quantitatively monitoring a biological effect, viz., bacterial mutagenicity (7). The role of enzymes with specific cofactors can be studied by either removal or addition of that cofactor. However, the relative importance of enzymes which depend on a common cofactor can only be monitored after purification to apparent homogeneity. The same applies to enzymes which do not require a particular cofactor such as epoxide hydrolase which merely adds the elements of water (9, 21). An additional point which is frequently not taken into account is that several reactive metabolites are often involved in a specific toxic effect. They can be toxicologically fundamentally different from one another and since they are reactive and shortlived their chemical quantification of characterization is in many cases difficult or even impossible.

In order to overcome this intrinsic problem, we monitored with various strains of bacteria, the potency of certain chemically synthesized reactive metabolites to cause mutation. When the K-region epoxide, BP-4, 5-oxide, was monitored with TA 98 and TA 1537, the strain 98 was somewhat more efficiently mutated (ration 1.6). After _in situ_ bioactivation of the synthetic BP-7, 8-dihydrodiol to the corresponding very short-lived bay-region BP-7, 8-dihydrodiol-9, 10-epoxide, the strain TA 98 was much more efficiently reverted than TA 1537 (ration 15). This ratio was again different from that observed after _in situ_ bioactivation of the BP-9, 10-dihydrodiol. Such characteristic ratios of how various strains of bacteria are reverted can distinguish individual reactive metabolites derived from the same precursor.

The importance of these considerations becomes clear from the fact that the effect of microsomal epoxide hydrolase was essentially opposite depending on the actual cytochrome P-450 isoenzyme pattern. This was investigated by adding apparently homogeneous epoxide hydrolase (22) to BP activated by liver microsomes. When these microsomes were from untreated mice the mutagenicity was decreased to 1 - 2% of the original rate. This small remaining mutagenicity could not be further reduced by addition of more purified microsomal epoxide hydrolase, being due to the epoxide hydrolase-resistant part of reactive metabolites. The ratios of the mutagenic potency towards various detector strains (see above) showed that the K-region BP-4, 5-oxide was predominantly responsible for the major, i.e., epoxide hydrolase-sensitive, portion of the mutagenic effect under these conditions.

However, the situation was changed fundamentally when liver microsomes from mice pretreated with 3-methylcholanthrene were used producing a different pattern of monooxygenase isoenzymes. The latter in turn generate a different pattern of primary reactive metabolites, and in this situation microsomal epoxide hydrolase has a weak but activating effect (10), with dihydrodiol epoxides being now predominantly responsible for the observed mutagenicity. The microsomes used to generate the metabolites contain endogenous epoxide hydrolase and can form dihydrodiol epoxides. Further addition of purified enzyme causes a small, but significant, increase of the mutagenicity, since more of these dihydrodiol epoxides are produced. If the amount of epoxide hydrolase is further increased, a small decrease and then a small increase of the mutagenicity is observed (10). The effect is multiphasic, since several metabolites are contributing to the mutagenicity. This situation - induction by 3-methylcholanthrene - is especially important, because BP as well as BA and other carcinogenic polycyclic aromatic hydrocarbons show their carcinogenic effects at doses of different inducers, which produce 3-methylcholanthrene-type monooxygenase induction. Thus, in the control of reactive metabolites and the biological effects produced by them, not only the quantity but even more the quality of the cytochromes P-450 represent a fundamental factor.

Enzymatic inactivation of reactive epoxides

Substitution patterns, steric hindrance and hydrophilic/lipophilic nature around the structural moiety to be enzymatically converted are important features often strongly determining relative activities of enzymes for given substrates. These factors have recently been investigated in our laboratory for a series of methylated styrene oxides using human liver cytosol and microsomes as sources for cytosolic and microsomal epoxide hydrolase (Table 1; 42). The obtained apparent K_M values and maximal turnover rates have been compared with those of the purified major isoenzymes of rat liver gluta-thione-S-transferase (Table 2; 43) which are closely related to human liver forms with respect to structural and biochemical properties (5). Apart from the fact, that only transferase isoenzymes composed of the subunits 3 and 4 were capable of metabolizing this class of reactive epoxides, and the striking preference for the least sterically hindered substrate, trans- - methylstyrene oxide, of all studied enzymes, there are characteristic requirements that exclude or favour the metabolism of differentially substituted oxiranes by either of the enzymes in question. While dialyzed cytosol, although at an about 10- to 40-fold lower rate, exhibits substrate specificities similar to those of microsomes, with cis − methylsubstituted styrene oxides being more efficiently metabolized than the corresponding trans-monosubstituted and β-disubstituted compounds, appear the glutathione transferases to prefer the trans-methylsubstituted styrene derivatives as substrates. Cis substituted oxides are slowly metabolized particularly in the presence of a cis- , -dimethyl substitution. This impressive example of the influence of structural properties on the detoxification of reactive intermediates demonstrates at the same time the complementary action of different enzyme systems on the inactivation of epoxides.

Among the enzymes capable of inactivating reactive epoxides derived from polycyclic aromatic hydrocarbons particularly microsomal epoxide hydrolase is

Table 1. Inactivation of different methylsubstituted styrene oxides by human liver microsomes and cytosol

Substrate	Microsomes		Cytosol	
	K_{mapp} [μM]	V_{max} [nmol.min^{-1}.mg^{-1}]	K_{mapp} [μM]	V_{max} [nmol.min^{-1}.mg^{-1}]
trans- β -methyl styrene oxide	–	86.5 X 10^3	–	2.1 X 10^3
cis- β-methyl styrene oxide	319	5.4	55	0.664
cis-α, β -dimethyl styrene oxide	83	0.665	89	0.034
trans-α, β -dimethyl styrene oxide	473	0.238	–	n.d.
β, β -dimethyl styrene oxide	229	0.130	358	0.122

n.d. not detectable

16

Table 2. Inactivation of different methylsubstituted styrene oxides by purified isoenzymes of rat liver glutathione-S-transferase (GST)

Substrate	GST A (3-3)*		GST C (3-4)		GST D (4-4)	
	K_{mapp} µM	V_{max} nmol.min^{-1}.mg^{-1}	K_{mapp} µM	V_{max} nmol.min^{-1}.mg^{-1}	K_{mapp} µM	V_{max} µmol.min^{-1}.mg^{-1}
trans- -Methyl styrene oxide	112	578 ± 139	800	575 ± 32	300	500 ± 33
cis- -methyl styrene oxide	330	7.7 ± 0.3	550	5.3 ± 0.8	250	0.7 ± 0.13
cis- , -dimethyl styrene oxide	500	0.26 ± 0.02	260	0.65 ± 0.18	250	0.80 ± 0.14
trans- , -dimethyl styrene oxide	2500	6.4 ± 0.3	1250	11.4 ± 1.0	2000	15.7 ± 4.0
, -dimethyl styrene oxide	600	43.0 ± 2.0	2000	$30,6 \pm 0.7$	3300	12.5 ± 3.4

*) for nomenclature see Mannervik (5)

much less efficient for epoxides which carry hydroxyl groups close to the
epoxide ring, whilst a hydrophobic center near the oxirane moiety is
generally favorable (21, 23, 24, 25). Therefore vicinal dihydrodiol epoxides
are, in the majority of cases investigated to date, the metabolites of
polycyclic hydrocarbons that are responsible for most of the carcinogenic and
mutagenic effects induced by these compounds (8, 26, 27, 28, 29, 30, 31, 32,
33). Mutagenicity and DNA-binding experiments with BP-7, 8-dihydrodiol 9,
10-epoxide indicate that some inactivation is caused in the presence of
glutathione (34, 35, 36) but not by microsomal epoxide hydrolase (10, 24, 37,
38). The latter finding was believed to have possibly resulted from the
short half-life of the dihydrodiol epoxide in an aqueous environment,
although this may not necessarily be the same in a biological membrane.
Since not all vicinal dihydrodiol epoxides are so unstable, we investigated
the non-bay-region vicinal dihydrodiol epoxide, BA-8, 9-dihydrodiol-10, 11-
epoxide, which has a half-life of many hours and is therefore especially
useful for metabolic studies. It is mutagenic (39, 40) and is often the
major DNA- binding species formed from BA in vivo and in vitro (32, 33, 41),
and it serves as a substitute model for less stable dihydrodiol epoxides. The
results reported below (Table 3) show that this vicinal dihydrodiol epoxide
can be metabolically inactivated by dihydrodiol dehydrogenase, but not by
microsomal or cytosolic epoxic hydrolase. If one assumes that the activities
of the investigated enzymes in vivo are comparable to those that are present
in our experiments in vitro, inactivation of the dihydrodiol epoxide by
dihydrodiol dehydrogenase would be slower than the rate of inactivation of
the K-region oxide by microsomal epoxide hydrolase, but still sufficiently
rapid to substantially affect dihydrodiol epoxide concentrations in mammalian
systems. To study their relative importance in metabolic inactivation, the
three enzymes (microsomal and cytosolic epoxide hydrolase and cytosolic dihydrodiol dehydrogenase) were purified and bacterial mutagenicity was used as
an indication of their effects on the metabolism of BA-8, 9-diol-10, 11-oxide
and the K-region epoxide BA-5, 6-oxide.

As expected from its substrate specificity (44, 45), microsomal epoxide
hydrolase readily inactivated BA-5, 6-oxide. No significant effect on the
mutagenicity of BA-8, 9-diol-10, 11-oxide was obtained even with a 100-fold
excess over that required for complete inactivation of BA-5, 6-oxide. BA-8,
9-diol-10, 11-oxide was not inactivated by cytosolic epoxide hydrolase
either, whilst the K-region oxide was inactivated. However, relatively large
amounts of cytosolic epoxide hydrolase were required to inactivate BA-5, 6-
oxide.

The vicinal dihydrodiol epoxide, on the other hand, was inactivated by
dihydrodiol dehydrogenase. Relatively high amounts of this enzyme were
needed for this inactivation, whereas a low amount of microsomal epoxide
hydrolase was sufficient for the inactivation of BA-5, 6-oxide. A 50%
inactivation of 1 ug BA-5, 6-oxide was achieved with either 0.4 units
purified rat microsomal epoxide hydrolase, equivalent to 1.3 mg liver, or
with 7 units purified rabbit cytosolic epoxide hydrolase, equivalent to 28 mg
liver. These are relatively small quantities of enzyme. Microsomal epoxide
hydrolase equivalent to 330 mg rat liver and cytosolic epoxide hydrolase
equivalent to 200 mg rabbit liver did not inactivate this mutagen, whereas
with dihydrodiol dehydrogenase an amount equivalent to 200 mg liver was
required to obtain a 50% inactivation. The effective but moderate rate of
inactivation of vicinal dihydrodiol epoxides by dihydrodiol dehydrogenase
suggests that differences in the activity of this enzyme between species,
organs, and physiological states are likely to be important contributing
factors to differences in the susceptibility to diol epoxide induced chemical
carcinogens.

More recent studies performed in our laboratory however strongly indicate
that dihydrodiol dehydrogenase may play an important protective role by meta-

Table 3. Role of various purified enzymes in the control of benz(a)anthracene 5, 6-oxide (a monofunctional epoxide) and benz(a)anthracene-8, 9-dihydrodiol 10, 11-oxide (a vicinal dihydrodiol epoxide)[a]

Enzyme	Enzyme concentration in liver[b] (μg/mg tissue)	BA 5, 6-oxide[c]		BA-8,9-diol 10,11-oxide[d]	
		μg/incubation	mg liver equivalents	μg/incubation	mg liver equivalents
Microsomal epoxide hydrolase	0.5	0.7	1.3	inactive (>>170)	>> 300[e]
Cytosolic epoxide hydrolase	0.16	4	30	inactive (>> 30)	>> 200[e]
Glutathione transferase A	0.5	0.11	0.2	110	200
Glutathione transferase B	2.2	0.5	0.2	130	60
Glutathione transferase C	1.1	0.02	0.017	30	20
Glutathione transferase X	0.25	0.003	0.011	6	20
Dihydrodiol dehydrogenase	0.45	inactive (>>3)	>> 1000	70	170

[a] Data taken from (Glatt et al., 25, 38)
[b] Values refer to untreated, adult males of the species from which the enzyme was purified
[c] BA 5, 6-oxide: benz(a)anthracene 5, 6-oxide
[d] BA-8 9-diol 10, 11-oxide: benz(a)anthracene-8, 9-dihydrodiol a0, 11-epoxide
[e] Only determinable as an upper limit from the experiment

bolizing dihydrodiols as the immediate precursors of the ultimately
carcinogenic dihydrodiol epoxides. We have shown that purified rat liver
dihydrodiol dehydrogenase is capable of oxidising certain dihydrodiols from
aromatic hydrocarbons which are known to be substrates for further cytochrome
P-450 dependent oxidation to dihydrodiol epoxides (Table 4). This observation deserves particular attention with respect to the detoxification of BA-
3, 4-dihydrodiol which has been reported a highly potent carcinogen (19, 20).
The physiological significance of the dihydrodiol dehydrogenase catalyzed
conversion of BA-3, 4-dihydrodiol to the corresponding catechol can be best
extrapolated by a comparison of the rate of metabolic formation of BA-3, 4-
dihydrodiol by rat liver microsomes with the kinetic constants reported in
Table 4.

BA is metabolized at a maximal rate of roughly 30 $nmol.min^{-1}.mg^{-1}$ cytochrome P-450 by microsomes from non-induced rats (15). The 3, 4-dihydrodiol
constitutes about 1.5 - 4% of all formed metabolites, i.e. 0.5 - 1.2
$nmol.min^{-1}.mg^{-1}$ cytochrome P-450. In microsomes from 3-methylcholanthrene

Table 4. Kinetic constants of the enzymatic oxidation of aromatic
dihydrodiols of diverse structure by purified rat liver dihydrodiol
dehydrogenase.[a]

Substrate	V_{max} [$nmol.min^{-1}.mg^{-1}$]	K_M [μM]	V_{max}/K_M [$nmol.min^{-1}.mg^{-1}$]
Benzene dihydrodiol	8.580	2.2	3.90
syn-Benzene dihydrodiol epoxide	3.660	7	0.52
anti-Benzene dihydrodiol epoxide	n.d.[b]	n.d.	0.15
Napthphalene-1, 2-dihydrodiol	n.d.	n.d.	0.03
Phenanthrene-1, 2-dihydrodiol	75.9	0.217	0.35
Phenanthrene-3, 4-dihydrodiol	27.8	0.251	0.11
Benz(a)anthracene-1, 2-dihydrodiol	28.0	0.095	0.29
Benz(a)anthracene-3, 4-dihydrodiol	20.0	0.021	0.95
Crysene-1, 2-dihydrodiol	n.d.	n.d.	0.39
Benz(a, h)anthracene-3, 4-dihydrodiol	n.d.	n.d.	0.35

[a] Data taken from Klein (6)
[b] n.d. not determined due to low solubility or high instability of the
substrate

treated rats, the rate of formation of the 3, 4-dihydrodiol is about 2.8
nmol.min^{-1}.mg^{-1} cytochrome P-450. This value is probably too high for an in
vivo situation because induction effects are certainly of lower significance
in vivo. The cytochrome P-450 isoenzyme c (P-450 IA1) is mainly responsible
for the enzymatic formation of the 3, 4-epoxide, the metabolic precursor of
the 3, 4-dihydrodiol. This particular isoenzyme is strongly induced by the
pretreatment of rats with 3-methylcholanthrene, but is almost undetectable in
microsomes from non-induced rats (46). Whole liver of adult rats contains
about 6 mg of total cytochrome P-450 (47) and therefore forms the BA-3, 4-
dihydrodiol at a rate of 1.8 - 5.4 nmol.min^{-1}. This rate of formation is in
the same order of magnitude as the oxidation of the 3, 4-hydrodiol by
cytosolic dehydrogenase. At a substrate concentration of 10 um which is
assumed not to be exceeded in vivo due to limited solubility, the 3, 4-
dihydrodiol is oxidized by dihydrodiol dehydrogenase at a rate of about 10
nmol.min^{-1}.mg^{-1}. Whole liver of adult rats contains about 1.5 mg dihydrodiol
dehydrogenase from which a potential conversion of the dihydrodiol to
catechol of 15 nmol per min can be calculated. Since the K_m of NADP$^+$ (with
benzene dihydrodiol as substrate) for dihydrodiol dehydrogenase has been
reported as 7.7 uM (48), the supply of NADP$^+$ should make near-maximal
reaction rates possible, since the concentration of NADP$^+$ in hepatic cytosol
has been measured as 20 - 60 uM (49, 50). Competing conjugating enzymes such
as sulfotransferases and UDP-glucuronosyl transferases have relatively low
turnover rates with dihydrodiols of polycyclic aromatic hydrocarbons (51, 52)
whereas the resulting catechol's appear to be rapidly conjugated and excreated
(1, 2, 53) rendering the whole reaction sequence basically irreversible.
Except, that the cytosolic location of dihydrodiol dehydrogenase may limit
the accessibility of substrate to a certain degree, dihydrodiol dehydrogenase
can be regarded as a very efficient enzyme in the metabolism of BA-3, 4-
dihydrodiol and similarly other dihydrodiols that may arise in vivo as precursors of ultimate chemical carcinogens (Table 4).

Interestingly, no enzymatic oxidation was observed with the non-pre-bay
region dihydrodiol compounds phenanthren-9, 10-dihydrodiol BA-5, 6-, -8, 9-
and -10, 11-dihydrodiol. Additionally, no turnover of the pre-bay syn- or
anti-BA-314-dihydrodiol-1, 2-epoxide could be detected confirming the suggested protective action against precursors of ultimate carcinogens.

The rat liver cytosolic fraction contains at least nine glutathione
transferases (5, 54). We have discovered and purified an additional enzyme
with distinct properties which we have termed glutathione transferase X (55).
This form and the glutathione transferases A (3 - 3), B (1 - 2) and C (3 -
4), which are the most abundant forms present in rat liver in terms of
amounts of protein (56), have been purified to apparent homogeneity and investigated for their abilities to inactivate the two prototype epoxides discussed above, BA-5, 6-oxide and BA-8, 9-diol-10, 11-oxide. Both prototype
epoxides, the K-region epoxide and the dihydrodiol-epoxide were inactivated
by glutathione transferases A, B, C, and X. About 1000-fold higher concentrations of glutathione transferase were required for inactivation of the
dihydriol-epoxide than for inactivation of the K-region epoxide. This was
independent of the enzyme form used. These similarities in the ratio of
inactivation of the two epoxides by the three enzymes were remarkable because
the enzymes differed substantially from each other in the efficiency with
which they inactivated the epoxides. Interestingly, the newly discovered
form X turned out to be the most efficient isoenzyme, followed by C $\frac{3}{8}$ A $\frac{3}{8}$ B.

Quantitive evaluation of the detoxification efficiencies of epoxide
hydrolase, dihydrodiol dehydrogenase and glutathione transferases on two
prototype epoxides

The amounts of the various purified enzymes required for a 50%
inactivation of the mutagenicity of the two prototype epoxides (BA 5, 6-

oxide, a K-region epoxide and BA-8, 9-diol 10, 11-oxide, a vicinal
dihydrodiol non-bay-region epoxide) are given in Table 3. When intrinsic
enzyme activities and the relative amounts of enzymes present in the liver
are considered, but subcellular compartmentalization is disregarded, the
glutathione transferases can play a more important role that dihydrodiol
dehydrogenase or the epoxide hydrolases, in the inactivation of both prototype epoxides, BA 5, 6-oxide and BA-8, 9-diol 10, 11-oxide, although the
hydrolases are specific for the hydrolysis of epoxides. Amongst the glutathione transferases, large differences in efficiency of detoxification occur
and this is even true for the immunologically closely related forms A, C and
X (55, 56), which differ in efficiency by more than one order of magnitude.
With regard to the enzymes present in rat liver, the forms C and X appear to
contribute most to the inactivation of the two epoxides examined here; form C
because of its quantitive abundance in rat liver, and form X because of its
high efficiency in inactivating these epoxides. Such an estimate is rather
crude, when different types of enzymes are compared, because of differences
in cofactor concentration, pH optima and in other environmental factors which
may lead to differences between enzyme activity in vivo and under our experimental conditions. For example, microsomal epoxide hydrolase was tested in
the mutagenicity experiments as the free purified enzyme, whereas in vivo it
is situated in the endoplasmic reticulum and in other cellular membranes
(57). This may be of advantage in comparison with cytosolic enzymes, since
it increases the opportunities for reaction with epoxides, which are generated in these membranes (57) and tend to stay there because of their relative
lipophilicity (34). A further factor to be taken into account is that
dihydrodiol dehydrogenase may not only inactivate dihydrodiol epoxides, but
as outlined above may also sequester precursor dihydrodiols, an effect that
has not been taken into account in the experiemtnal model, as only the effect
of the dihydrodiol epoxide was monitored. In spite of these limitations, the
data presented on the more efficient in vitro inactivation of epoxides by
glutathione transferasees indicate that glutathione transferases play an
important role in the in vivo inactivation of epoxides and that glutathione
transferases X is especially important. Such basic information is of crucial
importance for the extrapolation of experimental findings to the situation in
man.

ACKNOWLEDGEMENTS

The authors wish to thank Ms. S. Pollok for typing the manuscript. This
study was supported by the Deutsche Forschungsgemeinschaft (SFB 302).

REFERENCES

1 Boyland, E. (1986) Xenobiotica, 16, 899 - 913
2 Grover, P. L. (1986) Xenobiotica, 16, 915 - 931
3 Parke, D. V. (1987) Arch. Toxicol., 60, 5 - 15
4 Wixtrom, R. N. and Hammock, B. D. (91850 in Biochemical Pharmacology and
   Toxocology, Vol. I, Methodological Aspects of Drug Metabolizing Enzymes
   (Zakim, D. and Vessey, D. A. eds.) pp. 1 - 93. John Wiley and Sons, New
   York
5 Mannervik, B. (1985) Advances in Enzymology and Related Areas of
   Molecular Biology 57, 357 - 417
6 Lein, J. (1987) Ph.D. Thesis, University of Mainz, FRG.
7 Oesch, F. and Glatt, H. R. (1976) in Tests in Chemical Carcinogenesis.
   IARC Scientific Publications NO 12 (Montesano, R. ed.) p. 255.
   International Agency for Research on Cancer, Lyon
8 Levin, W., Wood, A. W., Wislocki, P. G., Chang, R. L., Kapitulnik, J.,
   Mah, H. D., Yagi, H., Jerina, D. M. and Conney, A. H. (1978) in
   Polycyclic Hydrocarbons and Cancer, Vol. 1. (Gelboin, H. V. and Ts'o,

P. O. P. eds.) p. 189. Academic Press, New York

9 Oesch, F. (1979) Arch. Toxicol., Supp. 2, 215 - 227

10 Bentley, P., Oesch, F. and Glatt, H. R. (1977) Arch. Toxicol. 39, 65 -75

11 Lehr, R. E. and Jerina, D. M. (1977) Arch. Toxicol. 39, 1 - 6

12 Slaga, T. J., Viaje, A., Bracken, W. M., Berry, D. L., Fischer, S. M., Miller, D. R. and Leclerc, S. M. (1977) Cancer Lett. 3, 23 - 30

13 Kapitulnik, J., Wislocki, P. G., Levin, W., Yagi H., Jerina, D. M. and Conney, A. H. (1978) Cancer Res. 38, 354 - 358

14 Kapitulnik, J., Wislocki, P. G., Levin, W., Yagi, H., Thakker, D. R., Akagi, H., Koreeda, M., Jerina, D. M. and Conney, A. H. (1978) Cancer Res. 38, 2661 - 2665

15 Thakker, D. R., Yagi, H., Levin, W., Wood, A. W., Conney, A.H. and Jerina D. M. (1985) in Bioactivation of Foreign Compounds (Anders, M. W. ed.) pp. 177 - 242. Academic Press, Washington

16 Sims, P. (1970) Biochem. Pharmacol. 19, 795 - 818

17 Tierney, B., Hewer, A. D., Mac Nicoll, A. D., Gervasi, P. G., Rattle, H., Walsh, C., Grover, P. L. and Sims (1978) Chem. Biol. Interact, 23, 243 - 257

18 Thakker, D. R., Levin, W., Yagi, H., Ryan, D., Thomas, P. E., Karle, J. M., Lehr, R. E., Jerina, D. M., and Conney, A.H. (1979) Mol. Pharmacol. 15, 138 - 153

19 Wood, A. W., Levin, W., Chang, R. L., Lehr, R. E., Schaefer-Ridder, M,. Karle, J. M., Jerina, D. M. and Conney, A. H. (1977) Proc. Natl. Acad. Sci. USA 74, 3176 - 3179

20 Wislocki, P. G., Kapitulnik, J., Levin, W., Lehr, R., Schaefer-Ridder, M., Karle, J. M., Jerina, D. M. and Conney, A. H. (1978) Cancer Res. 38, 693 - 696

21 Oesch, F. (1973) Xenobiotica 3, 305 - 340

22 Bentley, P. and Oesch, F. (1975) FEBS Lett. 59, 291 - 295

23 Oesch, F. (1974) Biochem. J. 139, 77 - 88

24 Wood, A. W., Levin, W., Lu, A. Y. H., Yagi, H., Hernandez, O., Jerina D. M. and Conney, A. H. (1976) J. Biol. Chem. 251, 4882 - 4890

25 Glatt, H. R., Friedberg, T., Grover, P. L., Sims, P. and Oesch, F. (1983) Cancer Res. 43, 5713 - 5717

26 Sims, P., Grover, P. L., Swaisland, A., Pal, K. and Hewer, A. (1974) Nature 25, 226 - 228

27 Humbermann, E., Sachs, L., Yang, S. K. and Gelboin, H. V. (1976) Proc. Natl. Acad. Sci. USA 73, 607 - 611

28 Newbold, R. F. and Brookes, P. (1976) Nature 261, 52 - 54

29 Slaga, T. J., Viaje, A, Berry, D. L., Bracken, W. M., Buty, S. G. and Scribner, J. D. (1976) Cancer Lett. 2, 115 - 122

30 Wislocki, P. G., Wood, A. W., Chang, R. L., Levin, W., Yagi. H., Hernandez, O., Jerina, D. M. and Conney, A. H. (1976) Biochem. Biophys. Res. Commun. 68, 1006 - 1012

31 Hecht, S. S., La Voie, E., Mazzorese, R., Amin, S., Bedenko, V. and Hoffmann, D. (1978) Cancer Res. 38, 2191 - 2198

32 Vigny, P., Kindts, M., Duquesne, M., Cooper, C. S., Grover, P. L. and Sims, P. (1980) Carcinogenesis 1, 33 - 41

33 MacNicoll, A. D., Cooper, C. S., Ribeiro, O., Pal, K., Hewer, A., Grover P. L. and Sims, P. (1981) Cancer Lett. 11, 243 - 249

34 Glatt, H. R. and Oesch, F. (1977) Arch. Toxicol. 39, 87- - 96

35 Glatt, H. R., Billings, R., Platt, K. L. and Oesch, F. (1981) Cancer Res. 41, 270 - 277

36 Guenthner, T. M., Jernstrom, B. and Orrenius, S. (1980) Carcinogenesis 1, 407 - 418

37 Glatt, H. R., Vogel, K., Bentley, P. and Oesch, F. (1979) Nature 277, 319 - 320

38 Glatt, H. R., Cooper, C. S., Grover, P. L., Sims, P., Bentley, P., Mertes, M., Waechter, F., Vogel, K., Guenthner, T. M. and Oesch, F. (1982) Science 215, 15071509

39 Malaveille, C., Kuroki, T., Sims, P., Grover, P. L. and Bartsch, H.

(1977) Mutat. Res. 44, 313 - 326

40 Wood, A. W., Chang, R. L., Levin, W., Lehr, R. E., Schaefer-Ridder, M., Karle, J. M., Jerina, D. M. and Conney, A. H. (1977) Proc. Natl. Acad. Sci. USA 74, 2746 - 2750

41 Cooper, C. S., MacNicoll, A. D., Riveiro, O., Gervasi, G. P., Hewer, A., Walsh, C., Pal, K. and Grover, P. L . (1980) Cancer Lett 9, 53 - 59

42 Schladt, L., Thomas, H., Hartmann, R. and Oesch, F. (1988) Eur. J. Biochem. in press

43 Milbert, U., Worner, W. and Oesch. F. (1986) in Primary Changes and Control Factors in Carcinogenesis (Friedberg, T. and Oesch, F. eds.) pp. 14 - 21. Deutscher Fachschriften-Verlag, Wiesbaden

44 Bentley, P., Shomassmann, H. U., Sims, P. and Oesch, F. (1976) Eur. J. Biochem. 69, 97 - 103

45 Jerina, D. M., Dansette, P. M., Lu, A. Y. H. and Levin, W. (1977) Mol. Pharmacol. 13, 342 - 351

46 Parkinson, A., Safe, S.H., Robertson, L. W., Thomas, P. E., Ryan, D. E., Reik, L. M. and Levin, W. 91983) J. Biol. Chem. 258, 5967 - 5976

47 Smith, M. T. and Orrenius, S. (1984) in Drug Metabolism and Drug Toxicity (Mitchell, J. R. and Horning, M. G. eds.) pp. 71 - 98. Raven Press, New York

48 Vogel, K., Bentley, P., Platt, K. L. and Oesch, F. (1980) J. Biol. Chem. 255, 9621 - 9625

49 Jacobsen, K. B. and Kaplan, N. O. (1957) J. Biol. Chem. 226, 603 - 613

50 Kalhorn, T. F., Thummel, K. E., Nelson, S. D. and Slattery, J. T. (1985) Anal. Biochem. 151, 343 - 347

51 Nemoto, N. and Gelboin, H. V. (1976) Biochem. Pharmacol. 25, 1221 - 1226

52 Nemoto, N., Takayama, S. and Gelboin, H. V. (1978) Chem. Biol. Interact. 23, 19 - 30

53 Bock, K. W., Lilienblum, W., Fischer, G., Schirmer, G. and Bock-Henning, B. S. (1987) Arch. Toxicol. 60, 22 - 29

54 Habig, W. H. Pabst, M. J. and Jakoby, W. B. (1974) J. Biol. Chem. 249, 7130 - 7139

55 Friedberg, T., Milbert, U., Bentley, P., Guenthner, T. M. and Oesch, F. (1983) Biochem, J. 215, 617 - 625

56 Jakoby, W. B., Ketley, J. N. and Habig, W. H. (1976) in Glutathione: Metabolism and Function, Vol. 6 (Arias, I. and Jakoby, W. eds.) p. 213. Raven Press, New York

57 Stasiecki, P., Oesch, F., Bruder, G., Jarasch, E. D. and Franke, W. W. (1980) Eur. J. Cell Biol. 21, 79 - 92

MECHANISMS OF CHEMICAL CARCINOGENESIS

Rudolf Preussman

Institute of Toxicology and Chemotherapy
German Cancer Research Centre
D-6900 Heidleberg FRG

ABSTRACT

Chemical carcinogenesis is a multistage process, consisting of different
steps leading ultimately to malignant tumors. Initiation is considered to be
the first stage, leading to a rapid and probably irreversible change in the
target cells. A generalized theory of the initiation step in chemical
carcinogenesis has been proposed by Miller & Miller (1976) which is now
generally accepted for so-called <u>genotoxic</u> carcinogens. Chemical carcinogens
are not normally reactive by themselves, but need metabolic activation to
form highly reactive, electrophilic species (ultimate carcinogens), interacting with nucleophilic sites in the cell, especially in DNA. Several
examples for typical genotoxic carcinogens will be given.

However, other chemicals are known which in animal experiments induce
tumors, but which probably do not form adducts with DNA and therefore are not
mutagenic in short-term tests. Some of these epigenetic carcinogens will be
mentioned and the available knowledge of reaction mechanisms discussed.

INTRODUCTION

There is little doubt nowadays that carcinogenesis in general and chemical
carcinogenesis specifically is usually a multistage process, comprising
different stages from the conversion of a normal somatic to a tumor cell
(transformation) and ultimately, after a long latency period, to a clinically
manifest malignant tumor.

While many of the later stages of carcinogenesis (promotion and progression) are little understood at present and are fields of intensive
research, especially from a biochemical and molecular-biological point of
view, the stage of initiation (ie the transformation of a normal to a potential tumor cell) has been extensively studied in the past two decades and has
finally resulted in a theory of chemical carcinogenesis that has been and is
widely accepted. Squeezed in between the metabolism of chemical carcinogens
and their interactions with nucleic acids in this meeting, I will concentrate
on some theories of initiation, especially the one by Miller and Miller
(1,2,3). It is based on the concept of somatic mutation (1) as the first
step in carcinogenesis.

Biochemistry of Chemical Carcinogenesis
Edited by R. Colin Garner and Jan Hradec
Plenum Press. New York. 1990

Any theory of carcinogenesis has to incorporate the fact that the initiation process leads to irreversible changes at the cellular level; thus, mechanistic theories must explain that the induced effect becomes permanent. The only plausible explanation at present is that the initiation step is a genetic change, probably a somatic mutation. Therefore, the direct interaction of the carcinogen or its activated metabolite/s with DNA as a promutational step has received much attention.

However, it should not be forgotten that reasonable models do exist, which discuss macromolecular cellular targets other than DNA (3). Thus, chemical carcinogenesis might also result from a primary alteration of a cellular RNA, since Temin (4) has shown that RNA's can be transcribed intracellularly as DNA and the resulting DNA can be intergrated into the host genome. Dickson and Robertson (5) discuss a mechanism of the potential regulatory role of specific RNA's in cellular development.

Alterations of specific proteins could also have a potential for permitting the development of cellular clones with altered genomes (3). A possible mechanism is the greatly increased error rates of certain DNA polymerases (6,7). Another possible molecular mechanism of carcinogenesis, not involving direct genomic change in the cells, is based on the repressor-derepressor systems that control the expression of genomes of viruses or bacteria. Monod and Jacob (8) and Pitot and Heidelberger (9) both have argued that loss or modification of protein repressors of parts of the genome that are expressed, could result in near stable states of cellular differentiation.

This historical "review" however should by no means distract from the fact that direct interaction of chemical carcinogens with DNA and the ensuing mutation is no doubt the most probable as well as the most plausible mechanism of initiation. The groups Lawley (10) and Brookes (11) first showed the reaction of direct alkylating carcinogens with nucleic acids and proteins, with covalent binding to these substrates. They also showed that DNA interaction correlated best with carcinogenicity data.

The indirectly acting carcinogen 4-dimethyl-aminobenzene (butter yellow) was first shown by Miller and Miller (12) to bind to proteins in 1947 and, later, by the same group, also to form adducts with nucleic acids (13,14).

Identification of formed adducts invariably showed that cellular nucleophiles were the sites to which binding occured, thus indicating that reactive intermediates of carcinogens were electrophilic chemical species.

The Miller and Miller Theory

These and similar results thus formed the basis of a more generally accepted theory of chemical carcinogenesis by Miller and Miller (1,2,3). It postulates that chemical carcinogens are either electrophiles (directly acting agents) or form electrophilic reactive metabolites in vivo from per se chemically nonreactive carcinogens (so-called procarcinogens). The reactive electrophilic species then interacts with cellular nucleophiles, especially in DNA. DNA adduct formation is then considered as a promutagenic step which after cell replication, may lead to a mutation and therefore to a "fixation" of the change. DNA interacting carcinogens are therefore also called geno-toxic carcinogens, they are usually mutagenic in appropriate test systems. To summarize: Genotoxic carcinogens bind covalently to DNA and tumor initiation is therefore the consequence of mutation(s) resulting from such interactions. This theory, based on new scientific results of carcinogenesis research, also lead to a revival of the somatic mutation hypothesis, first suggested by Boveri (15) in 1914 and later emphasized again by K.H.Bauer (16).

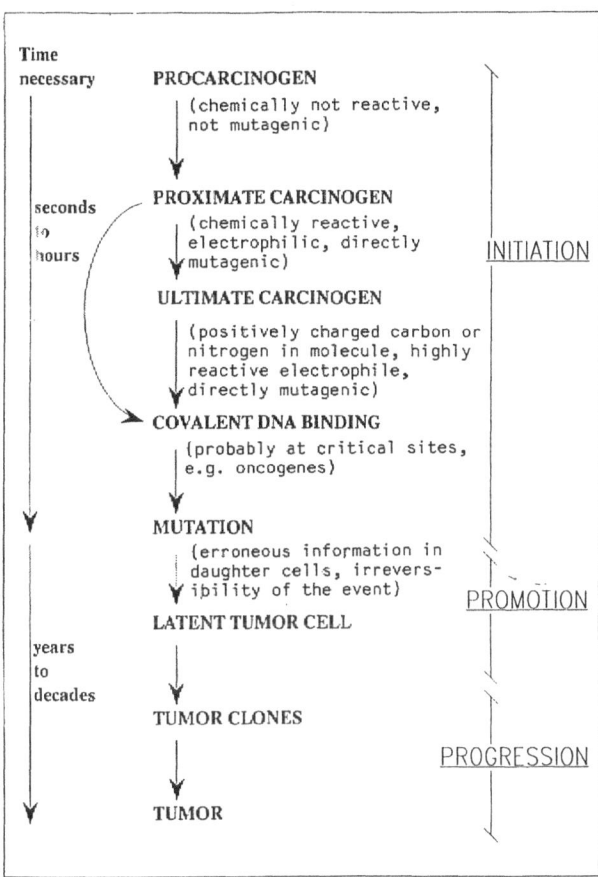

Figure 1. Scheme of mechanism of carcinogenesis according to Miller and Miller (1,2,3)

A generalized scheme of tumor development by genotoxic carcinogens is given in Figure 1. Figures 2-4 then give examples of the different activation steps for important genotoxic carcinogens.

In a critical discussion of the theory, of course, it must be strongly emphasized that it explains plausibly many of the known facts in carcinogenesis, but that it is an idealized scheme that does not explain all facts. It effectively explains the initiation mechanism of genotoxic carcinogens, but it certainly cannot easily explain carcinogenesis with proven human carcinogens, such as inorganic carcinogens (arsenic, nickel, chronium) and fibrous inorganics such as asbestos. Hormonal agents such as 17-β-estradiol or diethylstilboestrol are nonmutagenic, but carcinogenic in animal bioassays. Similarily many other chemical compounds are carcinogenic in certain animal bioassay systems, but are nonmutagenic and probably do not interact with DNA directly. It is now clearly evident that the title "Carcinogens are Mutagens" in the paper by B.Ames (17) is wishful thinking and in this generalized form wrong. I will therefore discuss now some carcinogens that formally do not fit into Miller's scheme.

Inorganic Carcinogens

Arsenicals, some nickel and chromium compounds are recognized human carcinogens and have also induced cancer in animal bioasssays. The usual short-term tests for assaying mutagenicity usually give negative results

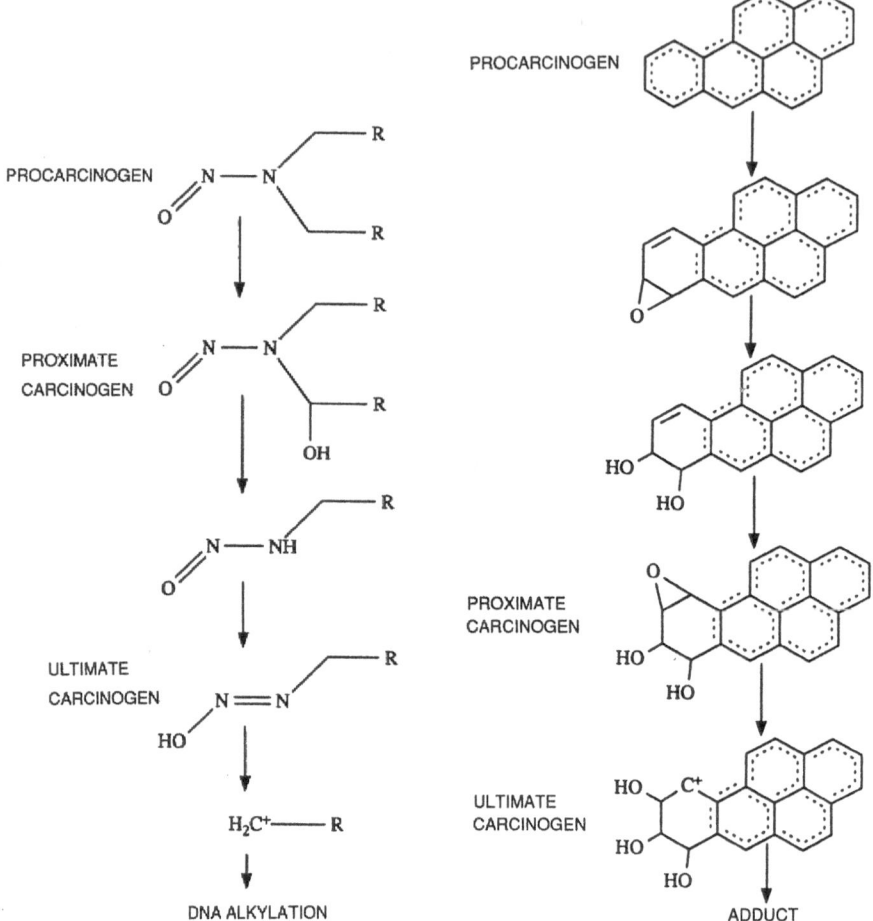

Figure 2. Activation pathway of
N-nitrosamines

Figure 3. Activation pathway of
benzo(a)pyrene

with inorganic compounds; however mammalian cell systems are better suited to test for DNA and chromosomal damages: Carcinogenic metal compounds sometimes but not always induce chromosomal aberrations in different test systems such as human lymphocytes or in Chinese hamster embryo cells where they also induce transformations. The topic has been reviewed by Gilman and Swierenza (18).

Asbestos and similar fibers, also negative in simple mutation test systems, induce chromosomal damage (19) and also transform Syrian hamster embryo cells (20,21).

Thus, for inorganic carcinogens there is some, but often not sufficient evidence for genotoxicity in a broad sense but they probably do not form DNA-adducts; their DNA interaction, mechanistically not understood, is of a more complex nature.

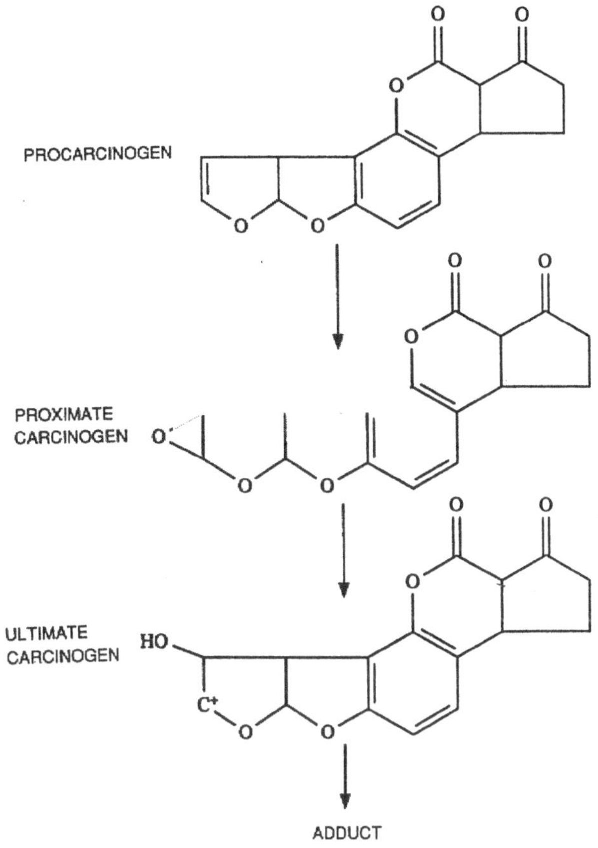

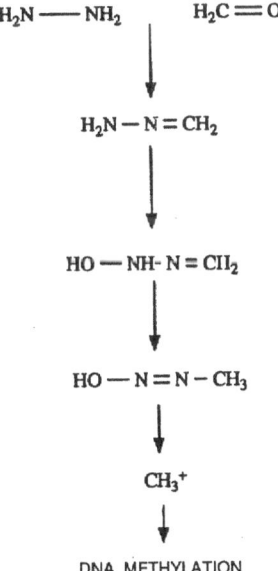

PROCARCINOGEN

PROXIMATE
CARCINOGEN

ULTIMATE
CARCINOGEN

ADDUCT

Figure 4.   Activation of aflatoxin B1

$H_2N \longrightarrow NH_2$ $H_2C = O$

$H_2N - N = CH_2$

$HO - NH - N = CH_2$

$HO - N = N - CH_3$

$CH_3{}^+$

DNA METHYLATION

Figure 5.   Proposed activation of hydrazine

Hydrazine

There is sufficient evidence for carcinogenicity of hydrazine in mice and rats (22) and it induces a variety of genetic changes in various test systems (22). Thus, it could be classified as a genotoxic agent. Shank and his group, however have demonstrated that hydrazine does not directly interact with DNA. Instead they clearly showed that methylation of DNA does occur (23) and N-7 as well as O-6-methylguanine have been identified as DNA adducts. Since hydrazine itself cannot methylate, a methylating intermediate must have been formed in vivo. Recent data support the formation of formaldehyde hydrazone as a condensation product of hydrazine and endogenous formaldehyde, which then is rapidly transformed, probably by catalase or catalase-like enzymes, to a methylating agent as an endogenously formed proximal carcinogen (24). The proposed mechanism is shown in Fig 5. It is interesting that the corresponding hydrazine from acetaldehyde does not lead to ethylation products in similiar test systems.

Hydrazine therefore is clearly a genotoxic carcinogen, but with an "indirect" mechanism of action.

Nonogentoxic (epigenetic) Carcinogens

By using the names of "nongenotoxic" or "epigenetic" for certain carcinogens producing tumors in some animal bioassay systems, we only express our ability to say anything about the mechanism of action of carcinogens; so we just define them by saying they act differently to gentotoxic carcinogens.

Nongenotoxic carcinogens are a very heterogenous group of chemicals (25) with obviously no common denominator. However, they seem to share at least the following characteristics:

1) By definition almost, they do not react directly or via metabolites with DNA and do not form stable, covalently bound DNA adducts.

2) They are negative in the majority of short-term tests based on mutagenic and/or clastogenic effects.

3) Tumors in animal carcinogenicity tests are often seen in only one organ of one species, sometimes only in one sex of this species (eg induction of mouse liver tumors).

4) Induced neoplasms very often, but not always, are benign; often a "spontaneous" tumor incidence is increased by nongenotoxic carcinogens.

5) Life-time application of high to very high doses, often in an acutely toxic range, are necessary to induce an often small incidence of tumors.

6) Dose-response studies usually reveal non-linear curves; the carcinogenic effect usually diminishes dramatically, when the exposure is reduced below the toxic level.

Table 1 compares such characteristic features between genotoxic and non-genotoxic carcinogens. Characteristic examples are given in Table 2.

Our present knowledge of the field has recently been summarized and reviewed at a Banbury meeting (25); for more detail, this review is recommended. I will here only pick out one example of a nongenotoxic carcinogen and discuss it in some detail.

Table 1. Feature of genotoxic and nongenotoxic carcinogens, adapted from (26)

Effects	Genotoxic carcinogens	Nongenotoxic carcinogens
DNA adduct formation	Form specific adducts	Do not form adducts
Short term-test based on Mutagenicity or clastogenicity	Majority of tests positive	Majority of tests negative
Carcinogenicity in experimental animals	Active in many different species	Often only active in one sensitive species, not in others
	Single dose exposure often active	Life-long high-dose exposure necessary
	Additive of synergistic effects proven or probable	Additivity uncertain
Dose-response relationship	Linear Low, non-toxic doses are effective	Non-linear Only high doses in or near the toxic range are carcinogenic

31

Table 2. Some epigenetic carcinogens, adapted from (26)

Compounds	Potential mechanism of action
Polyhalogenated compounds (DDT, TCDD)	Tumor promoter
Saccharin	Tumor promoter
Amitrole, estrogens	Hormone modifier
(Di-(2-ethylhexl)phthalate, clofibrate	Peroxisome proliferation
Nitrilotriacetic acid	cytotoxic

Di-(2-ethyhexyl)phthalate (DEHP)

This compound is widely used as plasticiser in many consumer plastic products and medical devices and also as lubricant. It is therefore considered as a widespread environmental synthetic chemical. In regard to carcinogenicity (27), no human data in form of case reports or epidemiological studies are available. Three early feeding studies in rats, with 4000-20000 ppm in the food, were negative (18,29,30) though not fully convincing because of low animal numbers in the experiment. An extensive NTP study (31) using 3000-12000 ppm in the food showed a very low incidence for hepatocellular carcinoma in the Fischer 349 rat with incidences of maximally 16%, while the mouse (B6C3F$_1$) was slightly more sensitive and had incidences of hepatocellular tumors of maximally 36% at 6000 ppm. Recent data from our Institute (32) showed no carcinogenicity in the Syrian Golden hamster after chronic i.p. administration (3 g/kg/week) or inhalation (continous exposure at 15µg/m^3).

A very large body of data on short-term tests in vitro and in vivo quite consistently show negative results in regard to mutagenicity/clastogenicity (27,32,33), even though one study claims induction of dominant lethal mutations in mice after systemic, but not after oral administration (39). There is no evidence of DNA interaction (35).

The most prominent biochemical effect of DEHP is a significant peroxisomal proliferation and perioxsome proliferation-associated oxidative stress has been correlated with the induction of liver tumors (36). A similarly acting compound, with similiar carcinogenicity profiles, is the hypolipidemic drug clofibrate. In favor of this hypothesis is the negative carcinogenicity experiment in hamster (32), where DEHP is only a weak peroxisome inducer (36). However Butterworth (33) when comparing exposure, liver tumors and peroxisome proliferation with DEHP and the corresponding Di-(2-ethylhexyl)-adipate, a perxisome-proliferator, but not carcinogenic in F344 rats, comes to the conclusion that there is no direct link between peroxisome proliferation and tumor induction.

Ward et al (38) have shown that DEHP has no initiating activity either; however, it demonstrated a weak second stage promoter activity. In the initiation/promotion model of liver carcinogenesis DEHP after diethylnitrosome initiation showed promoting activity in the mouse, but not in the rat.

In my summarizing evaluation of the carcinogenic activity of DEHP, it appears that no convincing evidence in regard to mechanisms of action are available. It is clear that it is not genotoxic, but neither its promoting

	MUTAGENICITY	DNA-BINDING	CARCINOGEN MOUSE SKIN
DMBA 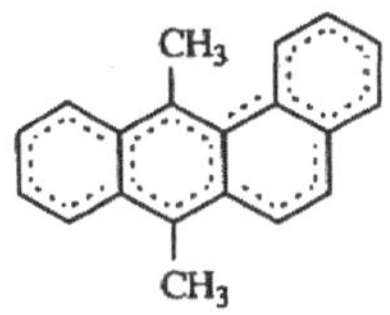	+	+	++
DMBA-5,6-IMINE 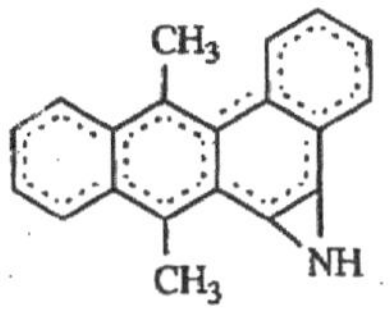	+	++	
9,10-DIME-ANTHRACEN 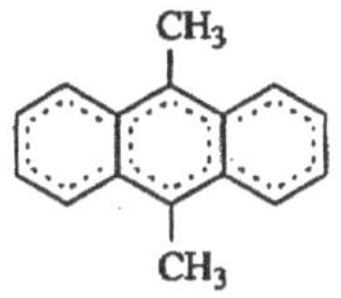	+	+	
ANTHRACEN--9-YL-METHYL-NITROSOUREA 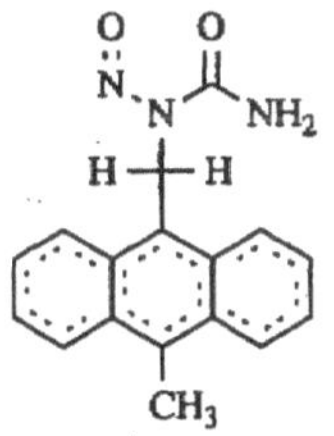	+	+++	

Figure 6. Mutagenicity, DNA-binding and carcinogenicity to mouse skin (initiation) of DMBA and some structural analogs.

activities nor its capacity to stimulate peroxisome proliferation at present seem enough to explain its biological activity for tumor induction.

Genotoxic Noncarcinogens

Recently groups at the German Cancer Research Center (Institute of Biochemistry: H, Friesel, E Hecker; Institute of Toxicology, University of Mainz (H.R. Glatt, F. Oesch) looked into correlations between DNA binding, mutagenicity and initiating activity in the mouse-skin two-stage carcinogenicity test system and reported (personal communication of unpublished data) highly interesting as well as unexpected data about DMBA as standard initiator and several structural analogs: The qualitative, preliminary data are summarized in Fig. 6.

Figure 6

The results show clearly that strong mutagenicity and strong DNA binding of structural analogs of the model carcinogens DMBA are not carcinogenic in the mouse skin model system. At present there is no explanation for this unexpected experimental result. Further validating additional tests will have to be performed to confirm such results. It would be especially desirable to have an additional and different test system in experimental animals to confirm lacking carcinogenicity. If repeatable, however, such results would then need additional thought in regard to the generalized and qualifiable theory to mechanistically explain chemical carcinogenesis. Any scientific theory is valid only as long as it can explain the majority of phenomena. If not, new hypotheses are required. We might be in such a situation in regard to the mechanisms of chemical carcinognesis.

REFERENCES

1   Miller, J.A. and Miller, E.C. (1966), Pharmacol. Rev. 18, 805-838.
2   Miller, J.A., (1970) Cancer Res. 30, 559-576.
3   Miller, J.A. and Miller, E.C. (1977) in Origins of Human Cancer (Hiatt
    et al. eds.) pp 605-627. Cold Spring Harbor Laboratory.
4   Temin, H.M. and Mizutani, S. (1970) Nature 226, 1211-1213.
5   Dickson, E. and Robertson, H.D. (1976) Cancer Res. 36. 3387-3393.
6   Loeb, L.A., Springgate, C.F. and Battula, N. (1974) Cancer Res. 34, 2311-
    2321.
7   Springgate, G.F. and Loeb, L.A. (1973) Proc. Natl. Acad. Sci. 70, 245-
    249.
8   Monod, I, and Jacob, F. (1962) Cold Spring Harbor Symp. Quant. Biol. 26,
    389-405.
9   Pitot, H.C. and Heidelberger, C. (1963) Cancer Res. 23, 1694-1706.
10 Lawley, P.D. and Wallick, A.M. (1957) Chem. Ind. 633-635.
11 Brookes, P.and Lawley, P.P. (1960) Biochem. J. 77, 478-481.
12 Miller, E.C. and Miller, J.A. (1947) Cancer Res. 7, 468-480.
13 Lin, J.K., Miller, J.A. and Miller, E.D. (1975) Cancer Res. 35, 844-850.
14 Lin, J.K., Schmall, B., Scharpe, I.D., Miaura, I., Miller, J.A. and
    Miller, E.c. (1975) Cancer Res. 35, 832-843.
15 Boveri, T.H. (1919) Zur Frage der Entstehung maligner Tumoren. S.
    Fischer-Verlag, Jena.
16 Bauer, K.H. (1949) Das Krebsproblem. Springer-Verlag Berlin Heidelberg.
17 Ames, B.N., Dursten, W.E., Yamasaki, E. and Lee, F.W. (1973) Proc. Natl.
    Acad. Sci. (USA) 70, 2281-2285.
18 Gilman, J.P.W. and Swierenga, S.H.H. (1984) in: Chemical Carcinogens, J.
    Ed., Am. Chem. Soc. Monograph No 182, p. 577-630.
19 Brown, R.C., Gromley, I.P., Chamberlain, M. and Davis, R. (1980) in: The
    in vitro effects of mineral dusts. Acadmemic Press, London.
20 DiPaolo, J.A., McMarino, S.J. and Doniger, J. (1983) Pharmacology 27, 65-
    73.
21 Barrett, J.C., Hesterberg, T.W., Oshimura, M. and Tjutsui, T. (1985) in:
    Carcinogenesis, Vol. ((J.C. Barrett and R.W. Tennant, Ed.) Raven Press,
    New York, p. 123-137.
22 IARC Monographs on the Evaluation of Carcinogenic Risks to Humans, Suppl.
    7 (1987) Lyon, France, p. 223-224.
23 Barrows, L.R. and Shank, R.C. (1978) Toxicol. Appl. Pharmacol. 60, 334-
    345.
24 Lambert, C.E. and Shank, R.C. (1988) Carcinogenesis 9, 65-70.
25 Banbury Report R. 25 (1987) Nongenotoxic mechanisms in carcinogenesis.
    (B.E. Butterworth and T.S. Slaga, eds.) Cold Spring Harbor Laboratory.
26 Williams, G.M. (1987), in Ref. 25, p. 339-350.
27 IARC Monographs on the evaluation of carcinogenic risk to humans, Vol.29
    (1982), p. 269-299.
28 Carpenter, C.P., Weil, C.S. and Smyth, H.F. (1953) Arch. Ind. Hyg. Vec.
    Med. 8, 219-226.
29 Harris, R.S., Hodge, H.C., Maynard, E.A. and Blanchet H.I. (1956) Arch.
    Ind. Health 13, 259-264.
30 Ganning, A.E., Brunk U. and Dallner, G. (1984) Hepatology 4, 541.
31 National Toxicology Program (1983). NTP Tech. Rep. Series No 217, NIH,
    Bethesda, Maryland.
32 Schmezer, P., Pool, B.L., Klein, R.G., Komitowski, D., and Schmahl, D.
    (1988) Carcinogenesis 9, 37-43.
33 Butterworth, B.E. (1987), in ref. 25, p.257-275.
34 Dillingham, E.O., and Autian, J. (1973) Environ. Health Persp. 3, 87-89.
35 Lutz, W.K. 91986) Environ Health Persp. 65, 267-269.

36 Reddy, J.K., Reddy, M.K., Usman, M.I., Lalwani, N.D., and Rao, M.S.
    (1986) Envir. Health Persp. 65, 317-327.
37 Lake, B.A., Gray, T.J.B., Foster, J.R., Stubberfield, C.R. and Gangolli,
    S.D. 91984) Toxicol. appl. Pharmacol. 72, 46-60.
38 Ward, I.M., Diwan, B.A., Oshima, M., Hu, H., Schuller, H.m., and Rice,
    J.M. (1986) Envir. Health Persp. 65, 279-292.

CHEMICAL CARCINOGENS PRESENT IN NIGERIAN (AFRICAN) FOODS AND DRINKS

Helen O. Kwanashie

Department of Pharmacology and Clinical Pharmacy, Ahmuda Bello
University, P. M. B. 1044, Zaria – Nigeria

ABSTRACT

Dietary association with chemical carcinogens is a well recognized
occurence the world over. This paper reviews the presence of some compounds
which have been detected and measured in foods and drinks that are indigenous
to Nigeria. Diverse staple food items are known to be heavily contaminated
with mycotoxins, particularly those producing the aflatoxin series. Polycyclic aromatic hydrocarbons such as benzpyrenes and anthracenes are present
in smoked fish and meat. Several locally brewed alcoholic beverages have
also been shown to contain nitrosatable amines. The widespread use of
pesticides for agricultural and disease control purposes ensures a generous
level of residues in foodstuffs. The presence of these and other chemical
carcinogens in Nigerian foods and drinks may be associated with environmental
factors, processing and handling practices and may account for the prevalence
of certain types of cancers in the population.

INTRODUCTION

It was Roe who stated that "it is almost impossible to conceive of an
environment that is totally free from carcinogens," (1), and Nigeria is no
exception. The environment (as opposed to genetic factors) is generally
believed to contribute 70–80% of all cancers. Part of this environmental
contribution is diet-related. Dietary association with chemical carcinogens
for example, has been demonstrated in many communities all over the world
(2–7). In Nigeria, a number of such compounds have also been shown to be
present in foods and drinks that are indigenous to the people. The most
important of these are mycotoxins, polycyclic aromatic hydrocarbons,
nitrosamines and chlorinated pesticides all of which are potent chemical
procarcinogens.

Mycotoxins

These are by far the most widely studied dietary contaminants in Nigeria,
presumably because of their preponderance. The relatively hot (20–35^{0}C) and
highly humid (>80%) atmospheric conditions prevalent in most parts of
tropical Nigeria, promote and guarantee the growth of a plethora of
mycotoxins (8). One species of these moulds, Aspergillus, is most notorious
for the production of toxic compounds. The well known mutagens and

carcinogens, the aflatoxins, are produced by <u>Aspergillus flavus</u>. The aflatoxins are ubiquitous and plentiful in the air and soil. They first attained limelight following the outbreak of the turkey "X" disease in poultry farms in England in 1960, due to their contamination of groundnuts used in the manufacture of animal feed. The groundnuts had been imported from Brazil and Africa.

In Nigeria as elsewhere, aflatoxins have been isolated from groundnuts and other nuts. However, mycotoxin contamination of Nigerian foodstuffs is not limited to nuts. Indeed a wide variety of staples have repeatedly been shown to support the growth of <u>A. flavus</u> and produce very high levels of aflatoxins of mostly the B and G series, (9-16). The implicated foodstuffs include dairy products (e.g. meat, poultry), other animal protein sources (e.g. fish), legumes (e.g. soya bean, black-eyed bean), cereals (e.g. millet, rice, corn, sorghum), oilseeds (e.g. ground nuts), root tubers (e.g. yam, cassave), vegetable oils (e.g. groundnut oil, cotton seed oil), fruits (e.g. cocoa, paw-paw, banana), bakery products (e.g. bread), etc. Beverages such as "burukutu" (millet beer) have also been shown to grow <u>A. flavus</u> and contain high levels of aflatoxins (17,18). Routine analysis by the various zonal laboratories of the Nigerian Stored Products Research Institute have confirmed again and again that many Nigerian foods and feeding stuffs are heavily contaminated with mycotoxins especially of the aflatoxin series, even shortly after harvesting, (19). Up to 90% of diverse food samples from local markets contained aflatoxins (20).

The levels of mycotoxin contamination are very often above permissible limits. For example it has been shown that the concentrations of aflatoxins in Nigerian sorghum varied between 15.8 and 211.2 μg/kg which is mostly above the maximum level of 30 μg/kg accepted by the United Nations Protein Advisory Group (1969) in food supplements for under-nourished children in developing countries, (14). Similarly, aflatoxin levels of as high as 5.0 mg/kg have been detected in local meat which is several times higher than the 0.12-0.51 mg/kg considered safe by the World Health Organization, (15).

Dietary contamination with aflatoxins affect both raw and processed food materials including prepared meals. Most of the food items that have been so investigated were in fact purchased from open markets. None of several on-the-table food samples analysed showed a zero value for aflatoxins (20). The average Nigerian no doubt ingests a reasonable dose of aflatoxin on a daily basis. This is borne out by the fact that normal "healthy" Nigerians, adult and children, regularly excrete aflatoxins in their urine (20,21). Excretion of aflatoxin has also been demonstrated in breast milk (personal communication). Aflatoxins have also been measured in sera of "normal" subjects, (20). It is pertinent to state that aflatoxins are generally not regarded as normal constituents of these body fluids.

Although , <u>A. flavus</u> is the most predominant contaminant of Nigerian diets, other toxin-producing moulds such as <u>A. niger</u>, <u>A. parasiticus</u> and <u>A. penicillin</u> are also present. Similarly, besides aflatoxins, ochratoxin, aspertoxin, sterimatocystin and O-methyl sterimatocystin are all hepato-toxic mycotoxins produced by fungal strains present in Nigerian foods.

Polycyclic Aromatic Hydrocarbons

These chemical carcinogens are known to be an important constituents of tar and tar products as well as smoke, such as liquid smoke, cigarette smoke and automobile exhaust fumes, (5,22). Elsewhere in the world, polycyclic aromatic hydrocarbons have been found as contaminants of charcoal broiled foods, (4,5). Charcoal broiling and food smoking are common processing and preservative practices in Nigeria and therefore smoked foods play a major role in the nutrition and dietary habits of Nigerians (23). Compared to

mycotoxins, studies regarding the presence of polycyclic aromatic hydrocarbons have demonstrated the presence of benzpyrene and anthracenes in
Nigerian smoked fish and meat (24,25). According to Lijinsky and Shubik
(1964) 8 µg/kg, which is the level of benzpyrene present in charcoal-broiled
steak, is equivalent to the levels that would be taken in from smoking, it is
definitely less than consumption of smoked foods, extrapolation from the
above would mean that the latter is the more significant from the point of
view of chemical carcinogens.

Nitrosamines

N-nitroso compounds such as nitrosamines (procarcinogens) and nitro-
samides (direct carcinogens) are well known as contaminants of foods and as
some of the most potent chemical carcinogens. Several fermented beverages in
Nigeria have been shown to contain such compounds, (26-28). The implicated
beverages are mostly local alcoholic drinks such as palm wine (fermented palm
sap), "ogogoro" (distilled gin from stale palm wine), "burukutu" (guinea corn
beer) and "pito" (millet beer). These beverages are frequently consumed in
large quantities by Nigerians for relaxation and for such festivities as
naming ceremonies, marriages and funerals. Non-alcoholic beverages such as
"nono" (a fermented milk product) are also contaminated (27). Both dimethy-
nitrosamine and diethyl nitrosamine are most often implicated and levels of
0.6-22 µg nitrosamine/litre have been reported (28).

Besides their occurence in drinks, nitrosamine precursors abound in
nature particularly in sea foods, air, soil, water and sewage. Acidic
conditions of the stomach is believed to provide a suitable environment for
dietary nitrite or nitrate to undergo N-nitrosation in vivo with nitrogen
containing compounds in food-stuffs. Sea foods such as shrimps, crabs and
fish are highly implicated and these form a substantial part of animal
protein intake of many Nigerians (23).

Pesticide Residues

Due to the problems of undernutrition in Nigeria, pesticides in diverse
forms such as herbicides, fungicides and insecticides are employed at various
levels of agriculture from planting to post-harvest storage in a bid to boost
food production. Insecticides also play an additional role in disease
control especially with regard to maleria (caused by the anopheles mosquito)
and trypanosomiasis (caused by the tse-tse fly). Thus pesticides are widely
used in Nigeria in establishments, farm settlements, private farms and homes.
Unfortunately, the available pesticides are not always of the newest and
safest categories. For example "Gamallin 20" and DDT (dichlorodiphenyl-
trichloro ethane), which are no longer in use in many countries because of
their toxicity are prevalent in our markets and readily available for use.
These class of pesticides i.e. the organochlorines unlike the safer
carbamates are notorious for their mutagenicity and chemical carcinogenicity.
The principal pesticides in use in Nigeria include lindane, aldrin and
dieldrin. Research results show that when these are applied as recommended
(e.g. 6.5-10 ppm lindane dust on sorghum and millet), toxic hazards to
consumers are well within acceptable limits (30). However, the possibilities
of non-uniformity of mixing, unwitting repetition of treatment and overdosing
by unsupervised, ignorant and illiterate peasant farmers arer very real. It
is not surprising therefore that pesticide residues have been found in our
treated foodstuffs and in the environment, (31-34).

Miscellaneous

Besides the four major contaminants of Nigerian foods and drinks discussed above, other chemical carcinogens are also associated with the
peoples' diets. The development of the food industry in recent years has led

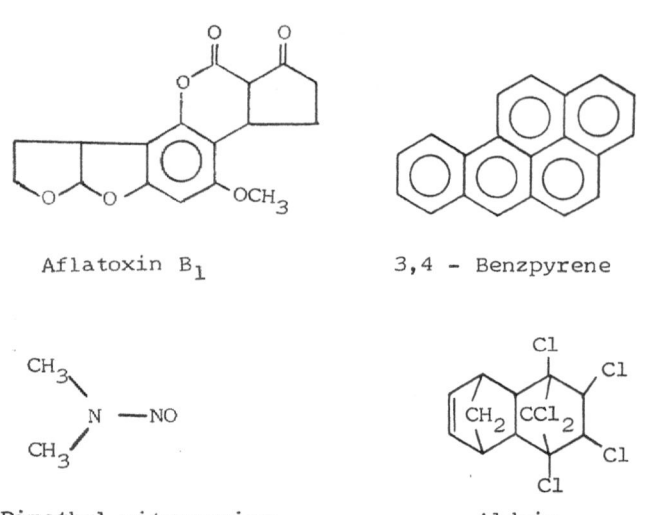

Aflatoxin B$_1$

3,4 - Benzpyrene

Dimethyl nitrosamine

Aldrin

Figure 1.   Some chemical carcinogenic contaminants of Nigerian foods and
drinks.

to an increase in consumption of additives which serves as colouring,
flavouring and preservative agents. Cyclamate and saccharin, banned in many
countries for their possible carcinogenic effects are widely used in Nigeria
to sweeten natural and artifical fruit drinks and squashes to the extent of
being openly listed in product labels. Many varieties of soup condiments are
available in the markets as cubes or salts of monsodium glutamate. A number
of dubious colouring agents of undeclared identity, particularly orange and
yellow are widely employed to enhance the appeal of soft drinks. Some food
colouring such as "butter yellow" (dimethylaminoaxobenzene) are well known as
chemical carcinogens. The miscellaneous compounds definitely make some
contribution to the carcinogenic level of Nigerian foods and drinks.

Incidence and Prevalence of Cancer in Nigeria and Possible Association with Dietary Chemical Carcinogens

Contrary to popular belief, cancer is not a disease restricted to the
developed industrialized world. Evidence now abounds to show that this
scurge of mankind is as much a problem in the poorly, less developed and less
industrialized nations like Nigeria. The apparent relative obscurity of
these conditions is due to the overwhelming health problems arising from the
poverty, ignorance and malnutrition that prevails in the tropics. As a
matter of fact, certain malignant tumors occur more frequently in tropical
countries than in any other part of the world (35). Examples of such cases
are the lymphomas (especially Burkett's), liver cell carcinoma (hepatoma) and
squamous carcinoma of the cervix. Doll (1969) had reported a high incidence
of different types of cancer for Nigeria including primary liver cancer,
(36). In this latter condition, Nigeria is ranked 3rd as the country of
greatest incidence in a study covering various countries in Africa, America,
Asia, U.S.S.R and Europe. It is perhaps significant that Mozambique and
Durban which came 1st and 2nd are also African. By 1978, Nigeria had
continued to be one of the countries of greatest incidence of primary liver
cancer, (37). In two surveys which covered the period between 1960 and 1970,
a cascading increase in cancer incidence in southern Nigeria was shown,
(38,39). There is no reason to believe that the situation is any different
today.

Going by the data emanating from the cancer registries in the south (Ibadan) and north (Zaria) of Nigeria, liver cell carcinoma is the prevalent cancer type, accounting for 7-19% of all malignancies, (35). According to Edington (1978) the upper range is one of the highest prevalence rate for any single type of human cancer the world over, (37). In a study in which American negroes and Nigerian Africans were compared, intestinal polyps were found in under 20s Nigerians whereas these occured in over 50s American negroes (40). These and other literature, which for want of space cannot be cited here, shown that primary liver cancer is an important disease condition in Nigeria. Other malignancies of the gastrointestinal tract, particularly those of the stomach, colon and oesphagus rank next to liver cancer in that order (35).

The prevalence of gastrointestinal carcinoma among Nigerians as against other forms of cancer suggest a dietary association. Such an association is not farfetched considering that the known chemical carcinogenic contaminants of Nigerian foods and drinks have the liver as their primary target organ (of attack) followed by the intestine.

The mycotoxins (aflatoxin) are responsible for the high incidence of primary liver cancer and other hepatic diseases in Nigeria and other parts of Africa, is no longer speculative. It has been shown that Nigerians with liver diseases had a higher level of urinary aflatoxin that normal subjects or patients without liver disease, (21). Furthermore, high levels of afla-toxins have been shown in the sera of patients with primary liver carcinoma, (41-43). Studies carried out outside Nigeria have also associated dietary aflatoxin with liver cancers in both animals and man, (44-47). This association was demonstrated as a linear correlation between the incidence of liver tumor and dietary aflatoxin intake over a range of 0.07-1.00 ppm in the rat and 4-200 ng/kg/day in man. One estimate had put the level of aflatoxin contamination of a consummable Nigerian food item at 0.5 ppm, (12). A number of human epidemiological studies in Africa and elsewhere had associated dietary aflatoxins with liver cancer, (48,49). These facts together point to dietary aflatoxin contamination as a possible dominant factor in the aetiology of cancer in Nigeria. Although this review had restricted itself to carcinogenicity, mycotoxins have been implicated in other lesions such as hepatic necrosis and kwashiorkor, the latter being a prevalent protein-energy-malnutrition syndrome in Nigerian children (20,50).

Polycyclic aromatic hydrocarbons, especially when associated with oral ingestion are known to cause neoplastic growths along the gastrointestinal tract, including hepatomas, (51). Studies have also shown that the organo-chlorine pesticides induce liver and other tumors in animals and possibly man, (52). Liver cancer and cancers of the oesphagus and kidney are also associated with nitrosamines, (7). Thus the presence of mycotoxins, polycyclic aromatic hydrocarbons, nitrosamines and pesticide residues in Nigerian foods and drinks may account for the prevalence of liver and other gastrointestinal cancers in the population.

Reducing Carcinogenic Contamination Risks

Dietary contamination by the above mentioned chemical carcinogens are associated with environmental factors, processing and handling practices. Like other environmental factors, the risks can be reduced by suitable measures some of which are mentioned here. Rapid drying of grains and other foodstuffs to a water content that will not support mycotoxins will definitely reduce the risk of aflatoxin contamination, (53). γ-Irradiation has also proved beneficial in inhibiting mycotoxin growth in Nigerian foods, (54,55). Good store hygiene and practices will not only reduce the high aflatoxin levels in foods stored for prolonged periods, (56), but also the need for pesticides. Possibilities of replacing or reducing smoking as a preservative method for fish and meat e.g. by massive refrigeration in cold houses, should be explored as this would reduce polycyclic aromatic hydrocarbon contamination, (24). Since the levels of nitrosamines are reduced upon fermentation of the palm sap, consumption after this stage may significantly reduce intake of these carcinogens, (29). Improved agricultural extension services and educational campaigns will no doubt help local farmers to apply pesticides correctly and thereby reduce incidence of residues in foods.

Such concerted efforts as these, will go a long way in reducing the levels of chemical carcinogens present in Nigerian food and drinks.

REFERENCES

1   Roe, H. (1972) Marie Curie Foundation Symposium on The Prevention of Cancer
2   Sargeant, K. et al. (1961) Nature, 192, 1096-1097
3   Forgacs, J. and Carll, W. T. (1962) Adv. Vet. Sci., 7, 273-403
4   Lijinsky, W. and Shubik, P. (1964) Science, 145, 53-55
5   Lijinsky, W. and Shubik, P. (1965) Toxicol. Appl. Pharmacol., 7, 337
6   Duggan, R. E. and Corneluissen, P. E. (1972) Pest. Monit. J., 5, 331
7   Preussmann, R. (1980) in Toxicology in the Tropics (Smith, R. L. and Bababunmi, E. A. eds.) pp. 108-123. Taylor & Francis Ltd., London
8   Hendrickse, R. G. (1985) Proc. Royal Coll. Phys, Edinburgh, 15, 138-156
9   Bassir, O. (1969) W. A. Journal Biol. Appl. Chem., 12, 3-6
10 Bassir, O. and Adekunle, A. (1972) Mycopath. Mycol. Applicata, 46, 241-6
11 Bababunmi, E. A. et al. (1976) Wld. Rev. Nut. Diet., 28, 188-209
12 Bababunmi, E. A. (1980) in Toxicology in the Tropics (Smith, R. L. and Bababunmi, E. A. eds.) pp. 93-107. Taylor & Francis Ltd. London
13 Emerole, G. O and Uwaifo, A. O. (1980) in Toxicology in the Tropics (Smith, R. L. and Bababunmi, E. A. eds.) Taylor & Francis Ltd., London
14 Uriah, N. and Ogbadu, G. (1980) Microbios Letters 14, 29-31
15 Abalaka, J. A. and Eronini, T. (1987) N. J. Sci. Res., 1, 17-19
16 MacDonald, D. and Harkness, C. (1967) Trop. Sci. 9, 148-161
17 Okoye, Z. S. C. and Ekpenyong, K. I. (1984) Trans. Royal Soc. Trop. Med. Hyg., 78, 417-418
18 Obasi, O. E. et al. (1987) Trans. Royal Soc. Trop. Med. Hyg., 81, 879
19 Nigerian Stored Products REsearch Institute, Technical Reports (1970-80)
20 Ibiam, G. E. N. (1988) The relationship of aflatoxin and kwashiokor: a study of underfives in Nigeria, M.Sc thesis, Ahmadu Belo University, Zaria, Nigeria
21 Bababunmi, E. A. (1976) in Detection and Prevention of Cancer (Nieburgs, H. E. ed.) Dekker, New York
22 Davis, B. R. et al. (1975) Br. J. Cancer, 31, 443-452
23 Osuji, F. N. C. (1976) Nig. Field, 41, 3-18
24 Akpobi, H. O. (1979) Detection of polycyclic aromatic hydrocarbons in Nigerian smoked fish, B.Sc Project, University of Ibadan, Nigeria
25 Emerole, G. O. (1980) Bull, Environ. Contam. Toxicol. 24, 641-646
26 Maduagwu, E. N. and Bassir, O. (1979a) J. Environ, Pathol. Toxicol., 2, 1183-1194
27 Maduagwu, E. N. and Bassir, O. (1979b) Toxicol. Lett., 4, 169-173
28 Maduagwu, E. N. et al. (1979) Trop, Geogr. Med., 31, 283-290
29 Maduagwu, E. N. and Bassir, O. (1979c) J. Agric. Food. Chem., 27, 60-63
30 Giles, P. H. (1964) Trop. Sci., 6, 113-121
31 Sands, W. A. (1959) Samaru Newsl., 5, 8-9
32 Taylor, T. A. (1968) J. W. Afric. Sci. Assoc., 13, 139-145
33 Halliday, D. and Kazaure, I. (1968) Report of the Nigerian Stored Products Research Institute, 45-52
34 Aduku, A. O. (1977) Samaru Agric. Newsl., 19, 97-102
35 Osunkoya, B. O. (1980) in Toxicology in the Tropics (Smith, R. L. and

Bababunmi, E. A. eds.) pp. 4-10, Francis & Taylor Ltd

36 Doll, R. (1969) Br. J. Cancer, 23, 1

37 Edington, G. M. (1978) Nig. Med. J. 8, 281-289

38 Edington, G. M. and Maclean, C. M. U (1965) Br. J. Cancer, 19, 471-481

39 Odebiyi, A. I. (1972) Demographic and socio-ecomonic aspects of cancer
    in the city of Ibadan. M.Sc thesis, University of Ibadan, Ibadan -Nigeria

40 Williams, A. O. et al. (1975) Br. J. Cancer, 31, 485-491

41 Onyemelukwe, G. et al. (1980) Trop. Geogr. Med. 32, 237-240

42 Onyemelukwe, G. C. et al. (1982a) Toxicol. Lett. 10. 309-312

43 Onyemelukwe, G. C. et al. (1982b) Mycotoxins and cancer, 5th Annual
    Conference of the Nigerian Cancer Society, Zaria - Nigeria Dec. 9 - 10

44 Lancaster, C. M. et al. (1961) Nature, 192, 1095-1096

45 Newberne, P. M, (1965) in Mycotoxins in Foodstuffs (Wogan, G. N. ed.) pp.
    187-208, Cambridge, mass., MIT Press

46 Schank, R. C. et al. (1972) Food Cosmet. Toxicol, 10, 71-84

47 Schank, R. C. et al. (1977) J. Toxicol. Environ. Health, 2, 1229-1244

48 Peers, F. G. and Linsell, C. A. (1973) Br. J. Cancer 27, 473-484

49 Peers, F. G. et al. (1976) Int. J. Cancer, 17, 167-176

50 Henderickse, R. G. et al. (1982) Br. Med. J., 285, 843-846

51 Klein, M. (1963) Cancer Res. 23, 1701

52 Cabral, J. R. P. (1980) in Toxicology in the Tropics (Smith, R. L. and
    Bababunmi, E. A. eds.) pp, 162-183, Taylor & Francis Ltd. London

53 Christensen, C. M. and IKaufman, C. M. (1974) in Storage of Cereal Grains
    and their Products (Christensen, C. M. ed.) pp. 158-192. American Cereal
    Chemists, Inc. St. Paul, Minn

54 Ogbadu, G. (1979) Microbios Letters, 10, 139-142

55 Ogbadu, G. (1980) Microbios, 27, 19-26

56 Nwokolo, C. and Okonkwo, P. (1978) Trans. Roy, Soc. Trop. Med. Hyg. 72,
    329-332

CONVERSION INDUCED BY TRANSFORMING GROWTH FACTORS α AND β :   THE ROLE OF

"WOUND HORMONES" IN MULTISTAGE SKIN CARCINOGENESIS

Friedrich Marks and Gerhard Fürstenberger

German Cancer Research Center
Institute of Biochemistry
Im Neuenheimer Feld 280
D - 6900 Heidelberg, F.R.G.

ABSTRACT

The induction of papillomas (and carcinomas) in mouse skin can be experimentally subdivided into the stages initiation, conversion and promotion
While initiation is most probably related to gene mutation, conversion is
required to render initiated skin sensitive to the tumor promoting effect of
agents which induce hyperplastic transformation of skin and sustained epidermal hyperplasia. Traditionally, conversion and promotion are achieved by
phorbol ester treatment. Conversion can also be affected by mechanical skin
wounding. It is shown that a combined injection of two putative "wound
hormones", i.e. TGFα and TGFβ, converts initiated mouse skin to a promotable
state. These results indicate a close relationship between the wound response and skin tumor development.

INTRODUCTION

Shortly after the first successful experimental induction of neoplastic
growth by coal-tar painting of rabbit skin, an accelerating effect of skin
wounding on skin tumor development was observed (1). Later on, mechanical
injury was shown to exert a promoting effect in multistage skin carcinogenesis, i.e. repeated wounding could stimulate tumor development in animal
skin upon initiation with a chemical carcinogen in the same way as it could
repeated application of a tumor-promoting agent such as croton oil or the
croton oil-derived phorbol ester TPA (2-4). An initiating effect of wounding
was not observed (5). Recently, the putative relationship between mechanisms
of wound repair and neoplastic development (6) has gained new attention by
the finding that the great majority of proto-oncogenes code for proteins
which are involved in the cellular transduction of growth-stimulatory signals
such as the polypeptide growth factors (7). Moreover, there is a steadily
growing body of evidence that one major route to neoplastic growth involves
the acquisition of cells for an autocrine production of such growth factors
many of which may play a physiological role as "wound hormones" (7,8). Among
the tissue models presently employed in experimental cancer research, skin
probably offers the best possibilities to study such relationships between
carcinogenesis and the wound response in detail.

Biochemistry of Chemical Carcinogenesis
Edited by R. Colin Garner and Jan Hradec
Plenum Press, New York, 1990

Hyperplastic transformation

A more detailed investigation of the biological and biochemical events
induced by a tumor promoter such as TPA in skin led to the conclusion that
these agents could directly interact with endogeneous mechanisms which are
normally involved in the skin's response to external irritation and injury.
This response has been called "hyperplastic transformation" because it is
characterized by a reversible transformation of epidermis into a state of
growth and differentiation which in many aspects resembles that of the neo-
natal tissue (9). Actually, the skin of newborn mice does not respond in
this way to phorbol ester treatment or mechanical injury but develops the
ability of hyperplastic transformation within the first 2-3 weeks after birth
when the adult morphology is developed (10,11). Moreover, hyperplastic
transformation has been shown to be a strictly controlled and highly specific
response to injury rather than to be simply the result of epidermal hyperproliferation (9).

The cellular and molecular events by which hyperplastic transformation is
characterized include:

- the rapid development of epidermal hyperplasia (within 1-2 days after
  stimulation)
- an inflammatory reaction
- the activation in epidermal cells of certain genes with c-fos, c-myc and
  the ornithine decarboxylase gene being the most prominent examples (12,13)
- a desensitization of epidermal cells for hormonal signals such as catecholamines (14), prostaglandin E (15) and epidermal chalone (16)
- an inhibition of intracellular communication in epidermis (17)
- a complex, oscillating activation pattern of arachidonic acid metabolism
  leading to the release of eicosanoid mediators of cell proliferation and
  tissue inflammation (see the following article).

The tumor promoting phorbol esters such as TPA rank among the most potent
inducers of hyperplastic transformation. The only biochemical effect of
these agents known so far is the stimulation of protein kinase C (18). This
key enzyme of intracellular signal transduction is physiologically activated
by the second messenger diacylglycerol (DAG) released together with
inositoltrisphosphate (IP_3) upon receptor-mediated activation of a membrane-
bound phosphatidylinositol-phosphodiesterase (19). Phorbol esters and other
potent skin tumor promoters act as strong DAG agonists (20). These
observations indicate an important role of the IP_3/DAG-cascade of transmem-
brane signalling in hyperplastic transformation although the physiological
activator of the cascade in skin has not yet been identified.

It must be emphasised that by certain non-irritating mechanical or
chemical manipulations of mouse skin a strong hyperproliferative response can
be evoked in epidermis which does not result in a rapid development of a
hyperplastic state and skin inflammation (21). When epidermal cell proliferation is induced in such a way, the cellular and biochemical events seen
in the course of hyperplastic transformation (as summarized above) are not
observed. This result clearly characterized hyperplastic transformation as a
specific response of skin irritation and injury. To the authors' knowledge
such a fundamental difference in the responses to external manipulation has
not yet been shown for any other tissue except skin. This fact places the
skin model in the front-line of research on in vivo growth control.

Multistage skin carcinogenesis

The fact that both wounding and application of tumor-promoting phorbol
esters induce the same sequences of cellular events which results in hyperplastic transformation indicates a close relationship between the induction

of tumor development and the wound response in skin. Actually, a clear-cut correlation exists between tumor-promoting efficacy and the ability to induce hyperplastic transformations at least within the phorbol ester series (22). One gets the impression that hyperplastic transformation is required to activate the "dormant" tumor cells generated by initiation. This situation is quite similar to the induction by skin injury of psoriatic eruptions or of wart growth in papilloma virus infected skin. During the wound response hyperplastic transformation may result in an activation of epidermal stem cells. Since initiation is generally believed to occur in the epidermal stem cell compartment, stem cell activation has been postulated also to be required for tumor development (23-25).

The multistage approach of experimental skin carcinogenesis allows a more detailed investigation of the mechanisms involved in tumor development. Today 4 stages are distinguished: initiation (26) and malignant progression (27) are most probably due to genotoxic events leading to irreversible genetic mutations, conversion and promotion seem to occur along epigenetic routes which are normally involved in the defense reactions of skin against irritation and wounding. Experimentally, initiation can be induced by local treatment of skin with a chemical carcinogen such as dimethylbenz[a]anthracene in a low ("subthreshold") dose, while conversion and promotion are generally achieved by phorbol ester application (Fig.1). For this purpose phorbol esters and related compounds with different effects on tumor development have been introduced. Thus, TPA exhibits both convertogenic and tumor-promoting efficacy, where as its unsaturated derivatives Ti8 (28) and retinoylphorbol-acetate (RPA) are almost devoid of convertogenic efficacy (29). Nevertheless, both Ti8 and RPA are as powerful as TPA in inducing hyperplastic transformation in skin.

The key observation which led to the discovery of the conversion stage of carcinogenesis is that the induction of chronic epidermal hyperplasia (by chronic RPA or Ti8 treatment) in initiated mouse skin results in tumor development only if it is preceeded by a short-term treatment with TPA carried out within a certain time period prior or after initiation (28-30, Fig.1).

This result clearly shows that chronic induction of hyperplastic transformation is a necessary but not a sufficient condition of papilloma growth in initiated skin.

Tumor promotion is generally understood as the clonal expansion of tumor cells in the course of a chronic hyperplastic response (31). This is exactly the effect which is induced by chronic RPA or Ti8 treatment. The short term TPA treatment has been called "conversion" because it is required to convert initiated skin into a state of promotability (30). According to this definition TPA is a convertogenic tumor promoter, whereas RPA, Ti8 and also mezerein (32) are non-convertogenic tumor promoters. While promotion may be understood to occur in the course of repeated induction of hyperplastic transformation carried out at such intervals that a chronic hyperplastic state develops (33) the mechanism of conversion is still largely unknown (for a detailed discussion of this point see ref. 34). Promotability is defined as sensitivity of initiated epidermal cells to the tumor-promoting effect of RPA, Ti8 or related compounds. It appears as if without conversion initiated cells are unable to respond to the promoter by hyperplastic transformation and clonal expansion, whereas the great majority of (non-initiated) epidermal cells exhibit the hyperplastic response independently of a pre-treatment with a convertogenic agent.

Conversion by wounding and "wound hormones"

As discussed above tumor promotion may be the result of hyperplastic

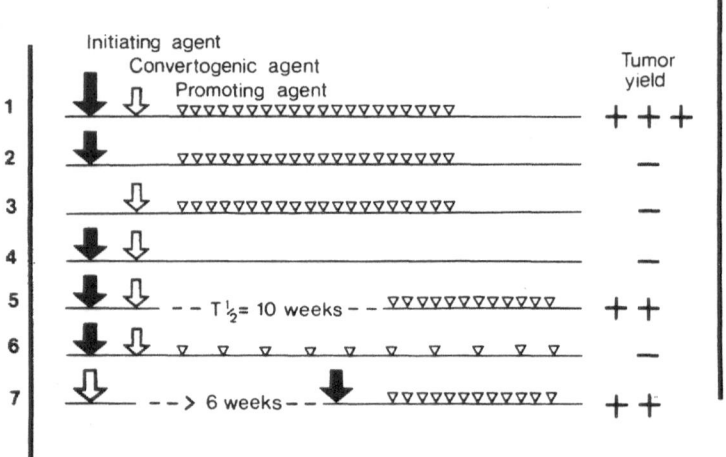

Figure 1. Scheme of the experimental phorbol of three-stage tumorigenesis (initiation-conversion-promotion) in mouse skin.
Papilloma development is induced by the combined topical treatment with an initiator (for instance DMBA, 100 nmol), a convertogenic agent (for instance 10 nmol TPA) and a promoter (for instance 10 nmol RPA once a week over a period of 20 weeks): line 1. No tumors develop when one of these treatments is omitted (line 2-4). While initiation is irreversible, the converted state is slowly reversible (half-life of 10 weeks in NMRI mice, see line 5 and ref.30). The promoting effect is lost when the time-interval between 2 subsequent promoter applications is extended from 1 to 2-3 weeks (line 6). Convertogenic treatment may also be carried out several weeks prior to initiation (line 7 and ref. 30). For application the substances are generally dissolved in 0.1 ml acetone.

transformation which is the general response of skin to external injury. conversion appears to be also closely related to the wound response. Such a conclusion is strongly supported by the fact that for the conversion of initiated skin into the state of promotability TPA treatment can be replaced by skin wounding (35). This result indicates that endogeneous factors such as inflammatory mediators and other "wound hormones" which are locally re- leased upon injury may be involved in the conversion stage of skin carcino- genesis.

Searching for such factors we initially came upon a platelet-derived peptide called Epstein-Barr-virus-inducing factor (EIF) because of its activity as an inducer of early antigen formation in latently Epstein-Barr- virus infected Raji cells (36). We considered EIF to be a candidate for a convertogenic agent since it exhibits synergistic effects with TPA in several in vitro systems including the stimulation of transformation of UV-initiated $C_3H10T1/2$ cells (37). Later on EIF was identified as β-type transforming growth factor (TGFβ, ref. 38 and G. Bauer, personal communication)).

EIF/TGF was tested for convertogenic efficacy by intracutaneous injec- tion into mouse skin initiated by local DMBA application 1 week before. The injection was then followed by tumor-promoting treatment, i.e. continuous application of the phorbol ester RPA once a week over a period of 20 weeks. This treatment was started 1 week after TGFβ-injection. Using this experi- mental protocol no convertogenic effect of TGFβ-injection could be observed (38,40). The reason may be that TGF is a potent inhibitor of epidermal

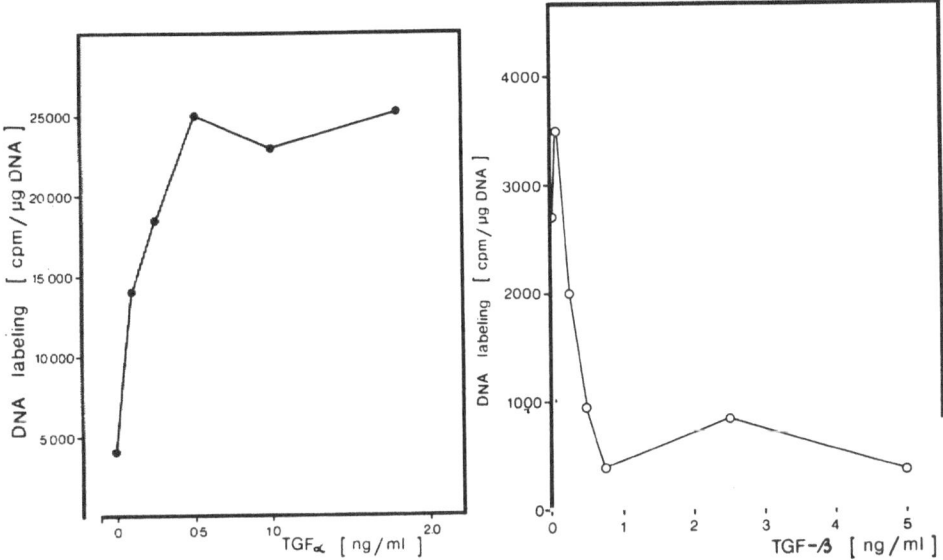

Figure 2. The effects of TGF. (left diagram) and TGF (right diagram) on
epidermal DNA synthesis in vitro.
Primary basal keratinocytes (2 X 10^6 cells) were prepared from newborn NMRI
mice according to ref.54 and grown in 4 X MEM medium containing 1.5% human
platelet-poor plasma. TGF β was added at the time of plating in the concentrations indicated. For experiments with TGFα the cells were grown in the
presence of 10% fetal calf serum instead of platelet-poor plasma and the
medium was changed to "low Ca^{2+} medium" (0.05 mM Ca^{2+}) plus 5% serum containing TGFα in the concentrations indicated after 4 hours. DNA synthesis
was determined by pulse-labelling with ^{3}H-thymidine (5μCi/ml medium) for 60
minutes carried out 30 hours (TGF experiment) or 48 hours (TGF experiment)
after plating.

cell proliferation in vitro (Fig.2). On the other hand, conversion had been
previously shown to depend critically on an induction of epidermal DNA
synthesis (39). We decided, therefore, to combine the injection of TGFβ with
a mitogenic stimulus which by itself does not exhibit a convertogenic effect.
For this purpose we treated initiated mouse skin by local RPA application
immediately after TGFβ-injection and started chronic RPA treatment for
promotion one week later (it should be noted that in this experiment RPA
application served for two purposes, i.e. to complete TGFβ-treatment for the
mitogenic component and to promote tumor development). As shown recently
(40), the combined TGFβ/RPA-treatment exhibited a distinct convertogenic
effect, i.e. rendered initiated mouse skin sensitive to RPA-induced tumor
promotion.

    In a skin would the mitogenic stimulus provided in the above mentioned
experiment by RPA-treatment must be necessarily derived from an additional
endogeneous factor. Since TGFα exhibits a strong mitogenic effect on keratinocytes in vitro (Fig.2), we tested whether TGFα could replace RPA as a
mitogenic agent in conversion. For this purpose a mixture of TGF - and TGFα
was intracutaneously injected into initiated mouse skin and promotion was
subsequently carried out by repeated RPA application starting 1 week after
growth factor injection. As shown in Fig.3, this treatment resulted indeed
in a 2-3 fold increase of tumor development as compared with control animals
which had received injections of isotonic saline instead of the growth
factors.

Conclusions and unresolved problems

The first conclusion which maybe drawn from our results is that the conversion of initiated skin into a promotable state is due to a complex process which consists of at least two components, a mitogenic one which is affected by TGFα (or RPA) and another one which is related to the effect of TGFβ on epidermis. The phorbol ester TPA exhibits both effects. Except for the observation that TGFβ inhibits epidermal cell proliferation in vitro (Fig.2 and ref.41), nothing is known about the physiological action of this factor in epidermis. Presently, any hypothesis on the role of TGFβ in conversion will be, therefore, entirely speculative.

Recently, TGFβ has been proposed to augment tumor development indirectly by immunosuppression and stimulation of stroma growth including formation of connective tissue and vascularisation (42). Whether such a hypothesis (which tries to connect carcinogenesis with elements of the wound response) can explain the convertogenic effect of TGFβ remains questionable since immunosuppression has been found to inhibit rather than to enhance skin tumor development (43,44) and the stroma reaction may be regarded to be a rather late process in tumorigenesis which probably does not play a role in the initial activation and clonal expansion of initiated cells. The finding that

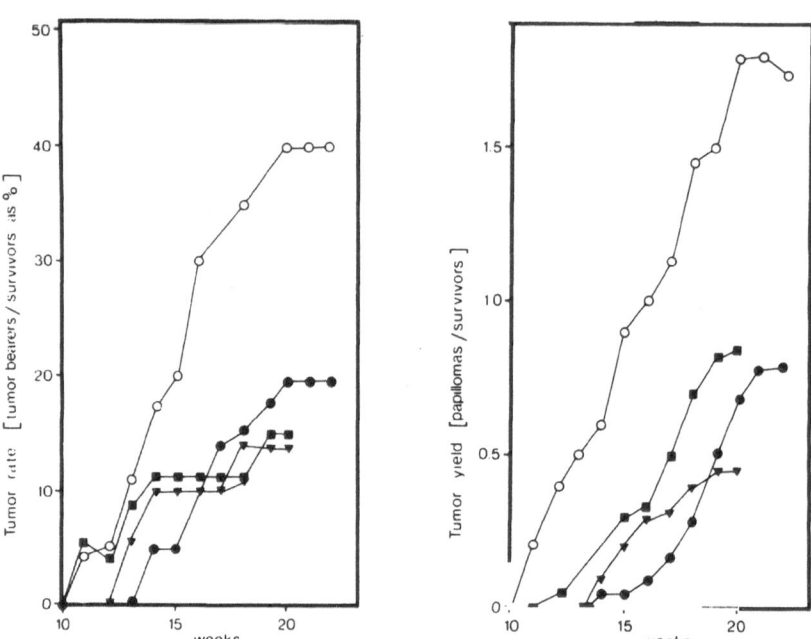

Figure 3. The conversion of initiated mouse skin into a promotable state by combined treatment with TGFα and TGFβ.
Female NMRI mice (age 7 weeks, groups of 20 animals) were initiated each with 100 nmol DMBA/0.01 ml acetone applied onto the shaved back skin at zero time. Conversion was achieved by a single intracutaneous injection of 12.5 µl TGFα, TGFβ or TGFα plus TGFβ (in phosphate-buffered saline) at 4 sites of the back skin one week after initiation. Control animals received only saline. Promotion was carried out by weekly applications of RPA (10 nmol/0.1 ml acetone) starting one week after conversion. The following TGF-doses were injected per site: TGFβ 10 ng (▼), TGFα 25 ng (■), TGFα (25 ng) plus TGFβ (10 ng) (○), none (●).

a combined treatment with TGFα and TGFβ converts initiated skin into a promotable state may nevertheless help to explain the convertogenic effect of skin wounding. However, before this conclusion can be drawn two questions have to be answered:

1. Are TGFα and TGFβ "wound hormones", i.e. endogenous factors which are locally released upon skin wounding and which act on the cells involved in the wound response?

2. Are the cellular actions of TGFα and TGFβ sufficient to explain the convertogenic effect of wounding or are additional factors involved?

As far as the first question is concerned the answer may turn out to be positive. Thus epidermal cells have been shown to be able to produce both TGFα(45,46) and TGFβ (47) as well as to express the corresponding receptors. In addition, transforming growth factors may be released from platelets (8, 42) the aggregation of which is one of the earliest responses to wounding. Finally, both factors have been shown to enhance wound-healing, TGFα by stimulating re-epithelization (48,49) and angiogenesis (50) and TGFβ by inducing stromal reactions (42).

Regarding the second question it has to be emphasized that the wound response in skin as well as conversion and tumor promotion are highly complex processes with other endogeneous signals being involved beside the transforming growth factors. These include first of all the so-called inflammatory mediators such as the kinins and the eicosanoids. Especially the latter have been shown to exert strong effects on epidermal cells and tumor development (see next chapter). In addition, interleukin 1 (51) which is also produced by epidermal cells (52) has been found to contribute to the induction of epidermal hyperproliferation.

Whether the production of such mediators is induced by the growth factors or occurs independently, remains an open question.

Another question regards the relationship between phorbol esters and growth factors. As already mentioned the only molecular action of phorbol esters known so far is the stimulatory effect on protein kinase C. Is this reaction sufficient for triggering the complex series of events involved in hyperplastic transformation, conversion and tumor promotion, or are there other cellular points of attack of phorbol esters which still have to be elucidated? To this question another problem is closely related: Does a phorbol ester somehow mimic the cellular effects of endogenous factors or are these factors locally released upon phorbol ester treatment thus mediating the stimulatory effect of the phorbol ester? After all, the synthesis of both TGFβ(47) and TGFα (53) in epidermis has been reported to be induced by phorbol ester treatment, probably via an activation of PKC, and the induction by phorbol esters of arachidonic acid metabolism is a well established fact (see next chapter).

This brief discussion may indicate that the demonstration of conversion by the transforming growth factors is only a first step towards an understanding of the wound effect on skin tumor development. Much more has to be learned about the complex interactions involved in the wound response before a well founded concept on the relationship between neoplastic growth and wound-healing can be formulated which is something more than just speculation. One of the most important problems which has to be resolved in this context is whether or not the effect of "wound hormones" on tumor development is restricted to mouse skin or represents a more general phenomenon which occurs also in other organs and may be even of significance for human cancer development.

Abbreviations

TPA, 12-0-tetradecanoylphorbol-13-acetate; RPA, 12-0-retinoylphorbol-13-acetate; Ti8, 12-tetradecatetre-2,4,6,8-enoylphorbol-13-acetate; DMBA, 7,12-dimethylbenz[a]anthracene; DAG, diacylglycerol; IP_3, inositol-1,4,5-trisphosphate; EIF, Epstein-Barr-virus-inducing factor; TGF, transforming growth factor.

REFERENCES

1   Deelman, H. T. (1924) Z. Krebsforsch., 21, 220-226
2   Hennings, H. and Boutwell, R. K. (1970) Cancer Res., 30, 312-325
3   Clark-Lewis, I. and Murray, A. W. (1978) Cancer Res., 38, 494-497
4   Argyris, R. S. (1986) CRC Crit. Rev. Toxicol., 14, 211-237
5   Hasegawa, R., St. John, M., Tibbels, S. and Cohen, S. (1987) J. Invest. Dermatol., 88, 652-656
6   Haddow, A. (1972) Adv. Cancer Res., 16, 181-234
7   Marks, R. (1988) Pathology Res. Pract., 182, 831-848
8   Sporn, M. B. and Roberts, A. B. (1986) J. Clin. Invest., 78, 329-332
9   Marks, F., Furstenberger, G., Ganss, M., Richter, K. H. and Seemann, D. (1983) Brit. J. Dermatol., 109, Suppl. 25, 18-21
10 Bertsch, S. and Marks, F. (1974) Cancer Res., 34, 3283-3288
11 Bertsch, S. and Marks, F. (1978) Cell Tissue Kinet., 11, 651-658
12 Gilmour, S. K., Verma, A.K., Madara, T. and O'Brien, T.G. (1987) Cancer Res., 47, 1221-1225
13 Rose-John, S., Furstenberger, G., Krieg, P., Besemfelder, E., Rincke, G. and Marks, F. (1988) Carcinogenesis,, 9, 831-835
14 Marks, F., Ganss, M. and Grimm, W. (1981) Biochem, Biophys. Acta, 678, 122-131
15 Marks, F. (1983) Carcinogenesis, 4, 1465-1470
16 Marks, F., Clauss, M., Tiegel, A. and Richter, K. H. (1986) in Biological Regulation of Cell Proliferation (Baserga, R., Foa, P., Metcalf, D., Polli, E. E., eds.) pp. 267-274. Raven Press, New York
17 Yamaski, H. (1987) in Concept and Theories in Carcinogenesis (Maskens, A. P., Ebbesen, P., Burny, A., eds.) pp. 117-133. Excerpta Medica, Amsterdam etc.
18 Castagna, M., Takai, Y., Kaibuchi, K., Sano, K., Kikkawa, V. and Nishizuka, Y. (1982) J. Biol. Chem., 257, 7847-7851
19 Berridge, M. J. (1987) Biochim. Biophys. Acta, 907, 33-46
20 Ashendel, C. L. (1985) Biochim. Biophys. Acta, 822, 219-242
21 Marks, F., Furstenberger, G., Ganss, M., Richter, K. H. and Seemann, D. (1983) in Psoriasis: Cell Proliferation (Wright, N. A. and Camplejohn, R. S., eds.) pp. 173-188. Churchill Livingstone, Edinburgh etc.
22 Slaga, T. J., Scribner, J. D., Thompson, S. and Viaje, A. (1978) J. Natl. Cancer Inst., 52, 1611-1618
23 McCutcheon, J. A., Bickenback, J. C. and Mackenzie, J. C. (1985) J. Dental. Res., 64, 298
24 Klein-Szanto, A. J. P. (1984) in Mechanisms of Tumor Promotion (Slaga, T. J., ed.) Vol 2, pp. 41-72. CRC Press, Boca Raton
25 Marks, F., Bertsch, S., Grimm, W. and Schweizer, J. (1978) in Mechanisms of Tumor Promotion and Cocarcinogenesis (Slaga, T. J., Sivak, A., Boutwell, R. K., eds.) pp. 97-116. Raven Press, New York
26 Balmain, A. and Brown, K. (1988) Adv. Cancer Res., in press
27 Hennings, H., Shores, R., Wenk, M. L., Spangler, E. F., Tarone, R. and Yuspa, S. J. (1983) Nature, 304, 67-69
28 Marks, F., Bertsch, S. and Furstenberger, G. (1979) Cancer Res., 38, 4183-4188
29 Furstenberger, G., Berry, D. D., Sorg, B. and Marks, F. (1981) Proc. Natl. Acad. Sci. USA, 78, 7722-7726
30 Furstenberger, G., Kinzel, V., Schwarz, M. and Marks, F. (1985) Science 230, 76-78

31 Yuspa, S. H. (1984) in Cellular Interactions by Environmental Tumor
    Promoters (Fujiki, H., Hecker, E., Moore, R. E., Sugimura, T., Weinstein
    I. B. eds.) pp. 315-326. Japan Scient. Soc. Press, Tokyo. VNU Science
    Press, Utrecht
32 Slaga, T. J., Fischer, S. M., Nelson, K. and Gleasson. G. L. (1980) Proc.
    Natl. Acad. Sci. USA, 77, 3659-3663
33 Sisskin, E. E., Gray, T. and Barrett, J. C. (1982) Carcinogenesis, 3,
    403-408
34 Kinzel, V., Furstenberger, G. and Marks, F. (1988) in Tumor Promoters:
    Biological Approaches for Mechanistic Studies and Assay Systems, in press
    Raven Press, New York
35 Furstenberger, G. and Marks, F. (1983) J. Invest. Dermatol., 81, 157s-
    161s
36 Bauer, G., Hofler, P. and Simon, H. J. (1982) J. Biol. Chem., 257,
    11405-11410, 11411-11415
37 Bauer, G., Hofler, P. and zur Hausen, H. (1982) Virology, 121, 184-194
38 Rogers, M., Furstenberger, G., Marks, F., Bauer, G. and Hofler, P. (1988)
    in Hormones and Cancer (Proc. of the 3rd Int. Congress on Hormones and
    Cancer, Hamburg 1987), in press. Raven Press, New York
39 Kinzel, V., Loehrke, H., Goerttler, K., Furstenberger, G. and Marks, F.
    (9184) Proc. Natl. acad. Sci. USA, 81, 5858-5862
40 Marks, F., Furstenberger, G., Gschwendt, M., Rogers, M., Schurich, B.,
    Kaina, B. and Bauer, G. (1988) in Models and Mechanisms in Chemical
    Carcinogenesis (Feo, F. et. al., eds.), in press. Plenum, New York
41 Coffey, R. J., Sipes, N. J., Bascom, C. C., Graves-Deal, R., Pennington,
    C. Y., Weissman, B. E. and Moses, H. L. (1988) Cancer Res., 48, 1596-
    1602.
42 Roberts, A. B., Thompson, N. L., Heine, U., Flanders, C. and Sporn, M. B.
    (1988) Brit. J. Cancer, 57, 594-600
43 Gschwendt, M., Kittstein, W. and Marks, F. 91987) Carcinogenesis, 8,
    203-207
44 Gschwendt, M., Kittstein, W. and Marks, F. (1987) Cancer Letters, 34,
    187-191
45 Coffey, R. J., Derynck, R., Wilcox, J. N., Bringman, T. S., Goustin,
    A. S., Moses, H. L. and Pttelkow, M. L. (1987) Nature, 328, 817-820
46 Gottlieb, A. B., Chang, K. C., Posnett, D. N., Fanelli, G. and Tam,
    P. J. (1988) J. Exp. Med., 167, 670-675
47 Akhurst, R. J., Fee, F. and Balmain, A. (1988) Nature, 331, 363-365
48 Sporn, M. B., Roberts, A. B., Shull, J. H., Smith, J. M., Ward, J. M. and
    Sodek, J. 91983) Science, 219, 1329-1331
49 Schultz, G. S., White, M., Mitchell, R., Brown, G., Lynch, J., Twardzik,
    D. R. andTodaro, G. J. (1987) Science, 235, 350-352
50 Schreiber, A. B., Winkler, M. and Derynck, R. (1986) Science, 232, 1250-
    1253
51 Ristow, H. J. (1987) Proc. Natl. Acad. Sci. USA, 84, 1940-1944
52 Luger, T. A., Kock, A., Danner, M., Colot, M. and Micksche, M. (1985)
    Brit, J. Dermatol., 113, Suppl. 28, 145-156
53 Pittelkow, M. R. and Coffey, R. J. (1988) Ann. N. Y. Acad. Sci., in Press
54 Furstenberger, G., Gross, M., Schweizer, J., Vogt, I. and Marks, F.
    (1986) Carcinogenesis, 7, 1745-1753

ARACHIDONIC ACID METABOLITES AS ENDOGENEOUS MEDIATORS OF CHROMOSOMAL

ALTERATIONS AND TUMOR DEVELOPMENT IN MOUSE SKIN IN VIVO

G Furstenberger, H Hagedorn, B Schurick, R T Petrusevska
N E Fusenig and F Marks

Institute of Biochemistry, German Cancer Research Center
Im Neunheimer Feld 280, D - 6900 Heidelberg, F.R.G.

ABSTRACT

In carcinogen-initiated NMRI mouse skin the process of tumor induction
can be subdivided into two stages: conversion induced by a single treatment
with the convertogenic tumor promoter TPA and promotion elicited by repetitive treatments with the irritant hyperplasiogenic phorbol ester RPA. The
induction by RPA or TPA of a stationary hyperplasia seems to be a sufficient
condition of promotion, cytogenic effects induced by TPA but not by RPA may
be critically involved in conversion. Prostaglandins E_2 and $F_{2\alpha}$ have been
found to be endogenous modulators of the phorbol esters' hyperplasiogenic
effects. Moreover, $PGF_{2\alpha}$ appears to be involved in a distinct step of the
conversion process. Phorbol esters induce an epidermal lipoxygenase activity
resulting in the formation of 8-HPETE/8-HETE. Inhibition of this pathway of
arachidonic acid metabolism prevents both the TPA-induced clastogenic activity in mouse keratinocytes in vitro and in vivo and conversion. These
results indicate that cytogenetic effects such as chromosomal alterations
induced via activation of epidermal arachidonic acid metabolism could play a
critical role in the conversion process.

INTRODUCTION

The generation of metabolites of arachidonic acid in a given tissue is
induced by distinct exogeneous and endogeneous stimuli. As biologically
potent locally acting agents they may be involved in the modulation of individual cell functions, cell interactions and tissue homeostasis. Skin
epidermis is a tissue exhibiting an active arachidonic acid metabolism (1)
and arachidonic acid metabolites such as prostaglandins leukotrienes and
hydroxyeicosatetraenoic acids have been shown to be involved in inflammatory
and immune processes (1-3), wound repair (4,5) and in proliferative skin
diseases such as psoriasis (6).

Epidermal arachidonic acid metabolism

Arachidonic acid released from cellular phospholipids by the action of
acyl hydrolases, predominantly phospholipase A_2, is metabolized via the
cyclooxygenase pathway to the endoperoxides prostaglandin G_2 (PGG_2) and PGH_2.
These products serve as precursors for the formation of PGE_2, PGD_2 and $PGF_{2\alpha}$,
the predominant prostaglandins in human, rat, mouse and guinea pig skin (for

review see 1). The lipoxygenases transform arachidonic acids to hydro-
peroxyeicosatetraenoid acids (HPETEs), which are further metabolized to
hydroxyeicosatetraenoic acids (HETEs) and leukotrienes. Among the epidermal
lipoxygenase products 12-S-HETE was shown to be the major metabolite of
arachidonic acid in human, mouse and guinea pig epidermis (1,2). In
psoriatic skin, elevated levels of the (R) enantiomer of 12-HETE were detected (7). Evidence for a functional 5-lipoxygenase was obtained in mouse and
human keratinocytes (8). Treatment of the latter with the Ca^{++} ionophore A
23187 induced the formation of 5-HETE and trace amounts of leukotriene D_4
(LTD_4). Leukotriene B_4 (LTB_4) was shown to be released from ionophore-
treated human keratinocytes (9). Treatment of skin in vivo and keratinocytes
in vitro both of various origins with irritant and hyperplasiogenic phorbol
esters such as 12-0-tetradecanoylphorbol-13-acetate (TPA) led to stimulation
of epidermal PGE_2 and $PGF_{2\alpha}$ formation (for review see ref. 10). Moreover,
TPA was shown to induce the generation of 8-hydroxyeicosatetraenoic acid (8-
HETE) in mouse epidermis in vivo (11). In addition to keratinocytes,
potential sources of arachidonic acid metabolites in normal and damaged skin
are macrophages, mast cells, fibroblasts and vascular endotheluim as well as
blood-born cells such as neutrophils, lymphocytes, monocytes, eosinophils and
platelets.

Functions of arachidonic acid metabolites in epidermis

Arachidonic acid metabolites were shown to exert a variety of functions
in epidermal cells in vivo and in vitro. In particular, PGE_2 has been found
to modulate growth of cultured human keratinocytes (12). Exogeneous addition
of PGE_2 to epidermal cells in vitro was shown to increase proliferation (13),
but to decrease epidermal DNA synthesis in skin explants (14). PGE_2 topically applied to skin has been shown either to increase epidermal DNA synthesis in hairless mice (15) or to have no effect on epidermal DNA synthesis and
mitotic activity in NMRI mice (10,16). 12-HETE as well as leukotrienes B_4,
C_4 and D_4 were reported to enhance human deratinocyte DNA synthesis in vitro
(17,18). Moreover, when topically applied to guinea pig skin in vivo, LTB_4
was shown to induce epidermal hyperproliferation (19). Both cyclooxygenase
and lipoxygenase metabolites of arachidonic acid were found to induce
symptoms of inflammation such as erthema and edema, and to be present at
elevated levels in inflamed skin (1,2). Metabolites of arachidonic acid
appear to participate also in the inflammatory and hyperproliferative
processes in psoriasis. Increased levels of 12-HETE were detected in kera-
tome slices of involved psoriatic epidermis, whereas the PGE_2 and $PGF_{2\alpha}$
levels were only slightly elevated (20). Extracts of psoriatic scales were
shown to contain LTB_4 and six different HETEs (21). Moreover, benoxaprofen,
an inhibitor of 5-lipoxygenase, thereby decreasing LTB_4 formation, markedly
improved psoriasis (22). This may be taken as an indication that LTB_4 may be
critical to the inflammatory and proliferative processes in psoriasis.

Prostaglandins and phorbol ester-induced hyperplastic transformation of mouse epidermis

Additional evidence for a role of arachidonic acid metabolites, in particular PGE_2, as mediators of epidermal hyperproliferation came from studies
with the irritant and hyperplasiogenic phorbol esters 12-0-retinoylphorbol-
13-acetate (RPA) and TPA or the calcium ionophore A 23187 in mice (strain
NMRI; for a review see ref. 23). Treatment of mouse skin with these agents
was shown to stimulate an early increase of the epidermal PGE_2 content
(24,25) which turned out to be a necessary event for the subsequent induced
of ornithine decarboxylase activity, DNA-synthesis and mitotic activity.
Prevention of PGE_2-synthesis by cyclooxygenase inhibitors such as indomethacin resulted in an inhibition of TPA-induced ODC-activity and hyperplasia.
This inhibition could be specifically reversed by PGE_2 in mice (23,25), by
both PGE_2 and $PGF_{2\alpha}$ in guinea pigs (26). In the absence of phorbol ester or

ionophore, PGE_2 or $PGF_{2\alpha}$ were found to be inactive as mitogens in NMRI mouse and guinea pig skin in vivo (10).

Thus, the induction of hyperplastic transformation seems to due to a synergistic action of an exogeneous stimulus and endogeneous PGE_2. The mechanism of this synergistic effect remains to be elucidated. Experimental evidence has been provided that the synergistic effect of PGE_2 on the induction of epidermal hyperproliferation is not mediated by cyclic AMP (27). An involvement of prostaglandins in the induction of epidermal hyperplasia was also observed after mechanical stimulation of mouse epidermis, such as sand paper rubbing (28) and full skin wounding (5). This indicates that PGE_2-mediated induction of epidermal hyperplasia is part of a general response of skin to irritation and damage called hyperplastic transformation (see article of Marks & Furstenberger, this volume).

Endogeneous prostaglandin synthesis was also shown to be involved in the induction of DNA synthesis and mitotic activity in liver after partial hepatectomy (29) and in estrogen-induced DNA-synthesis in rat uterus (30).

Prostaglandins and phorbol ester-induced tumor formation in initiated NMRI mouse skin

The hyperplasia-inducing efficacy of phorbol esters such as TPA is closely related to another property of these agents, i.e. the ability to induce tumor formation to initiated skin upon repetitive treatments (31; see also the chapter by Marks & Furstenberger, this volume). Moreover, the introduction of certain diterpene esters such as mezerein (32) and 12-O-retinoylphorbol-13-acetate (RPA; 33) has demonstrated tumor induction experiments in initiated mouse skin to be composed of two steps called conversion and promotion. Promotion is generally understood as clonal hyperplasia induced by chronic RPA or mezerein treatment (31,34). Conversion achieved by one or two TPA applications prior to promotion is required to render initiated skin promotable. According to this, TPA presents a tumor promoter with convertogenic activity, whereas RPA is a tumor promoting agent almost devoid of convertogenic activities. A more detailed discussion of multistage skin tumorigenesis is presented by Marks & Furstenberger (this volume).

Evidence that prostaglandins are involved in multistage skin carcinogenesis (NMRI mouse) is based upon the following observations: 1. TPA-induced-epidermal DNA-synthesis is necessary but not sufficient component of the conversion stage (23,25); 2. A stationary hyperplasia is a necessary and probably sufficient condition of promotion (31). As shown above, phorbol ester-induced epidermal hyperproliferation is a PGE_2-mediated event. Thus, an inhibition of both the conversion and the promotion stage by the cyclooxygenase inhibitor idomethacin may be expected and was indeed observed (Table 1 and 2).

An unexpected observation was that the inhibition of the promotion stage by indomethacin was specifically reversed by $PGF_{2\alpha}$ instead of by PGE_2 (Table 1).

An explanation for this discrepancy could be that PGE_2 specifically mediates the initial induction of epidermal hyprplasia, whereas the maintenance of the hyperplastic state achieved by repeated phorbol ester treatment and required for tumor promotion (31,34) depends on $PGF_{2\alpha}$.

Conversion is also inhibited by indomethacin treatment. Again, the inhibition could be overcome by $PGF_{2\alpha}$ rather than - as expected - by PGE_2 (Table 2). Moreover, the time course of this inhibition differed from that of the inhibition by indomethacin of epidermal DNA-synthesis: when applied 1 hour prior to TPA, indomethacin caused maximal inhibition of epidermal DNA

Table 1. Effects of indomethacin alone or in combination with PGE_2 or PGF_2 on the promotion stage of skin carcinogenesis.

No. of experiments	Treatment	Tumor response after 18 weeks	
		Tumor rate (%)	Tumor yield (pap./surv.)
1.	acetone	66	3.1
2.	indomethacin	33	1.2
3.	indomethacin + PGE_2	27	1.2
4.	indomethacin + PGE_{2d}	66	2.6
5.	controls	13	0.5

Female NMRI mice at the age of 7 weeks were assigned by random distribution to experimental groups of 16 animals. For initiation, DMBA (100 nmol/0.1 ml acetone) was topically applied to the shaved back skin. For conversion animals received two topical applications of TPA (10 nmol/0.1 ml acetone) at days 7 and 10 after initiation. Control animals received acetone (0.1 ml) instead of TPA (exp. 5). In the promotion stage the treatments with acetone (0.1 ml; exp. 1), indomethacin (550 nmol/0.1 ml acetone + 28 nmol/0.1 ml acetone, exp. 3 and 4) were carried out 60 min. prior to each RPA application. The tumor response was read as tumor rate (percentage of tumor bearers per survivors) and tumor yield (papillomas/survivors) after 17 weeks of RPA treatment. At the end of the experiment $\geq$ 95% of the animals were alive.

synthesis (23,24) but only a moderate inhibitory effect on conversion, whereas maximal inhibition of conversion but no effect on DNA synthesis (23,24) was observed with indomethacin applied 3 hrs after TPA (Table 2). When applied at this time point, indomethacin inhibited the second wave of $PGF_{2\alpha}$ induced by the phorbol ester in mouse skin in vivo (36).

It is concluded that epidermal $PGF_{2\alpha}$ formation plays a critical role in the conversion process induced by TPA. This conclusion is supported by the observation that the non-convertogenic promoter RPA did not evoke an accumulation of $PGF_{2\alpha}$ 3 hrs after treatment (36). On the other hand, $PGF_{2\alpha}$ applied 3 hrs after RPA treatment could not confer convertogenic activity to the latter indicating a more specific synergism between TPA and $PGF_{2\alpha}$ rather than $PGF_{2\alpha}$ being a mediator of the phorbol ester's convertogenic efficacy (Table 2).

Involvement of lipoxygenase-derived arachidonic acid metabolites in conversion

In addition to prostaglandins products of lipoxygenase pathways of arachidonic acid metabolism such as 5-HETE, or 12-HETE or LTB_4 have been identified in normal and ionophore A 23187 treated mouse skin (see above). Moreover, the phorbol ester TPA was found to induce a cytosolic 8-lipoxygenase activity in mouse epidermis in vivo (Fig 1). 8-HETE formation started after a lag phase of 3 hrs, reached a first maximum of activity 18 hrs and a second maximum around 48 hrs after phorbol ester treatment. The production of 5-HETE was only slightly increased, the amount of 12-HETE essentially remained unchanged in epidermal cytosol obtained 17 hrs after TPA treatment.

Table 2. Effects of indomethacin alone or in combination with PGE_2 and PGF_2 applied 3 hrs after TPA application on the conversion stage of skin carcinogenesis

No. of experiments	Treatment	Tumor response after 18 weeks	
		Tumor rate (%)	Tumor yield (pap./surv.)
1.	acetone	63	3.3
2.	indomethacin	31	1.1
3.	indomethacin + PGF_2	56	2.7
4.	indomethacin + PGE_2	31	1.3
5.	controls	13	0.7
6.	PGF_2 + RPA	13	0.3

The experiments were carried out as described in Table 1 except that, for conversion, the treatment with acetone (0.1 ml; exp. 1), indomethacin (550 nmol/0.1 ml acetone; exp. 2) or indomethacin + prostaglandin (550 nml/0.1 ml acetone + 28 nmol/0.1 ml acetone; exp. 3 and 4) was carried out 3 hrs after TPA application (10 nmol/0.1 ml acetone). Controls were treated with acetone (0.1 ml; exp. 5) instead of TPA. In experiment 6 RPA (10 nmol/0.1 ml acetone) was applied instead of TPA and PGF_2 (28 nmol/0.1 ml acetone) was given 3 hrs later. Promotion was performed by twice weekly application of RPA (10 nmol/0.1 ml acetone).

Lipoxygenase inhibitors such as quercetin, rottlerin and eicosa-5,8,11,14-tetraynoic acid (ETYA) suppressed 5-, 8- and 12-HETE formation in adose dependent manner. Both the convertogenic tumore promoter TPA and the non-convertogenic promoter RPA induced 8-lipoxygenase activity to a similar degree. This result seems to argue against a critical role of this lipoxygenase activity in the conversion process. On the other hand, ETYA was found to inhibit both the cytosolic lipoxygenase activity and conversion in a dose dependent manner (37). Moreover, ETYA exhibited its maximal inhibitory activity when applied either 1 hr prior to or 18 hrs after TPA treatment. Inbetween skin was found to be refractory to ETYA. Thus, ETYA may impair two different events in the conversion process. With the ETYA doses used, TPA-induced prostaglandin synthesis and epidermal DNA synthesis were not inhibited (38). This result indicates that lipoxygenase metabolites of arachidonic acid are critically involved in the process of conversion.

A likely candidate for such a metabolite could be 8-HETE, as the predominant TPA-induced product. However, other still unknown metabolites cannot yet be exluded. They could be formed either by alternative pathways of arachidonic acid metabolism in skin or could be intermediates or by-products of 8-HETE formation such as 8-HPETE or cleavage products thereof, oxygen radicals and lipid peroxides. The latter aspect is particularly intriguing for two reasons: 1) Free radical generating compounds have been shown to elicit tumor growth in initiated skin (39) and 2) cyclooxygenase and lipoxygenase derived arachidonic acid metabolites were shown to be involved in the chromosome damaging (clastogenic) action of phorbol esters in white blood cells, a reaction, which has been speculated also to be involved in tumor promotion (40).

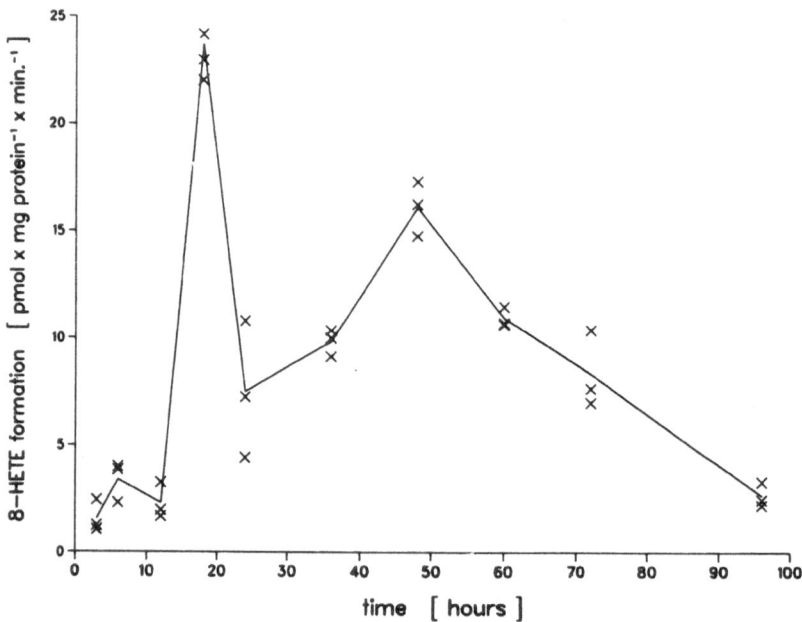

Fig 1. Kinetics of the induction by TPA of an epidermal 8-lipoxygenase
activity in mouse skin in vivo. Female NMRI mouse were treated with a single
topical applicatiron of TPA (10 nmol/0.1 ml acetone) and were killed by
cervical dislocation at the time points indicated. Epidermal cytosolic
fractions (1 mg protein/ml) were incubated at 37^0 with 35 uM 1-^{14}C-arachi-
donic acid for 5 minutes in 250 ul buffer (50 mM Tris, 2 mM EDTA, pH 8.0).
Extraction and analysis of 8-HETE was performed as described in reference 11.

Lipoxygenase products, chromosomal alterations and conversion

      The clastogenic effect of TPA treatment is indeed not restricted to leu-
kocytes but also found in mouse keratinocyte cell lines (41,42) and in normal
primary mouse keratinocytes (38).

      In Table 3 the frequency and type of such structural chromosomal aberr-
ations found upon TPA treatment is presented. They vary from simple breaks
and centromeric splitting to more severe alterations such as chromosomal exchanges, ring chromosomes, chromosomal rearrangements and fragmentations.
Chromosome rearrangements include translocations, duplications and deletions.
The total rate of metaphases with structural chromosomal aberrations was enhanced ninefold after a single TPA treatment. Interestingly, ETYA, the above
mentioned inhibitor of arachidonic acid metabolism and conversion caused a
70% reduction of TPA-induced structural chromosomal aberrations. This result
indicates that lipoxygenase-derived arachidonic acid metabolites are involved
in both TPA-induced clastogenicity and conversion (see above). This data
gained strong support by the observation that the non-convertogenic tumor
promoter RPA did not induce chromosomal aberrations (Table 4). To find out
whether or not similar clastogenic effects could be induced by TPA in mouse
skin in vivo, mice were treated with TPA (10 nmol/0.1ml acetone) for 48 hrs
and epidermal basal cells isolated from the treated skin area were analysed
for chromosomal aberrations. The results of such experiments are summarized
in Table 5. They show a distinct clastogenic effect of TPA in vivo, whereas
the same dose of RPA, applied also for 48 hrs was much less efficient in
enhancing the frequency of structural chromosomal aberrations. Thus, only the
convertogenic efficacy seems to correlate with clastogenic efficacy.

Table 3. Inhibition of TPA-induced structural chromosomal aberrations by ETYA in primary mouse epidermal cells in vitro

Treatment	No. of meta-phases (exp.)	% of metaphases with structural chromosomal aberrations				
		breaks[a]	rings	QR + TR[b]	chromosomal rearrangements[c] (t, dp, df)	total
acetone (0.1%)	165(5)	3	–	–	–	3
ETYA (10^{-6}M)	100(1)	2	–	–	–	.2
TPA (10^{-8}M)	280(5)	21	2.5	2	1.5	27
ETYA (10^{-6}M) TPA (10^{-8}M)	75(2)	7	–	1	–	8

[a]preferentially of isochromatid type; [b]chromosomal exchanges of triradial and quadriradial type; [c]chromosomal rearrangements: t = translocation, dp = duplication; d = deficiency (deletion)
Primary cultures from newborn C_3H mice were plated centrally at high density (5×10^5 attached cells/cm^2) in 35 mm Petri dishes and were cultivated at 37^0C in minimal essential medium with a fourfold concentration of amino acids and vitamins (4 X MEM) supplemented with 17% heat inactivated fetal calf serum. Cultures were washed 24 h after plating and treated the next day with phrobol esters and/or simultaneously with ETYA for two rounds of division. TPA, RPA and ETYA were dissolved in acetone (final concentration 0.05%). For the harvesting procedure and the analysis of structural aberrations see ref. 38.

Table 4. Induction of structural chromosomal aberrations by phorbol esters in primary mouse epidermal cells in vitro

Treatment	No. of meta-phases (exp.)	% of metaphases with structural chromosomal aberrations				
		breaks	rings	QR + TR	chromosomal rearrange-ments[c] (t, dp, df)	total
acetone (0.1%)	165(5)	3	–	–	–	3
TPA 10^{-8}M	280(5)	21	2.5	2	1.5	27
RPA 10^{-8}M	205(5)	4	–	0.5	–	4.5

[a]preferentially of isochromatid type; [b]chromosomal exchanges of triradial and quadriradial type; [c]chromosomal rearrangements: t = translocation, dp = duplications; d = deficiency (deletion). For experimental details, see Table 3 and reference 38.

Table 5. Induction of structural chromosomal aberrations by phorbol esters in NMRI mouse skin in vivo

Treatment	No. of meta-phases (exp.)	% of metaphases with structural chromosomal aberrations				
		breaks	QR + TR	chromosomal rearrange-ments (t, df)	total	TMC*
untreated	107(1)	2.8	-	-	2.8	-
acetone (0.1 ml)	160(2)	3.8	-	-	3.8	-
TPA (10 nmol/ 0.1 ml acetone	235(3)	20.0	1.2	1.2	21.9	1.5
RPA (0/1 ml/ 0.1 ml acetone	232(3)	3.8	-	1.5	5.3	-

*TMC: to many aberrations to count

Female NMRI mice were treated by topical applications of TPA (10 nmol/0.1 ml acetone), RPA (10 nmol/0.1 ml acetone) or acetone (0.1 ml) onto the shaved back skin and killed by cervical dislocation 48 hrs later. The skin was dissected and mouse epidermis was separated from dermis in thin-split sections by trypsinization (for details see reference 43). From the isolated keratinocytes the basal cell fraction was obtained using discontinuous Percoll density gradient centrifugation. $4.5 \times 5.0 \times 10^6$ basal keratinocytes were plated into 35 mm collagen-coated Falcon dishes. The cells were maintained in minimal essential medium with a fourfold concentration of amino acids and vitamins with 10% fetal calf serum at 34^0C. Subconfluent cultures were harvested by incubation with EDTA/trypsin (0.1%/0.1%) containing 1 ug/ml colcemid for 20 minutes. After detachement by vigourous pipetting, the cells were sollen in 75 mM KC1 for 20 minutes and fixed three times in methanol/ glacial acetic acid (3:1). G-banding was achieved with trypsin (0.0125%) treatment for 45-60 seconds, followed by staining of the slides with 3% buffered Giemsa for 8 to 10 minutes (see also reference 38).

It may be speculated that a critical or even a causal relationship exists between these two effects with a TPA-induced arachidonic acid metabolite as a common mediator, which may be produced in the course of a 8-lipoxygenation reaction (see above). Such hypothesis apparently is in conflict with the fact that RPA is as potent as TPA in inducing 8-HETE formation (37). One possible explanation for this discrepancy could be that the deficiency of RPA as a convertogenic and clastogenic agent could be due to the specific unsaturated structure of the molecule. Interaction of this structure with biologically active intermediates could lead to an inactivation of the putative mediator of both clastogenic and convertogenic effects.

It has indeed been shown that HPETEs are constituents of the clastogenic factors released by phorbol ester treated macrophages (44) and that HPETEs do exert clastogenic effects in $C_3H10T1/2$ cells (45). Experiments to test the convertogenic activity of HPETEs in the multistage carcinogenesis approach of mouse skin are underway.

Provided the assumption of a relationship between cytogenetic effects and conversion is correct, the question arises by which mechanisms chromosomal alterations could stimulate tumor induction in intitiated skin. Are the observed chromosomal alterations indicative for specific effects at the gene level or for more unspecific effects related to cell death and tissue damage? If the clastogenic effects are proposed to lead to cell death on may assume that cell loss could evoke a would response in the tissue. This may in turn lead to an activation of an epidermal stem cell compartment which is most probably the target of initiation. This hypothesis gains support by the observation that skin wounding (31) like TPA exerts a convertogenic activity in initiated NMRI mouse skin. If, on the other hand, specific genetic changes are critical for conversion, they must be different form those thought to be involved in initiation. The reversibility of the converted state (46), indeed, argues strongly against mutagenic events or irreversible gene rearrangements being involved in conversion. Thus, more transient effects such as reversible alterations of gene activity have to be considered critical for the conversion process.

ACKNOWLEDGEMENTS

We thank D Kucher, B Steinbauer and A Schrodersecker for expert technical asssistence with the tumorigenesis experiments. The support of the work by the Deutsche Forschungsgemeinschaft is gratefully acknowledged.

REFERENCES

1   Ruzicka, Th. and Printz, M. P. (1984) Rev. Physiol. Biochem. Pharmacol. 100, 121-160
2   Bonta, J. L. and Parnham, M. J. (1978) Biochem. Pharmacol. 27, 1611-1623
3   Samuelson, B. (1983) Science 220, 568-575
4   Anggard, E. and Jonsson, C. F. (1972) in Prostaglandins in cellular biology (Ramwell, P. W. and Harris, BB. eds) Vol. 1, pp. 269-291, Plenum Press New York
5   Bertsch, S. and Marks, F. (1982) Cell Tissue Kinet 15, 81-87
6   Voorhees, J. J. (1983) Arch. Dermatol. 119, 541-547
7   Woollard, P. M. (1986) Biochem. Biophys. Res. commun. 136, 169-176
8   Ziboh, V. A., Casebolt, T. L., Marcelo, C. I. and Voorhees, J. J. (1984) J. Invest. Dermatol. 83, 426-430
9   Brain, S. D., Camp, R. D. R., Leigh, J. M., Ford-Hutchinson, A. W. (1982) J. Invest. Dermatol. 78, 328
10 Furstenberger, G. and Marks, F. (1985) in Arachidonic acid metabolism and tumor promotion (Fischer, S. M. and Slaga, T. J., eds) pp. 49-72, Martinus Nijhoff Publishing, Boston
11 Gschwendt, M., Furstenberger, G., Kittstein, W., Besemfelder, E, Hull, W. E., Hagedorn, H., Opferkuch, H. J. and Marks, F. (1986) Carcinogenesis 7, 449-455
12 Pentland, A. P. and Needleman, P. (1986) J. Clin. Invest. 77, 246-254
13 Bem, J. L and Greaves, M. W. (1974) Arch. Dermatol. Forsch. 251, 35-41
14 Harper, R. A. (1976) Prostaglandins 12, 1019-1025
15 Lowe, N. J. and Stoughton, R. B. (1977) J. Invest. Dermatol. 68, 134-137
16 Furstenberger, G. and Marks, F. (1978) Biochem. Biophys. Res. Commun. 84, 1103-1108
17 Kragballe, K. and Fallon, J. D. (1986) Arch. Dermatol. 278, 449-453
18 Kragballe, K., Desjarlais, I. and Vorhees, J. J. (1985) J. Dermatol. 113, 43-52
19 Chan, C. D., Duhamel, L. and Ford-Hutchinson, A. 91985) J. Invest. Dermatol
20 Hammarstrom, S., Hamber, M., Samuellson, B., Duell, E. A., Stawiski, M. and Vorhees, J. J. (1975) Proc. Natl. Acad. Sci. USA 72, 5130-5134
21 Camp, R. D., Mallet, A. J., Wollard, P. M., Brain, S. D., Kobza-Black

A., Greaves, M. W. (1983) Prostaglandins, 26, 431-447

22 Kragballe, K. and Herlin, (1983) J. Invest. Dermatol. 119, 548-

23 Furstenberger, G. Gross, M. and Marks, F. (1984) in Eicosanoids and Cancer (Thaler-Dao, H., Crastes de Paulet, A. and Paoletti, R., eds) pp. 91-100, Raven Press, New York

24 Furstenberger, G. and Marks, F. (1980) Biochem. Biophys. Res. Commun. 92, 749-756

25 Marks, F., Furstenberger, G. and Kownatzki, E. (1981) Cancer Res. 41, 696-702

26 Delescluse, C., Furstenberger, G., Marks, F. and Prunieras, M. (1982) Cancer Res. 42, 1975-1979

27 Marks, F. (1983) Carcinogenesis 4, 1465-1470

28 Furstenberger, G., de Bravo, M., Bertsch, S. and Marks, F. (1979) Res. Commun. Chem. Pathol. Pharmacol. 24, 533-541

29 Rixon, R. H. and Whitfield, J. F. (1982) J. Cell Physiol. 113, 281-288

30 Stewart, P. J., Zalondek, C. J., Murphy, J. M. and Webster, R. A. (1983) Life Sci., 33, 2349-2356

31 Furstenberger, G. and Marks, F. (1983) J. Invest. Dermatol., 81, 157s-161s

32 Slaga, T. J., Fischer, S. M., Nelson, K. and Gleason, G. L. (1980) Proc. Natl. Acad. Sci. USA, 77, 3659-3663

33 Furstenberger, G., Berry, D. L., Sorg, B. and Marks, F. (1981) Proc. Natl. Acad. sci. USA, 78, 7722-7726

34 Yuspa, S. H., Hennings, H., Kulez-Martin, M. and Lichti, U. (1982) in cocarcinogenesis and Biological Effects of Tumor Promoters (Hecker, E., Fusenig, N. E., Kunz, W., Marks, F. and Thielmann, H. W., eds.) Carcinogenesis, a comprehensive survey, Vol 7, pp. 217-230, Raven Press, New York

35 Kinzel, V., Loehrke, H., Goerttler, K., Furstenberger, G. and Marks, F. (1984) Proc. natl. Acad. Sci. USA, 81, 5858-5862

36 Furstenberger, G., Gross, M, and Marks, F. (1988) Carcinogenesis, in press

37 Furstenberger, G., Gschwendt, M., Hagedorn, H. and Marks, F. (1987) in Prostaglandins in Cancer REsearch (Garaci, E., Paoletti, R., Santoro, M. G., eds.) pp. 48-61. Springer Verlag, Berlin-Heidelberg

38 Petrusevska, R. T., Furstenberger, G., Marks, F. and Fusenig, N. E. (1988) Carcinogenesis, 9, 1207-1215

39 Slaga, T. J., Klein-Szanto, A. J. P., Triplett, L. L. and Yotti, L. P. (1981) Science, 213, 1023-1025

40 Emerit, J. and Cerutti, P. 91984) in Icosanoids and Cancer (Thaler-Dao H., Crastes de Paulet, A. and Paoletti, R., eds.) pp. 79-90. Raven Press, New York

41 Dutton, D. R. and Bowden, G. T. (1985) Carcinogenesis, 6, 1279-1284

42 Dzarlieva-Petruvska, R. T. and Fusenig, N. E. (1985) Carcinogenesis, 6, 1447-1456

43 Gross, M., Furstenberger, G. and Marks, F, (1987) Exp. Cell. Res., 171, 460-474

44 Lozumbo, W. J., Muehlematter, D., Jorg, A., Emerit, J. and Cerutti, P. A. (1987) Carcinogenesis, 8, 321-326

45 Ochi, E. and Cerutti, P. (1987) Proc. Natl. Acad. Sci. USA., 84, 990-994

46 Furstenberger, G., Kinzel, V., Schwarz, M. and Marks, F. (1985) Science 230, 76-78

STUDIES ON THE MECHANISM OF CONVERSION EFFECTED BY THE TUMOR PROMOTER TPA

V. Kinzel[1], B. Färber[1], M. Kaszkin[1], R. T. Petrusevska[2]
N. E. Fusenig[2], F. Marks[2] and G. Fürstenberger[2]

Institute of Experimental Pathology[1]
Institute of Biochemistry[2]
German Cancer Research Center
Im Neunheimer Feld 280
D-6900 Heidelberg, FRG

ABSTRACT

Tumorigenesis can be effected in mouse skin in a number of steps: by initiation with DMBA, by conversion with one or two applications of TPA and by
promotion with repeated RPA treatment. The conversion step is characterized
by its half-life of 10 to 12 weeks (in NMRI mice) and its independence from
initiation. For conversion the mitogenic capacity of TPA appears to be
necessary in addition to a specific effect of TPA (by which TPA differs
qualitatively from RPA) which consists most probably of the clastogenic activity detected. From a number of inhibition studies it is concluded that the
specific convertogenic effect of TPA precedes the mitogenic effect and thus
takes place rather rapidly. In support of these considerations are studies
in a model cell. It can be demonstrated that TPA mediated clastogenesis accompanied by radiomimetic cell cycle delays is detectable at the earliest
measurable point. Studies on the inhibition of conversion and of clastogenesis as well as data from metabolic studies in culture seem to indicate
that arachidonic acid metabolites play an important role in these processes.

INTRODUCTION

Tumorigenesis in mouse skin can be induced by a subthreshold dose of a
carcinogen (initiation) followed by repetitive treatment with a non-
carcinogen tumor promoter. The process of tumor promotion has been subdivided by Boutwell (1) into two operationally defined stages I ("conversion") and II ("propagation") using Croton oil and turpentine, respectively.
The model systems have been further developed by Fürstenberger et al. (3)
using NMRI mice and by Slaga et al. (2) using Sencar mice. In both animal
systems TPA (12-0-tetradecanoylphorbol-13-acetate) was used to effect stage
I; for eliciting stage II in Sencar mice mezereine and in NMRI mice RPA (12-
0-retinoylphorbol-13-acetate) was applied repeatedly.

Abbreviations

DMBA = 7,12 dimethylbenz[a]anthracene, ETYA = eicosatetraynoic acid,
RPA = 12-0-retinoylphorbol-13-acetate, TPA = 12-0-tetradecanoylphorbol-13-
acetate.

The effectiveness of TPA in stage I has a number of important characteristics: (i) TPA needs to be applied only once (2), (ii) stage II treatment
with repeated doses of RPA does not need to follow TPA treatment immediately
but can be delayed for several weeks (4) indicating that TPA has relatively
long lasting effects, (iii) unlike initiation (which is irreversible) the
action of TPA on NMRI mouse skin exhibits a half-life of 10-12 weeks (5); the
half-life of the action of TPA or RPA in stage II appears to be smaller than
two weeks, (iv) these effects of TPA do not depend on initiation since TPA
can be given several weeks prior to DMBA. Taken together these results are
indicative of a unique mechanism by which TPA elicits this independent step
of multistage tumorigenesis. According to a proposal by Boutwell (1) this
particular step received the name conversion, while stage II is called
promotion (5). Accordingly, TPA is a promoter with an additional convertogenic potency, whereas the promoter RPA is almost ineffective in this
respect.

Part of the strategy of elucidating the mechanism of conversion by TPA
consists of a comparison of the biological activity of TPA with that of RPA.
Both phorbolesters are equally active as irritants and skin mitogens and also
the biochemical and biological responses elicited by TPA and RPA in many
other cellular systems appear to be almost identical (6). The difference in
the convertogenic efficacies of TPA and RPA, therefore, must lie in different
yet undiscovered cellular effects. The convertogenic action is most closely
correlated with cytogenetic effects elicited by TPA but not by RPA (7). In
order to elucidate the significance and the mechanism of the TPA-induced
chromosomal aberrations, a precise timing of the occurrence of conversion as
well as of the chromosomal events is required. In the first part of this
contribution we discuss the possible timing of the conversion taking place in
mouse skin.

In the second part of this contribution the influence of TPA on chromosomes is studied in a model cell - the HeLa cell line - in connection (i)
with the radiomimetic effect of TPA in cell cycle (8) in order to elucidate
sensitive cycle phases and the kinetics of induction of chromosome lesions,
(ii) with alterations of the phospholipid metabolism in order to search for
metabolites mediating cytogenetic and cell cycle changes.

Materials and Methods

The phorbol ester TPA was a generous gift from Dr. E. Hecker (German
Cancer Research Center), RPA was kindly supplied by Dr. B. Sorg. The phorbol
esters were dissolved in analytical grade acetone from Merck (Darmstadt) and
kept at -70^0 in dark bottles until use. Solutions were protected from direct
light during handling.

The performance of the animal experiments by using DMBA as the initiator,
TPA as the convertogenic agent and RPA as the promoter has been described in
detail elsewhere (3,9). Details of the studies using various inhibitors of
conversion have been published with respect to hydroxyurea (9,10), to ETYA
and indomethacin (11) and to cycloheximide (12) as indicated.

HeLa cells were cultivated routinely as monolayers in Eagle's minimal
essential medium containing Earle's salts supplemented with 10% calf serum.
Cells were kept in plastic bottles gassed with 95% air and with 5% CO_2 in a
humidified incubator at 37^0C. For experiments, HeLa cultures were established in appropriate plastic ware for approx. 16hrs prior to experiments.
Phorbol esters were added in 0.2% acetone. Cell cycle analysis by time lapse
photography of prophase cells was done as described (13). Analysis of the
release of the arachidonic acid from prelabeled cultures as well as the
extraction of cellular lipids has been described elsewhere (14). Determination of cellular diacylglycerol was done by thin layer chromatography

(diethylether:petroleather:acetic acid 60:40:1 v/v/v) on silica gel plates
from Merck (Darmstadt) using 1,2- and 1,3-sn-dioleylglycerol as the
standards. Radioactivity (^{14}C) was determined by means of a Linear Analyzer
(Berthold, Wildbad, FRG).

For chromosome analysis the cultures were harvested by incubation in EDTA
(0.2% in PBS) containing 1-2 µg/ml colcemid for 15 min, followed by EDTA/
Trypsin (0.1% each) for additional 5 min. The cells were removed by
pipetting and swollen in hypotonic buffer (0.02 M HEPES, pH 7.4; 0.04 M KCL;
0.5×10^{-3} M EGTA) for 15 min at 37^0C and fixed 3 times with methanol/glacial
acetic acid (3:1). The slides were stained with Giemsa stain (3%) for
approximately 5 min. The metaphases were analyzed for the following structural aberrations; gaps, breaks (chromosome as well as chromatid type),
deletions, rings, centric breaks. For each group 100 metaphases were
evaluated.

RESULTS AND DISCUSSION

When does conversion take place?

The conversion step in multistep tumorigenesis in mouse skin can be
effected by a single application of TPA but not by RPA (2). The fact that
both agents are equally mitogenic and irritant supports an earlier notion
according to which both biological properties are necessary but not sufficient conditions which tumor promoters have to fulfill. If TPA induced DNA
synthesis, which becomes evident by a peak 18 hours after application, is inhibited in a non-toxic fashion by hydroxyurea, conversion is largely aboli-
ished indicating that the induction of cell replication plays an important
role in conversion (9,10). Coincidentally, the inhibitor of enzymes of the
arachidonic acid metabolism ETYA (11) and the inhibitor of protein synthesis
cycloheximide applied 18 hrs after TPA also suppress conversion to various
degrees (for a summary see Figure 1). The reason for these effects is
largely unknown; however, one may speculate that the specific effect of TPA
(not caused by RPA) may somehow become "fixed" by at least one round of
subsequent cell replication.

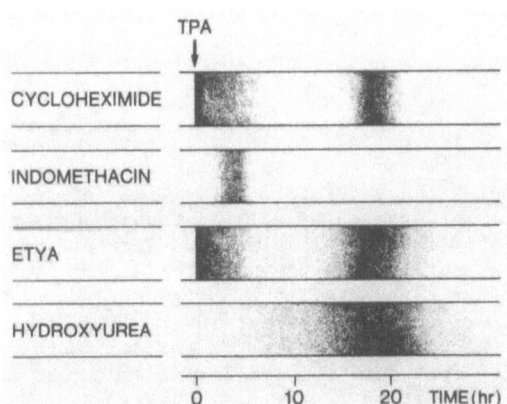

Figure 1. Inhibition of TPA-effected conversion. Mice initiated with DMBA
were treated with TPA at zero time. The inhibitors indicated were applied
once to different groups of animals various times after TPA as described in
detail elsewhere (for ETYA and indomethacine see ref. 11, for hydroxyurea see
ref. 9,10, for cycloheximide see ref. 12). Subsequent promotion was with RPA
for 18-20 weeks. Time and degree of inhibition of conversion is given by the
degree and period of shading.

Therefore, the effect of TPA which is critical for conversion (presumably structural chromosome aberrations; 7) should be effected **prior** to the onset of DNA replication, i.e. within 0-12 hours after treatment with TPA. Experiments with inhibitors of conversion indicate that in addition to a sensitive phase around 18 hours after TPA application a second sensitive phase exists at earlier time points, which is not related to mitogenesis (Figure 1). Thus, indomethacine given 3 hours after TPA inhibits conversion without abolishment of TPA induced DNA replication (11). Similarly, if ETYA is applied shortly prior or after TPA it inhibits conversion without inhibition of TPA induced DNA replication. This early inhibition of conversion by ETYA and by indomethacine (11) may indicate that the conversion-specific effect of TPA occurs around this time. In this respect it should be noted that ETYA has indeed been shown to inhibit TPA effected clastogenesis in epidermal cell cultures completely (7). Between the first and the second inhibition of conversion, i.e. at about 10-12 hours after TPA application, the system becomes refractory to the inhibitory action of ETYA (Fig. 1). ETYA, therefore, appears to interfere with conversion at 2 mechanistically distinct steps. It may be proposed that arachidonic acid metabolites, the formation of which is inhibited by ETYA, are involved as endogeneous mediators of clastogenesis and conversion.

A number of authors have hypothesized that inflammatory cells such as polymorphonuclear leucocytes may be the source of such mediators (for review see 15). However, in the above mentioned studies in epidermal cell cultures no leucocytes were present, indicating that the keratinocytes themselves can be stimulated by TPA to release endogenous clastogenic factors. Moreover, if TPA induced clastogenesis in mouse skin indeed occurs around that time at which the first inhibition maximum is observed with various inhibitors, inflammatory cells are not likely to play a key role since the first signs of inflammation appear later.

Under which circumstances does conversion take place?

In the case of the action of clastogens such as X-irradiation the susceptibility of cells and the type of chromosome aberrations elicited depends on the cell cycle staging. Intimately connected with the action of X-rays are changes, mostly delays, in the cell cycle traverse. To what extend the generation of lesions in the chromosomes and/or their repair are causally linked to the cell cycle delays are unknown. The fact that cell cycle alterations are elicited by a number of radiomimetic chemicals may indicate a mutual relation between cell cycle changes and chromosome aberrations.

The early effect of TPA on cell cycle parameters in mouse skin between 0 and 8 hours after application (16) can be interpreted as a delayed traverse of cells through S-phase (reduced rate of thymidine incorporation into a normal number of cells in S-phase) as well as through G2-phase (rapid decrease of mitotic activity and a simultaneous increase of cells with the DNA content to 2 n i.e. cells in G2-phase). These cell cycle changes resemble those seen after X-irradiation of cell cultures (17). Events in mouse skin beyond 8 hours start to reflect the results of the mitogenic action of TPA. From these considerations it follows that the specific interaction of TPA with the mouse epidermis seems to occur at the same time at which the replicating portion of the basal cells respond to TPA with an apparently radiomimetic response.

Use of a model cell - the HeLa cell

In order to study the early events effected by TPA - apart from events elicited by the mitogenic action of this compound - in detail, HeLa cells, a replicating cell culture system was used. This cell line has been shown earlier to respond by changes of phospholipid metabolism to a large number of phorbolesters according to their biological activity in mouse skin (18). The HeLa cell system, moreover, exhibits changes in the cell cycle traverse after application of TPA which resemble those in mouse epidermis as well as those seen after X-irradiation in culture. These changes comprise a delayed traverse through S-phase as well as a temporary inhibition in G2-phase (8). The latter response has been used to establish dose response relationships. It could be shown that TPA is an order of magnitude more potent in delaying cells in G2 than RPA (13,19). This cell line therefore can discriminate between TPA and RPA. Moreover, pilot experiments indicate that TPA but practically not RPA may cause chromosome alterations in these cells (for details see below).

Taken together, these circumstances indicate that the HeLa cell system may represent a suitable model cell for the mechanistic evaluation of TPA-induced chromosome lesions in relation to the cell cycle alterations particularly under kinetic aspects. Moreover, the use of HeLa cells may facilitate the search for endogenous mediators of clastogenesis.

The action of phorbolesters in HeLa cells

On application of TPA, the HeLa cell exhibits a number of changes in the cell cycle traverse from which the cells recover in the presence of the phorbolester. As an example the temporary inhibition of cells in G2-phase may be discussed in more detail. At a minimum of 10 min after addition of TPA HeLa cells stop entering mitosis (Fig. 2), i.e. they are blocked in the preceeding G2-phase. An increase of the TPA concentration from 10^{-8} to 10^{-7} M increases the G2-delay from approximately 134 min to 240 min. However, an increase from 10^{-7} to 10^{-6} M does not result in a further increase of the G2-delay (13). After a certain period of time they recover from inhibition - at high concentrations of phorbolesters, indicating that phorbolesters act indirectly either by mimiking and/or eliciting cellular mediators. After a certain period of time, the cells become refactory to the action of the phorbolester.

Treatment of HeLa cells with various concentrations of TPA for 24 hrs and analysis of structural chromosome aberrations (including gaps and breaks)

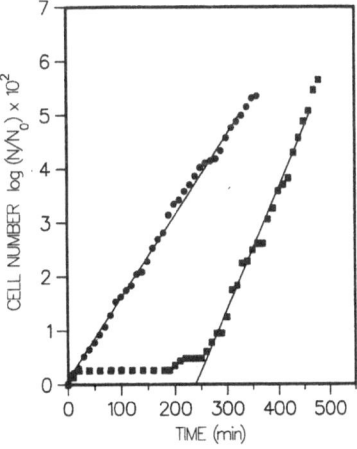

Figure 2. TPA effected inhibition of HeLa cells in G2 phase. Given are cumulative numbers of prophase cells in presence of 10^{-7} M TPA (■) or solvent; 0.2% acetone (●). In parts of the curves every second value ommited. For details see ref. 13.

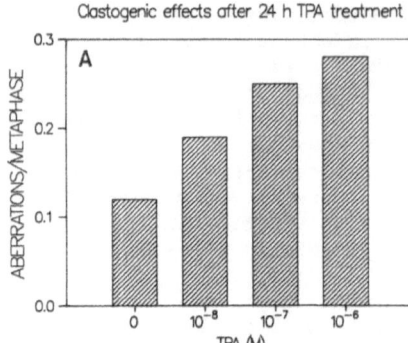
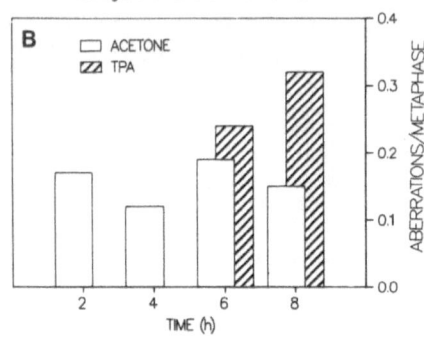

Figure 3. A) Clastogenic effect of various concentrations of TPA and of solvent (0.2% acetone) after 24 hrs treatment of HeLa cells. Each value was obtained from 100 metaphases.
B) Clastogenic effect of TPA (10^{-7} M) and of acetone (0.2%) after various periods of time as indicated. Each value was obtained from 100 metaphases.

also reveal a dose-dependent clastogenic activity of TPA in HeLa cells (Fig. 3A). At TPA concentrations of 10^{-7} to 10^{-6} M the control values obtained in the presence of solvent (acetone 0.2%) are at least doubled. The values obtained with the different concentrations of RPA (10^{-8} to 10^{-7} M) were usually equal to the acetone control, in a few cases, however, a slight elevation of the values was observed which was always much smaller than that obtained with TPA (data not shown).

Chromatid aberrations were observed much more frequently than chromosome type aberrations, thus indicating that aberrations were probably induced during S-phase or later. This assumption was supported by another experiment (Figure 3B) in which HeLa cells were treated for 2, 4, 6 and 8 hrs with 10^{-7} M TPA or solvent. At 2 and 4 hrs a chromosome analysis was only possible from control cultures, whereas the TPA treated cultures were devoid of mitotic activity as expected from the G2-block which lasts approximately 4 hrs (see Figure 2). As soon as the cells, however, overcame the G2-inhibition and entered mitosis, TPA induced structural chromosome aberrations became detectable and they are fully evident after 8 hrs (Fig. 3B). For kinetic reasons, TPA, therefore, most probably exerted the clastogenic effect during late S-phase or during G2-phase. The kinetics of the induction of structural chromatid aberrations may occur rather rapidly even though it takes a number of hours before they can be visualized within the condensed chromosome during metaphase. From these results one can derive the working hypothesis that TPA may induce chromosome lesions in mouse epidermis within a few hours after application. If this is the case, the early inhibition of conversion in mouse skin seen with ETYA, indomethacin and cyclcoheximide (see above) may indicate an interference of these inhibitors with TPA-induced clastogenicity. The exact kinetics of the induction of chromosome lesions by TPA have yet to be worked out by the use of inhibitors. Such studies may help to obtain correlations and possibly causal relationships between metabolic events such as arachidonic acid metabolism on the one hand and the induction of chromosomal aberrations and the convertogenic effect on the other.

As mentioned above, ETYA, an inhibitor of enzymes of the arachidonic acid cascade, inhibits TPA-induced chromosome lesions in mouse keratinocyte cultures (7) as well as the convertogenic activity of TPA in skin carcinogenesis (11) indicating that mediators responsible for the clastogenic and the convertogenic activity of TPA may consist of arachidonic metabolites. TPA also induces the release of arachidonic acid from HeLa cells, probably by

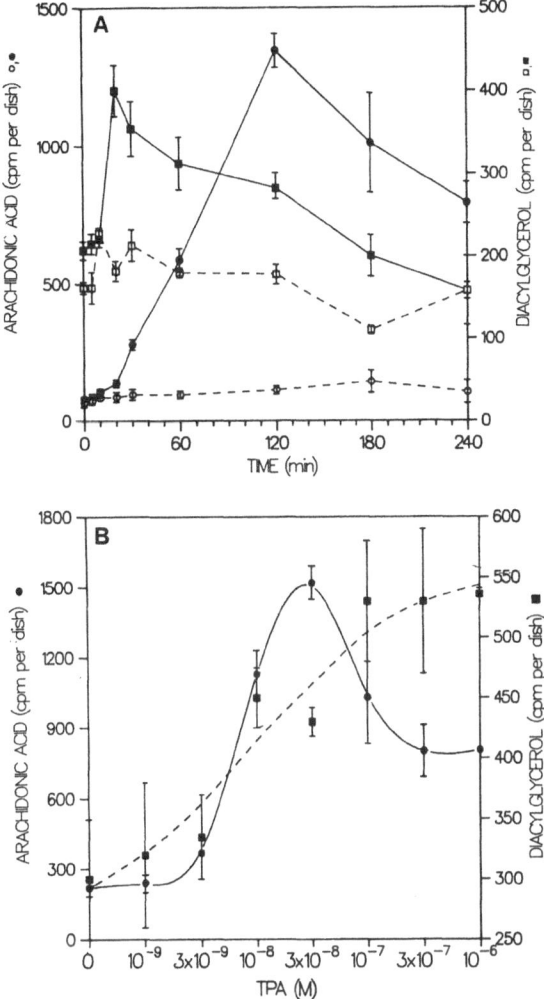

Figure 4. Appearance of cellular 1,2-diacylglycerol (■, □) and of extra-cellular arachidonic acid (●, ○) A.) at different periods after treatment of HeLa cells with 10^{-7} M TPA (■, ●) or acetone (□, ○) and B.) at different concentrations of TPA measured in case of diacylglycerol (■) after 20 min treatment and in case of arachidonic acid (●) after 120 min treatment. These experiments have been carried out in presence of bovine serum albumin (0.5%) instead of calf serum. For details see Materials and Methods. Each value represents the mean of 3 dishes + deviations

activation of phospholipase A_2 (20), with a maximum at 2 hours (Figure 4A). We have not yet approached the analysis of TPA-induced clastogenic factors in HeLa cells directly, however, the following correlation should be noted. The dose response relation for TPA-induced release of arachidonic acid goes through a maximum at 3×10^{-8} M TPA (Figure 4B). Several experiments on the clastogenic activity have shown that 3×10^{-8} M TPA was more effective in that respect than 10^{-8} M or 10^{-7} M TPA thus following the dose response of the arachidonic acid release. These results further support the conclusion that arachidonic acid metabolites may be involved in TPA-induced clasto-genesis in this cellular system.

Arachidonic acid, however, could be excluded as a mediator of the TPA-induced G2 inhibition even though arachidonic acid was capable of inducing a temporary G2-block (21). In a number of biological systems TPA has been shown to stimulate a phospholipase C (22,23) resulting in the generation of 1,2-diacylglycerol. Comparable results have been obtained with HeLa cells as shown in Figure 4. The appearance of diacylglycerol preceeds the release of arachidonic acid (Fig. 4A) and it exhibits a monophasic dose response relation (Fig. 4B). A number of lines of evidence including those from studies with epidermal growth factor as well as synthetic 1,2-diacylglycerol-agonist and -inducer may be responsible for the temporary blockage of cells in G2.

The data obtained so far on the action of TPA in HeLa cells indicate that certain radiomimetic cellular responses such as chromosome aberrations and the temporary G2 inhibition are elicited by TPA via different metabolic pathways in order to obtain insight into relationships between cell cycle changes and chromosome aberrations.

Acknowledgement

We thank G. Bonheim and J. Richards for expert technical assistance and A. Lampe-Gegenheimer for excellent secretarial assistance. The work was supported by the Deutsche Forschungsgemeinschaft.

References

1 Boutwell, R.K. (1964) Progr. Exp. Tumor Res., 4, 207-250
2 Fürstenberger, G., Berry, D.L., Sorg, B. and Marks, F. (1981) Proc. Natl. Acad. Sci. USA, 78, 7722-7726
3 Slaga, T.J., Fischer, S.M., Nelson, K. and Gleason, G.L. (1980) Proc. Natl. Acad. Sci. U|SA, 77. 3659-3663
4 Fürstenberger, G., Sorg, B. and Marks, F. (1983) Science, 220: 89-91.
5 Fürstenberger, G., Kinzel, V., Schwarz, M. and Marks, F. (1985)
6 Marks, F. and Fürstenberger, G. (1984) in Cellular interactions by environmental tumor promoters (Fujiki, H., Hecker, E., Moore, R.E., Sugimura, T. and Weinstein, J.B. eds.) pp. 273-287, VNU Science Press BV, Utrecht, The Netherlands
7 Petrusevska, R.T., Fürstenberger, G., Marks, F. and Fusenig, N.E. (1988) Carcinogenesis, in press.
8 Kinzel, V., Richards, J., and Stohr, M. (1980) Science 210, 429-431
9 Kinzel, V., Loehrke, H., Goerttler, K., Fürstenberger, G., and Marks, F. (1984) Proc. Natl. Acad. Sci. USA. 81, 5858-5862
10 Kinzel, V., Fürstenberger. G., Loehrke, H. and Marks, F. (1986) Carcino-genesis, 7, 779-782
11 Fürstenberger, G., Gschwendt, M., Hagedorn, H. and Marks, F. (1986) in Prostaglandins in Cancer Research (Garaci, E., Paoletti, R. and Santoro, M.G. eds.) pp. 48-61. Springer Verlag, Heidelberg, Berlin
12 Kinzel, V., Fürstenberger., G. and Marks, F. (1988) in: Tumor Promoters: Biological Approaches for Mechanistic Studies and Assay Systems. Raven Press, in press
13 Kinzel, V., Bonheim, G. and Richards, J. (1988) Cancer Res., 48, 1759-1762
14 Espe, U., Fürstenberger, G., Marks, F., Kaszkin, M. and Kinzel, V. (1987) J. Cancer Res. Clin. Oncol., 113, 137-144
15 Birnboim, H. (1983) in Radioprotectors and Anticarcinogens (Nygaard, O.F. and Simic, M.G. eds.) pp. 539-556. Academic Press, Inc., New York
16 Astrup, E.G. and Iversen, O.H. (1981) Carcinogenesis 2, 999-1006
17 Elkind, M.M. and Whitmore, G.F. (1967) The Radiation Biology of Cultured Mammalian Cells. Gordon and Breach, New York
18 Kinzel, V., Kreibich, G., Hecker, E. and Suss, R. (1979) Cancer Res., 39, 2743-2750

19 Kinzel, V., Richards, J., Marks, F. and Fürstenberger, G. (1984) Cancer Res., 44, 139-143

20 Levine, L. and Hassid, A. (1977) Biochem. Biophys. Res. Commun. 79, 477-484

21 Kinzel, V., Kaszkin, M., Espe, U., Richards, J. and Fürstenberger, G. (1987) Exp. Cell Res., 173, 305-310

22 Daniel, L.W., Waite, M. and Wykle, R.L. (1986) J. Biol. Chem. 261, 9128-9132

23 Bestermann, J.M., Duronio, V. and Cuatrecasas, P. (1986) Proc. Natl. Acad. Sci. USA 83, 6785-6789

GENOTOXIC EFFECT OF NATURALLY OCCURRING AND COOKED-FOOD-RELATED MUTAGENS

M Börzsonyi

National Institute of Hygiene
Budapest

SUMMARY

Epidemiological data have shown that diets and life-styles are closely related to human cancer. Many immigrant studies have revealed the importance of dietary habits in inducing cancers of the digestive tract. Foods contain various types of mutagens and carcinogens, and contain both initiators and promoters of carcinogenesis. In general mutagens and carcinogens in food may be A. naturally occurring constituents especially in edible plants or spices; B. heterocyclic amines and polycyclic aromatic hydrocarbons, mainly pyrolysis products of amino-acids and proteins (IQ, MeIQ, MeIQx, Trp-P-1, Trp-P-2, Glu-P-1, Glu-P-2, A C, MeA C etc.) C. mutagenic dicarbonyl compounds produced by heating carbohydrates or by fermentation; D. mutagens formed by (aminocarbonyl reactions) browning reaction; E. naturally occurring tumor promoters: phorbolesters, dehydroteleocidin B. teleocidin, lynygbyatoxin, aplysiatoxin.

Naturally occurring mutagens and carcinogens

Pyrrolizidine alkaloids (Hirono, 1986) are present in thousands of plant species. About 30 alkaloids are recognized as being hepatotoxic. Some of them are ingested by humans, particulary in herbs and herbal teas and occasionally in honey and can cause lung and liver lesions (Mattocks, 1986). Other pure alkaloids have been found to hepatocarcinogenic in rats and some others have also been to be carcinogenic in different experimental animals (IARC Monogr., lo, 333- 1976). Particularly, diester and cyclic diester pyrrolizidine alkaloids which contain necine, retronecine, and otonecine, are strong animal hepatocarcinogens. Among the most widespread of the known naturally occurring mutagens of plant origin are the flavonoids (Knudsen, 1982). Among the flavonoids quercetin, kaempferol and galangin have been shown to be mutagenic (Fujiki et al. 1986). Quercetin occurs in conjugated or free forms in many edible plant products, including fruits, vegetables, (e.g. onion) tea, red wine, dillweed, sumac, and bracken fern. They are carcinogenic for the jejunal and bladder epithelium of the rats (Pamukcu et al. 1980; Sugimura, 1982).

Bracken fern is consumed in certain areas of the world as a food delicacy and salad green. Recently (Matoba et al. 1987), bracken carcinogen ptaquiloside was successfully isolated from bracken fern by following the active principle with a carcinogenicity test. All ptaquiloside treated rats had

Biochemistry of Chemical Carcinogenesis
Edited by R. Colin Garner and Jan Hradec
Plenum Press, New York, 1990

mammary cancer. Multiple ileal adenocarcinomas and urinary bladder tumors
were also observed. Van der Hoeven et al. (1983) isolated from bracken fern
a new mutagenic compound which has the same planar structure as ptaquiloside
and named it Aquilide A.

Numerous alkylbenzene derivatives occur in oils from a wide variety of
plants (Ames, 1983). Among these naturally occurring compounds safrol,
estragol, methyl eugenol, isosafrol and beta asarone have weak or moderate
hepatocarcinogenic activity in mice and rats. These agents occur in mixed
human diet at very low levels (e.g. in black pepper). Human intake of black
pepper is over 2 mg/kg/day.

Betel nut has been considered as one of the important causative factors
in the high incidence of human oral cancer in many Asian countries (Hirono,
1985). In animal experiment mouse skin was painted with an extract of
typical betel (tobacco quid), and squamous cell carcinomas and papillomas
occurred in the painted area (Muir and Kirk, 1960). Several authors suggested that tobacco contains materials which although not in themselves carcinogenic, can enhance the carcinogenic effect of substances present in betel
nut. Indian scientists suggested (Jusawalla and Deshpande, 1971) that the
habit of chewing betel, leads to development of cancer in the oral cavity.
Animal experimental data suggest the possibility that betel quid ingredients
have some carcinogenic or tumor promoting activities in the liver as well as
the upper digestive tract. It is well known that the betel nut contains
several pyriding alkaloids such as arecoline, guvacoline arecaidine and guva-
cine (Hirono, 1985). Very recently it has been reported saliva of chewers
contains the areca derived N-nitrosoguvacoline, and other tobacco specific
nitrosamines when tobacco is added to the quid (Prokopczyk et al. 1987).

Most hydrazines that have been tested are carcinogens and mutagens, and
large numbers of carcinogenic hydrazines are present in edible mushrooms. The
widely eaten Gyromitra esculenta contains 11 hydrazines, three of which are
known carcinogens. One of these, N-methyl-N-formylhydrazine, causes lung and
blood vessels tumors in mice. The most common mushroom, Agaricus bisporus
contains agaritine, derivative of the mutagenic 4-hydrocymethyl-phenyl-hydra-
zine. Some agaritine is metabolized by the mushroom to a diazonium derava-
tive which is a very potent carcinogen. Gyromytrin can be converted at low pH
to methylhydrazine which is an indirect mutagen in Ames' test and a potent
colon carcinogen in mice (Ames, 1983; Toth, 1977, 1984).

Quinones and their phenol precursors (Irons and Sawahata, 1985) are widespread in human diet. Mutagenic anthraquinone derivatives are found in
plants such as rhubarb and mould toxins. Dihydroxylated phenols such as
catechol, resorcinol, caffeic acid and trihydroxylated phenols such as pyro-
gallol and gallic acid have chromosome damaging potential in CHO cells and
are mutagenic in Ames test. Catechol, for example, is a tumor promoter of
animal carcinogenesis (Carmella et al. 1982), and and inducer of DNA damage.

Fava bean (Vicia faba) a common food. It contains the toxins vicine,
convicine and divicine. The latter is a hydrolyzed form of vicine. But faba
bean also contains indols such as 4-chloro-6-methoxyindol which can be converted to a strong mutagenic compound in the presence of nitrite (Wakabayashi
et al. 1987). A nitrosochloroindol, has been identified a very potent mutagen after nitrosating fava beans (Correa, 1987). It seems likely that there
are many tumor initiators in our environment. Meanwhile, the two (or multi-)
step carcinogenesis model has been widely accepted by experimental and
clinical oncologists as well as epidemiologists. The crucial step in the
development of human cancer may be the second stage, that of promotion, not
the first, tumor initiation stage. Phorbol esters isolated from croton oil
have been most intensively studied as tumor promoters. They may have been a
cause of cancer in China and esophageal cancer in Curacao (Hecker, 1981). For

a long time, the phorbol ester TPA (12-0-tetradecanoyl-phorbol-13 acetate) has been used in experimental cancer research as a typical tumor promoter (Hecker, 1981; Hirayama and Ito, 1981; Van Duuren, 1969). In Sugimura's laboratory, a survey was successfully made to detect other tumor promoters that were as effective as TPA. As a result dehydroteleocidin B, teleocidin, and lyngbyatoxin A were found to induce ODC and exert various biological activities in vitro (Fujiki and Sugimura, 1987).

Teleocidin is a product of streptomyces and dihydroteleocidin B, is a catalytically dehydrogentated derivative of teleocidin B. Lyngbyatoxin A is produced by the blue-green alga lyngbya majascula. Teleocidin and lyngbya-toxin A are indol alkaloids (Fujiki et al. 1983). Later aplysiatoxin and debromoaplysiatoxin were also isolated from another variety of the blue-green alga. They have a strong promoter effect on carcinogenesis (Sugimura, 1982).

Heterocyclic amines and related compounds

After the overlapping of mutagens and carcinogens was established mainly from data obtained by using the Ames test, it becomes possible to detect the presence of probable carcinogens in food by detecting mutagenic activity of food. The formation of mutagens upon broiling dried fish and ground meat was first noticed by Japanese authors. These observations led them to identify mutagens formed during cooking. These compounds could probably be produced from creatinine, aldehydes, and Maillard reaction products. First they (Kasai et al. 1981a, 1981b; Spingarn et al. 1980) identified from broiled dried sardines and beef IQ, MeIQ MeIQx and later 3, 4, 8 DiMeIQx. All compounds have a common 2-amino imidazol structure. The former two are quinolin congeners while the others are quinoxalin congeners (Felton et al. 1983; Hatch and Felton, 1986; Sugimura, 1982).

IQ and MeIQ bear a structural similarity the potent colon, breast and prostate carcinogen DMAB (3, 2'-dimethyl-4-aminobiphenyl). Other types of heterocyclyc amines were isolated from pyrolysate of amino acids and proteins. They are Trp-P-1 and Trp-P-2 from tryptophan pyrolysate, and Glu-P-1 and Glu-P-2 from glutamic acid pyrolysate. Amino- -carboline (A C) and methyl-amino- -carboline (MeA C) form a pyrolysate of soybean globulin. They have a common 2-amino-pyridine structure (Hatch and Felton, 1986; Sugimura et al. 1979, 1981; Sugimura and Sato, 1983).

Compounds in the third group (Har, NorHaR and Lys-P-1) are heterocyclic imino compounds. These heterocyclic amines are highly mutagenic toward S. typhimurium TA98 with S9 mix (Fujiki and Sugimura, 1983; Buolo and Schuessler, 1985).

The pyrolysis products induced diphtheria toxin and ouabain resistance, chromosomal aberration, SCE and in vitro transformation in mammalian cells. They also induced 8-azaguanin resistance, chromosomal damage and SCE in human cells (Furihata an Matsushima, 1986). In carcinogenicity studies, IQ induced predominantly hepatomas, forestomach carcinomas and lung tumors in mice. MeIQx proved to be non carcinogenic. Trp-P-1 and Trp-P-2 induced hepatocarcinomas, predominantly in female mice. Recently Ishikawa et al. 1985, observed the activation of Ha-ras oncongene in a rat hepatoma which was induced by IQ feeding. They also found an activation of raf oncogene in another rat liver tumor induced by IQ. In a Glu-P-2 induced adenocarcinoma in the small intestine, the activation of N-ras oncogen was detected.

Mutagens other than heterocyclic compounds formed during cooking

Kinouchi et al. (1986) and Ohnishi et al. (1985) observed the formation of 1-nitropyrene and dinitropyrenes in grilled chicken. Nitropyrenes are some of the most potent mutagens detected widely as environmental pollutants.

Their widespread occurrence is not surprising because NPs are readily formed by exposure of pyrene to NO_2. In addition, if pyrene, formed in incomplete combustion of fats in foods is exposed to NO_2 in burning urban gas, mutagenic nitro derivatives would be readily induced. NPs are carcinogenic.

REFERENCES

1 Hirono, I. (1986) Genetic Toxicology of the Diet, Alan R. Liss, Inc., pp. 45-53
2 Mattocks, A.R. (1986) in Chemistry and toxicology of pyrrolizidine alkaloids p.232. Academic Press, London
3 IARC Monographs (1976) Volume lo. Some Naturally Occurring Substances, Lyon
4 Knudsen, I. (1982) in Mutagens in our environment oo. 315-326. Alan R. Liss, Inc.
5 Fujiki, H., Horiuchi, T., Yamashite, K., Hakii, A., Hirata, Y., Sugimura, T. (1986) In Plant Flavonoids biology and medicine: Biochemical, pharmalogical, and sturcture-activity relationships, pp. 429-440. Alan R. Liss, Inc.
6 Pamukcu, A.M., Yalciner, S., Hatcher, J.F., Bryan, G.T. (1980) Cancer Research, 40, 3468-3472
7 Sugimura, T. (1982) Cancer, 49, 1970-1984
8 Matoba, M., Saito, E., Saito, K., Koyama, K., Natori, S., Matsushima, T., Takimoto, M. (1987) Mutagenesis, 2, 419-423
9 Van Der Hoeven, J.C.M., Lagerweij, W.J., Posthumus, M.A., Van veldhuizen, A., Holterman, H.A.J. (1983) Carcinogenesis 4, 1587-1590
10 Ames, B.N. (1983) Science, 221, 1256-1264
11 Hirono, I. (1985) J. Environ. Sci. Health, C3(2), 145-187
12 Muir, C.S., Kirk, R. (1960) Br. J. Cancer, 14, 597-608
13 Jussawalla, D.J., Deshpande, V.A. (1971) Cancer, 28, 244-252
14 Prokopczyk, B., Rivenson, A., Bertinato, P., Brunnemann, K.D., Hoffman, D. (1987) Cancer Res. 47, 467-471
15 Toth B. (1977) Cancer, 40, 2427-2431
16 Toth B. (1984) J. Environ. Sci. Health, 1, 51-102
17 Morgan, R.W., Hoffman, G.R. (1983) Mutation Research, 114, 19-58
18 Hoffmann, G.R., Morgan, R.W. (1984) Environmental Mutagenesis, 6, 103-116
19 Irons, R.D., Sawahata, T. (1985) In Bioactivation of Foreign Compounds, pp. 259-281. Academic Press., Inc.
20 Carmella, S.G., Lavoie, E.J., Hecht, S.S. (1982) Food Chem. Toxicol., 20, 587-590
21 Tahira, T., Nakayasu, M., Ohgaki, H., Takayama, S., Sugimura, T. (1987) IARC Sci. Publ., 84, 287-291
22 Correa, P. (1987) IARC Sci. Publ., 84, 485-491
23 Hecker, E. (1981) J. Cancer Res. Clin. Oncol., 99, 103-124
24 Hirayama, T., Ito, Y. (1981) Prev. Med., lo, 614-622
25 Van Duuren, B.L. (1969) Prog. Exp. Tumor. Res., 11, 31-68
26 Fujiki, H., Sugimura, T. (1987) Cancer Research, 49, 223-264
27 Fujiki, H., Suganuma, M., Sugimura, T., Moore, R.E. (1983) in Human Carcinogenesis, pp. 303-324. Acad. Press Inc.
28 Sugimura, T. (1982) Gann, 73, 499-507
29 Kasai, H., Shiomi, T., Sugimura, T., Nishimura, S. (1981a) Chem. Lett., 675-578
30 Kasai, H., Yamaizumi, Z., Shiomi, T., Yokoyama, S., Miyazawa, T., Wakabayashi, K., Nagao, M., Sugimura, T., Nishimura, S. (1981b) Chem. Lett., 485-488
31 Spingarn, N.E., Kasai, H., Vuolo, L.L, Nishimura, S., Yamaizumi, Z., Sugimura, T., Matsushima, T., Weisburger, J.H. (1980) Cancer Lett., 9, 177-183
32 Felton, J.S., Hatch, F.T., Knize, M.G., Bjeldanes, L.F. (1983) Diet, Nutrition, and Cancer: From Basic Research to Policy Implications, pp.

177-194. Alan R. Liss, Inc.

33 Hatch, F.T., Felton, J.S. (1986) Genetic Toxicology of the Diet, 109-131

34 Sugimura, T., Nagao, M. (1979) CRC Critical Reviews in Toxicology, pp. 189-209. CRC Press.

35 Sugimura, T., Kawachi, T., Nagao, M., Yahagi, T. (1981) Nutrition and Cancer: Etiology and Treatment, pp. 59-71. Raven Press.

36 Sugimura, T., Sato, S. (1983) Cancer Res., (Suppl.), 43, 2415S-2421S

37 Fujiki, H., Sugimura, T. (1983) Genes and proteins in ocogenesis 111-123

38 Vuolo, L.L, Schuessler, G.J. (1985) Environmental Mutagenesis, 7, 577-598

39 Furihata, C., Matsushima, T. (1986) Ann. Rev. Nutr., 6, 67-94

40 Ishikawa, F., Takaku, F., Nagao, M., Ochiai, M., Hayashi, K., Takayama, S., Sugimura, T. (1985) Jpn. J. Cancer Res., 76, 425-418

41 Kinouchi, T., Tsutsui, H., Ohnishi, Y. (1986) Mutation Research, 171, 105-113

42 Ohnishi, Y., Kinouchi, T., Tsutsui, H., Uejima, M., Nishifuji, K. (1986) Diet, Nutrition and Cancer, Japan Scientific Societies Press, Tokyo/VNU/ Science Press B.V., Utrecht

BIOCHEMSISTRY OF DNA REPAIR PATCH SYNTHESIS IN UV-IRRADIATED PERMEABLE

DIPLOID HUMAN FIBROBLASTS

Steven L. Dresler, Kevin Sean Kimbo, Mark G. Frattini and
Rona M. Robinson-Hill

Department of Pathology, Washington University School
of Medicine,
St Louis, MO 63110 USA

ABSTRACT

Using a well-characterized permeable cell technique and several inhibitors of mammalian DNA polymerases, we have identified polymerase delta
as the enzyme responsible for repair patch synthesis in UV-irradiated
human fibroblasts. UV-induced repair synthesis appears to be mediated by
polymerase delta regardless of (i) the dose of UV administered, (ii) whether
the damaged cells are growing or growth-arrested, or (iii) whether repair
synthesis is studied immediately or at late times (14 hours or more) after
irradiation. The permeable cell technique has also been used to study de-
oxyribonucleoside triphosphate dNTP) concentration dependences of UV-induced
repair synthesis. Apparent K_m values for dCTP, dGTP, and dTTP for repair
synthesis are 0.11 µM, 0.11 µM, and 0.44 µM, respectively, for AG1518 fibroblasts, and 0.06 µM, 0.07µM, and 0.24 µM, respectively, for IMR-90 fibroblasts. These values are an order of magnitude lower than the K_m values for
DNA replication. They are also much lower than the K_m values for isolated
polymerase delta (2.0 µM for dGTP and 3.5 µM for dTTP), suggesting that when
the polymerase functions in DNA repair, its characteristics are altered
either by association with accessory proteins or by post-translational
modification. Also, the K_m values for repair synthesis are 5 to 80-fold lower
than dNTP concentrations found in intact human fibroblasts, indicating that
the "free" dNTP pools of the cell are probably adequate to support high rates
of DNA repair in vivo.

INTRODUCTION

The existence of specific inhibitors has facilitated the identification
of DNA polymerases involved in DNA repair patch synthesis in mammalian cells.
The best studied example is repair of DNA damage induced by UV irradiation.
Numerous studies (1-11) have established that repair synthesis induced by
high doses of UV in non-growing mammalian cells is inhibited by aphidicolin
and thus is mediated by one or both of the aphidicolin-sensitive DNA polymerases (ie DNA polymerase alpha and/or DNA polymerase delta; see references
12 and 13). Subsequent work in permeable human fibroblasts using two
additional inhibitors, butylphenyl-deoxyguanosine triphosphate (BuPh-dGTP)
and dideoxythymidine triphosphate (ddTTP), has specifically implicated
polymerase delta as the enzyme responsible for repair synthesis under these

conditions (14-16). This conclusion is supported by biochemical complement-
ation studies, also performed with human fibroblasts (17). Recently, we have
used our permeable cell system to explore DNA polymerase involvment in UV-
induced DNA repair synthesis in a variety of different situations. Our
results, which are described below, implicate DNA polymerase delta as the
enzyme responsible for DNA repair patch synthesis in UV-irradiated human
fibroblasts under all physiologic conditions. We have also used the
permeable cell system to characterize enzymologically the repair synthesis
complex active in UV-irradiated cells. Our data indicate that the
characteristics of polymerase delta are altered substantially when the repair
complex is formed.

MATERIALS AND METHODS

    Diploid human fibroblasts (AG1518 or IMR-90; Cornell Institute for
Medical Research) were grown in glass roller bottles or plastic culture
dishes and prelabelled with [^{14}C]dThd as described (8,18). For some studies,
intact confluent cells were irradiated with UV while attached to culture
dishes, incubated at 37°C with culture medium for the indicated times,
harvested, and made permeable as described (8). In other cases, the cells
were harvested, made permeable, and irradiated with UV as a cell suspension
in a plastic dish on ice (8). For repair synthesis measurements, unless
otherwise indicated, permeable cells were incubated for 15 min at 37° C with
reaction mix (8) containing 3 uM dATP, 3 uM dCTP, 3 uM dGTP, and 3 uM dTTP
(with one of the nucleotides [^{32}P]-labelled in the alpha position) and repair
synthesis was determined by taking the difference between acid-precipitable
nucleotide incorporation in corresponding damaged and undamaged samples (8).
For all studies of repair synthesis in growing cells, and for studies of
repair synthesis at late times after UV irradiation in confluent cells, dTTP
was replaced by BrdUTP and DNA was isolated from each sample and analyzed by
isopycnic centrifugation in alkaline CsCl as described (8). For measurements
of DNA replication, growing cells were harvested, made permeable, incubated
for 5 min at 37°C with reaction mix (18) containing the indicated concentrations of nucleotides, and acid-precipitable nucleotide incorporation was
determined (18).

RESULTS AND DISCUSSION

    The effect of aphidicolin on repair synthesis induced in confluent cells
by high doses of UV is shown in Figure 1. Whether repair is examined at
early (20 min) or late (14 hr) times after irradiation, a major portion of
the UV-induced repair synthesis is strongly inhibited by aphidicolin. As has
been reported by several investigators (11,19), however, approximately 20% of
UV-induced repair synthesis in confluent cells is resistant to aphidicolin.
These data indicate that the great bulk of DNA repair patch synthesis induced
by UV in confluent cells is mediated by one of the aphidicolin-sensitive DNA
polymerases (ie polymerase alpha or polymerase delta; see references 12 and
13) but that, at least in cells treated with aphidicolin, a small portion is
mediated by an aphidicolin-resisitant DNA polymerase, probably polymerase
beta.

    To determine whether polymerase beta is involved in UV-induced DNA
repair synthesis in unperturbed confluent cells, the sensitivity of such
repair synthesis to dideoxythymidine triphosphate was examined (Figure 2).
DNA polymerase beta is substantially more sensitive to ddTTP than either
of the aphidocolin-sensitive DNA polymerase (13, 21). Repair synthesis
induced in confluent cells by 12 J/m^2 UV was, however, completely resistant
to a concentration of ddTTP which inhibited polymerase beta by over 50% (Fig
2). Furthermore, the ddTTP inhibition curve for repair synthesis lay between

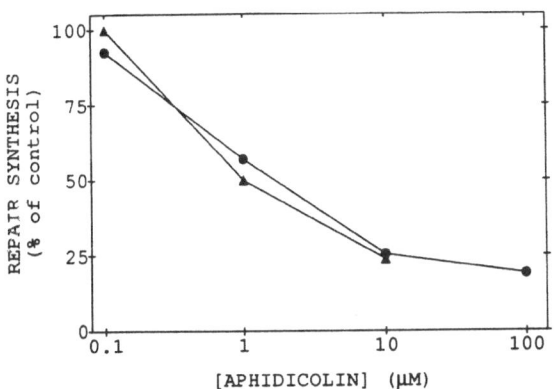

Figure 1: Aphidicolin sensitivity of DNA repair synthesis at early and late times after UV irradiation. Culture plates of confluent AG1518 fibroblasts were either irradiated with 12 J/m^2 UV and incubated for 20 min at 37°C with culture medium (●) or irradiated with 30 J/m^2 UV and incubated for 14 hr at 37°C with culture medium (▲). Cells were then harvested and made permeable, and repair synthesis was determined in the presence of the indicated concentrations of aphidicolin (with dCTP at 0.3 µM). Data are expressed as percentages of the repair synthesis seen in control samples incubated without aphidicolin. Data for repair at 14 hr after irradiation are taken from reference 20.

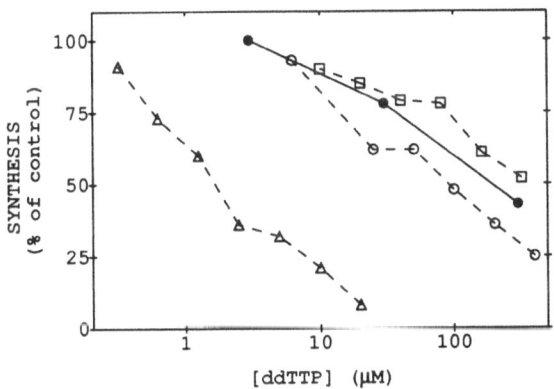

Figure 2. Dideoxythymide triphosphate sensitivity of UV-induced DNA repair synthesis. Culture plates of confluent AG1518 cells were irradiated with 12 J/m^2 UV and incubated for 20 min at 37° C with culture medium. Cells were harvested and made permeable, and repair synthesis (●) was determined in the presence of the indicated concentrations of ddTTP (with dTTP at 1 µM). DNA polymerase alpha (□), beta (△), and delta (○) were also assayed under repair synthesis conditions as described (15) in the presence of the indicated concentratiions of ddTTP (with dTTP a 1 µM). Data are expressed as percentages of the synthesis seen in control samples incubated without ddTTP. Data for polymerases alpha and beta are taken from reference 15.

the curves for the aphidicolin-sensitive polymerase, alpha and delta. Thus, in cells which have not been treated with aphidicolin, there does not appear to be any UV-induced DNA repair synthesis which can be ascribed to DNA polymerase beta. Evidently, polymerase beta becomes involved in UV-induced repair synthesis only when the aphidicolin-sensitive DNA polymerases are inhibited.

In previous studies of intact human fibroblasts which used the bromodeoxyuridine density shift technique, we and others found that [3H]dThd incorporation induced by low doses (2 J/m^2 or less) of UV was largely resistant to aphidicolin (11,19). Subsequently, we have obtained results suggesting that aphidocolin can induce changes in cellular nucleotide metabolism which, in some circumstances, obscure the inhibitory effects of the drug on UV-induced repair synthesis in intact cells (20). For this reason, we have re-examined the aphidicolin sensitivity of repair synthesis induced by low doses of UV using the permeable cell technique, which is unaffected by alterations of cellular nucleotide metabolism (22). When examined in this way UV-induced repair synthesis is highly sensitive to aphidicolin even at UV doses as low as 0.5 J/m^2 (Table I). The aphidicolin sensitivity of repair patch synthesis induced by low dose UV has been confirmed in intact cells using the DNA strand break accumulation technique (Dresler, S.L., Robinson-Hill, R.M., Gowans, B.J. and Hunting, D.J. submitted), which is also insensitive to changes in cellular nucleotide metabolism.

Previous studies, conducted in intact cell systems, have yielded conflicting estimates of the aphidicolin sensitivity of UV-induced DNA repair synthesis in growing mammalian cells (3,6,23-28). Our studies suggesting that aphidicolin can induce changes in nucleotide metabolism which obscure its inhibitory effects on repair synthesis (20) provide a possible explanation for these conflicts. To avoid the effects of changes in nucleotide metabolism, we studied the aphidicolin sensitivity of repair synthesis in growing cells using the permeable cell technique. As seen in Figure 3, UV-induced DNA repair synthesis in permeable growth-phase fibroblasts is completely inhibited by 100 uM aphidicolin. These permeable cell results have been confirmed in intact human fibroblasts using the DNA strand break accumulation technique (Hunting, D.J., Gowans, B.J. and Dresler, S.L., submitted).

In all the situations described above, UV-induced DNA repair synthesis is sensitive to aphidocolin, indicating that it involves either DNA polymerase alpha or DNA polymerase delta. Using the nucleotide analog BuPh-dGTP (29),

Table 1. Aphidicolin sensitivity pf DNA repair synthesis induced by low doses of UV.

UV dose (J/m^2)	Inhibition by aphidicolin (% of control)
0.5	90
1.0	90
2.0	83
8.0	87

Confluent AG1518 cells were irradiated with the indicated doses of UV, incubated at 37°C with culture medium for 20 min, harvested, and made permeable. Repair synthesis at each UV dose was determined without inhibitor and in the presence of 60 uM aphidicolin and the inhibition by aphidicolin was calculated.

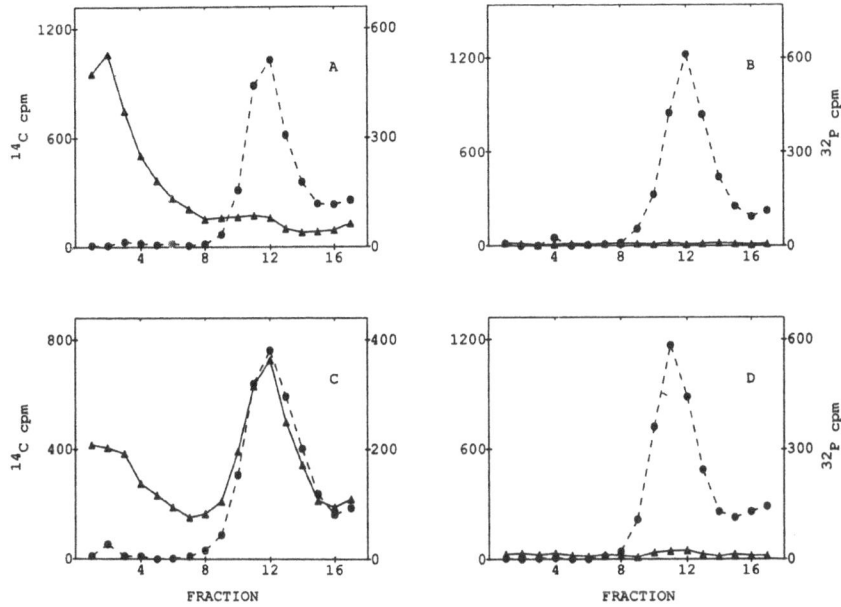

Figure 3. Aphidicolin sensitivity of UV-induced DNA repair synthesis in growing human fibroblasts. Growing AG1518 cells, prelabelled with [^{14}C]dThd, were harvested and made permeable. Portions of permeable cell suspension, either unirradiated (A,B) or irradiated with 100 J/m^2 UV (C,D), were incubated with reaction mixes containing BrdUTP in place of dTTP and either 100 µM aphidicolin (B,D) or no inhibitator (A,C). Following the incubations, DNA was isolated from each sample and analyzed by CsC1 density gradient centrifugation. The gradients were fractionated, and ^{32}P (▲) and ^{14}C (●) radioactivity were determined. The bottom of each gradient is to the left. Nucleotides incorporated by DNA repair (^{32}P-labelled) band at the same density as the parental DNA (^{14}C-labelled). Nucleotides incorporated by DNA relication (also ^{32}P-labelled) are denser than the parental DNA and thus band closer to the bottom of the gradient.

which inhibits polymerase alpha strongly and polymerase delta weakly (30-32), we have differentiated between these two possibilities (Figure 4). Regardless of the dose of UV employed, the growth state of the cells used, or the time after damage at which repair was studied, the BuPh-dGTP sensitivity of UV-induced repair synthesis is essentially·identical to that of DNA polymerase delta. We conclude that all DNA repair synthesis in UV-irradiated diploid human fibroblasts is mediated by DNA polymerase delta.

Using the permeable cell technique one can also characterize enzymologically the various forms of cellular DNA synthesis (18). This opportunity arises because it is possible to control precisely the concentrations of substrates such as the deoxyribonucleotide triphosphates (dNTPs) in the permeable cell reaction mix. Using this approach, we have measured the K_m values of both replication and UV-induced repair synthesis for dCTP, dGTP, and dTTP in diploid human fibroblasts of two different types (Table 2). In all cases, the K_m values for repair synthesis are remarkably low, and those of replication are higher by at least an order of magnitude. This dramatic difference in K_m values is of particular interest because of the possibility that polymerase delta is involved not only in UV-induced DNA repair synthesis

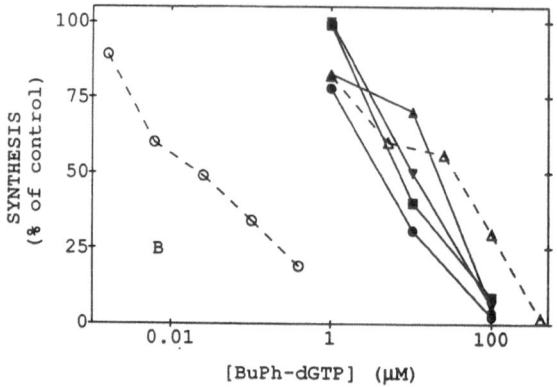

Figure 4. BuPh-dGTP sensitivity of UV-induced DNA repair synthesis. Confluent AG1518 cells were irradiated with either 1 J/^{2} (▲) or J/m^2 UV (●), incubated for 20 min with culture medium, and then harvested and made permeable, or irradiated with 30 J/m^2 UV, incubated for 14 hr with culture medium, and then harvested and made permeable (■). Growing AG1518 cells were harvested, made permeable, and irradiated with 100 J/m^2 UV (▼). Permeable cells were incubated with reaction mixes containing the indicated concentrations of BuPh-dGTP (with dGTP at 0.3 µM) and repair synthesis was determined. DNA polymerases alpha (○) and delta (△) were assayed in the presence of the indicated concentrations of BuPh-dGPT. Data are expressed as percentages of the synthesis seen in control samples incubated without BuPh-dGTP. Data for DNA polymerases alpha and delta are from reference 20.

(14-17), but also in DNA replication (36-38). The K_m values of UV-induced repair synthesis are also much lower than those of isolated DNA polymerase delta (Table 2), suggesting that when polymerase functions in DNA repair, its characteristics are substantially altered either by post-translational modification or by association with specific accessory proteins.

The dNTP K_m values in Table 2 can also be used to gain insight into the mechanisms by which substrate dNTPs are supplied to cellular sites of DNA repair. Snyder (39) has previously shown that UV-induced DNA repair synthesis in diploid human fibroblasts utilize dATP and dGTP synthesized by "salvage" pathways from exogenous deoxyadenosine and deoxyguanosine as efficiently as it utilizes dATP and dGTP synthesized de novo. One explanation for this finding is that dNTPs required for repair synthesis are derived directly from "free" cellular dNTPs pools, without specialized compartmentation or "channelling". We examined this hypothesis by comparing the dNTP K_m values which we measured for UV-induced DNA repair synthesis (Table 2) with dNTP levels which are likely to be present in intact cells. Taking the dNTP pool sizes which have been measured for cultured diploid human fibroblasts (40) and assuming a mean cell volume of 2 picoliters (41), one obtains the following dNTP concentrations: for unsynchronized, exponentially growing cells, dCTP - 5.7 µM, dGTP - 3.2 µM, dTTP - 24.7 µM; for synchronized S-phase cells, dCTP - 10.0 µM, dGTP - 4.0 µM, dTTP - 40.8 µM; and for confluent, growth-arrested cells, dCTP - 4.8 µM, dGTP - 2.6 µM, dTTP - 2.2 µM. Thus, the dNTP K_m values for UV-induced DNA repair synthesis are 29 to 103-fold lower than the dNTP concentrations found in growing cells and 5 to 80-fold lower than the concentrations found in confluent cells. In both growing and non-growing cells, high rates of DNA repair could be supported by the "free" dNTP pools of the cell, without the need for specialized compartmentation to generate higher local dNTP concentrations at sites of repair. In DNA

Table 2. Apparent K_m values for dCTP, dGTP, and dTTP of UV-induced DNA repair synthesis, DNA replication, and DNA polymerase delta

	Apparent K_m (μM)				
	Repair Synthesis		Replication		DNA Polymerase Delta
Nucleotide	AG1518	IMR-90	AG1518	IMR-90	
dCTP	0.11	0.06	1.8	1.7	---
dGTP	0.11	0.07	1.7	1.2	2.0
dTTP	0.44	0.24	2.9	2.7	3.5

For studies of repair synthesis, confluent AG1518 or IMR-90 cells were harvested, made permeable, and irradiated with 100 J/m2 UV. For studies of replication, growing cells were cells harvested and made permeable. Portions of permeable cell suspension were incubated with reaction mixes containing various concentrations of the indicated dTNP with the other dTNPs at 100 μM, and repair synthesis or replication was determined. DNA polymerase delta, isolated from calf thymus (32), was assayed using activated DNA as template-primer (33) with various concentrations of the indicated nucleotide and the other nucleotides at 100 μM. K_m values were calculated by fitting to each set of data a curve of the form described by the Michaelis-Menten equation (34). Each value in the table represents the mean of either three or four determinations. The data are taken from reference 35.

replication, on the other hand "channelling" of dNTPs is probably required to attain the higher rates of synthesis seen in vitro (42).

CONCLUSION

    Permeable cells have proven very useful for the biochemical analysis of DNA repair synthesis in human cells damaged by UV radiation. A single enzyme system has been described which appears to be responsible for U-V-induced DNA repair patch synthesis in all situations. We are currently analyzing, in a similar fashion, the DNA repair synthesis which follows damage by alkylating agents, bulky aromatic hydrocarbons, and drugs which produce DNA strand breaks. Preliminary data suggest the existence of additional repair synthesis systems which operate at different types of damage sites.

ACKNOWLEDGEMENTS

    This work was supported by USPHS Grant CA37261 from the National Cancer Institute, by a grant from the Life and Health Insurance Medical Research Fund, and by Brown and Williamson Tobacco Corporation, Phillip Morris, Incorporated, R.J. Reynolds Tobacco Company, and the United States Tobacco Company. We thank Brenda Jo Mengeling and Joseph A. DiGiuseppe for their helpful comments on the manuscript.

REFERENCES

1 Berger, N.A., Kurohara,K.K., Petzold,S.J. and Sikorski,G. (1979) Biochem. Biophys. Res. Commun., 89, 218-225.
2 Ciarrocchi,G., Jose,J.G. and Linn,S. (1979) Nucleic Acids Res., 7, 1205-1219

3   Hanaoka,F., Kato,H., Ikegami,S., Ohashi,M. and Yamada,M. (1979) Biochem. Biophys. Res. Commun., 87, 575-580
4   Waters,R. (1981) Carcinogenisis, 2, 795-797
5   Snyder,R.D. and Regan,J.D. (1981) Biochem. Biophys. Res. Commun., 99, 1088-1094
6   Snyder,R.D. and Regan,J.D. (1982) Biochem. Biophys. Acta, 697, 229-234
7   Collins,A.R.S., Squires,S. and Johnson,R.T. (1982) Nucleic Acids Res., 10, 1203-1213
8   Dresler,S.L., Roberts,J.D. and Lieberman,M.W. (1982) Biochemistry, 26, 2664-2668
9   Miller,M.R. and Chinault,D.N. (1982) J. Biol. Chem., 257, 46-49
10 Miller,M.R. and Chinault,D.N. (1982) J. Biol. Chem., 257, 10204-10209
11 Dresler,S.L. and Lieberman,M.W. (1983) J. Biol. Chem., 258, 9990-9994
12 Huberman,J.A. (1981) Cell, 23, 647-648
13 Lee,M.Y.W.T., Tan,C.-K., Downey,K.M. and So,A. (1981) Prog. Nucleic_ Acids Res. Mol. Biol., 26, 83-96
14 Dresler,S.L. and Frattini,M.G. (1986) Nucleic Acids Res., 14, 7093-7102
15 Dresler,S.L. and Kimbro,K.S. (1987) Biochemistry, 26, 2664-2668
16 Dresler,S.L. and Frattini,M.G. (1988) Biochem. Pharmacol., 37, 1033-1037
17 Nishida,C., Reinhard,P. and Linn,S. (1988) J. Biol. Chem., 263, 501-510
18 Dresler,S.L. (1984) J. Biol. Chem., 259, 13947-13952
19 Smith,C.A. and Okumoto,D.S. (1984) Biochemistry,23, 1383-1391
20 Dresler,S.L., Gowans,B.L., Robinson-Hill,R.M. and Hunting,D.J. (1988) Biochemistry, in press
21 Kornberg,A. (1980) DNA Replication, p. 204, W.H. Freeman, San Francisco
22 Hunting,D.J. and Dresler,S.L. (1985) Carcinogenesis, 6, 1525-1528
23 Pedrali-Noy,G. and Spadari,S. (1980) Mutat. Res., 70, 389-394
24 Hardt,N., Pedrali-Noy,G., Focher,F. and Spadari,S. (1981) Biochem. J., 199, 453-455
25 Morita,T., Tsutsui,T., Nishiyama,Y., Nakamura,H. and Yoshida,S.(1982) Int. J. Radiat. Biol., 42, 471-480
26 Collins,A. (1983) Biochim. Biophys. Acta, 741, 341-347
27 Seki,S., Hosogi,N. and Oda,T. (1984) Acta Med. Okayama, 38, 227-237
28 Yamada,K., Hanaoka,F. and Yamada,M. (1985) J. Biol. Chem, 260, 10412-10417
29 Kahn,N.N., Wright,G.E., Dudycz,L.W. and Brown,N.C. (1984) Nucleic Acids Res.,12, 3695-3706
30 Byrnes,J.J. (1985) Biochem. Biophys. Res. Commun., 132, 628-634
31 Lee,M.Y.W.T., Toomey,N.L. and Wright,G.W. (1985) Nucleic Acids Res., 13, 8623-8630
32 Crute,J.J., Wahl,A.F. and Bambara,R.A. (1986) Biochemistry, 25, 26-36
33 Wahl,A.F., Crute,J.J., Bodner,J.B., Marraccino,R.L., Harwell,L.W., Lord,E.M. and Bambara,R.A. Biochemistry, 25, 7821-7827
34 Segel,I.H. (1975) Enzyme Kinetics, pp. 18-22, John Wiley, New York
35 Dresler,S.L., Frattini,M.F. and Robinson-Hill,R.M. (1988) Biochemistry in press
36 Prelich,G., Kostura,M., Marslak,D.R., Mathews,M.B. and Stillman,B. (1987) Nature, 326, 471-475
37 Prelich,G. and Stillman,B. (1988) Cell, 53, 117-126
38 Downey,K.M., Tan,C.-K., Andrews,D.M., Li,X. and So,A. (1988) in Cancer Cells, Vol. 6, Eukaryotic DNA Replication (Kelly,T. and Stillman,B., eds.) Cold Spring Harbor Laboratory, Cold Spring Harbor, New York, in press
39 Snyder,R.D. (1984) Mutat. Res., 131, 163-172
40 Snyder,R.D. (1984) Biochem. Pharmacol., 33, 1515-1518
41 Williams,J.I. and Friedberg,E.C. (1982) Photochem. Photobiol., 36, 423-427
42 Mathews,C.K. and Slabaugh,M.B. (1986) Exp. Cell Res., 162, 285-295

INTERACTIONS OF CARCINOGENS WITH NUCLEIC ACIDS

R. Colin Garner

Cancer Research Unit
University of York
Heslington
York   YO1 5DD
United Kingdom

ABSTRACT

Since the discovery in the 1960s that chemical carcinogens can covalently react with nucleic acids, there have been a multitude of publications describing these interactions, identifying the chemical structures of the DNA adducts, studying their enzymic removal, the biological consequences and correlating extents of adduction with carcinogenic response. The majority of these studies have been centred on experimental models and only a few on man, in real carcinogen exposure situations. Up until recently we still did not have direct evidence that chemical carcinogens interact with human DNA in the target organ. This position has become clearer with the advent of various methods to measure carcinogen-DNA adducts in human tissues. Our laboratory has been particularly involved in developing immunological methods to both concentrate and quantify two potential human carcinogens, aflatoxin B_1 (AFB_1) and benzo(a)pyrene (BP). We have used immunoaffinity concentration prior to immunoassay of purified DNA and have found strong evidence that BP adducts are present in human lung tissue and AFB_1 adducts in human liver. Our most recent results indicate that combining immunological procedures with physico-chemical methods, such as P^{32}-postlabelling, enables us to measure adduct concentrations as low as 1 adduct per 10^{10} bases. The implications of these findings will be discussed in relation to risk assessment.

INTRODUCTION

The aetiological agents responsible for the majority of human cancers are largely unknown. Historically, observations as early as the 18th century suggested that industrial exposure to soot can give rise to scrotal cancer; snuff-taking was identified as a causative agent of nasal cancer. Since this time, occupational exposures to a variety of industrial chemicals or processes has been linked with cancers of the lung, bladder, liver, skin, lymphatic system, possibly the brain, kidney, bone etc, so that all organs which can be exposed to carcinogens are, in certain circumstances, cancer-susceptible [1]. However, epidemiological evidence is highly suggestive of an environmental aetiology for up to 80 percent of human cancers; occupational cancers probably account for no more than 5 percent of human cancer [2].

Biochemistry of Chemical Carcinogenesis
Edited by R. Colin Garner and Jan Hradec
Plenum Press, New York, 1990

The first experimental recognition that chemicals can induce cancer were the experiments reported from Japan, in 1915, on the painting of coal tar on rabbits' ears. In the 1930's some of the carcinogenic components of soots, tars and oils were identified as polycyclic aromatic hydrocarbons (PAH's) and studies on these compounds have been continued to this time. In the 1930's and 1940's, came the recognition that certain aromatic amines and dyestuffs are animal carcinogens and, subsequent to this, the recognition of other structural groups, such as the nitrosamines and nitrosoureas, the halogenated hydrocarbons, ethylenic and vinyl compounds and alkylating agents. The diversity of these compounds rendered any sensible hypothesis regarding their mode of action as irresolveable until investigations by the Millers and their collaborators (3) and Brookes and Lawley (4) came to the conclusion that the majority of organic chemical carcinogens were activated in vivo to electrophilic species, or that the chemical was already reactive per se. Figure 1 represents a hypothetical model of an organic chemical containing all the known functional groups which can give rise to electrophilic species.

A fundamental property of electrophiles is that they are reactive with nucleophiles forming stable covalently-bound molecules. If, therefore, during the process of phase 1 or phase 2 foreign compound metabolism a chemical is metabolised in the body to an electrophile, it will form co-valent adducts with cellular nucleophiles. There are a variety of biological molecules within the cell which have nucleophilic centres. These centres are the sites at which covalently-bonded carcinogen molecules can be found. In this short review I will concern myself with reactions of carcinogens with DNA since it is thought that these interactions are of critical importance in carcinogenesis.

Chemical Reactivity of Carcinogen Electrophiles as a Determinant of Carcinogenicity

The realisation that electrophilic activation was essential for organic chemical carcinogenicity has led to a body of research linking chemical reactivity with DNA and carcinogenicity. Research, particularly by Lawley and colleagues (reviewed in reference 5) and Dipple and his colleagues (6), has enabled a number of general statements to be made.

1) that the reactivity of the electrophile can be related to the Swain-Scott factor.

2) that nucleophiles can be divided into 'hard' and 'soft' Lewis bases; hard acids react with hard bases.

3) that electrophilic reactivity proceeds through an SN1 or an SN2 mechanism, or something between the two.

Whilst it has not been possible to apply these rules to all organic chemical carcinogens, considerable progress has been made in relating chemical and biological activity for the alkylating agents, particularly the methylating and ethylating agents. Thus a correlation has been demonstrated between thymic lymphomas in mice and the extent of the O^6-alkylation of gua-nine in thymus DNA in vivo (7). O^6-alkylation had been proposed by Loveless to be a promutagenic lesion, interfering with the normal base-pairing of de-oxyguanosine in DNA (8). Interestingly, the number of dose-dependent hits required to cause thymic lymphomas was one for ethylnitrosourea, two for ethylmethane sulphonate and three for methylnitrosourea. Perhaps the single hit required resulted in a dominant mutation such as has been demonstrated with the ras oncogene.

Whilst good correlations have been observed between chemical reactivity and carcinogenicity for the simple alkylating agents, the correlation appears to break down when bulky aromatic carcinogens are considered. Dipple has

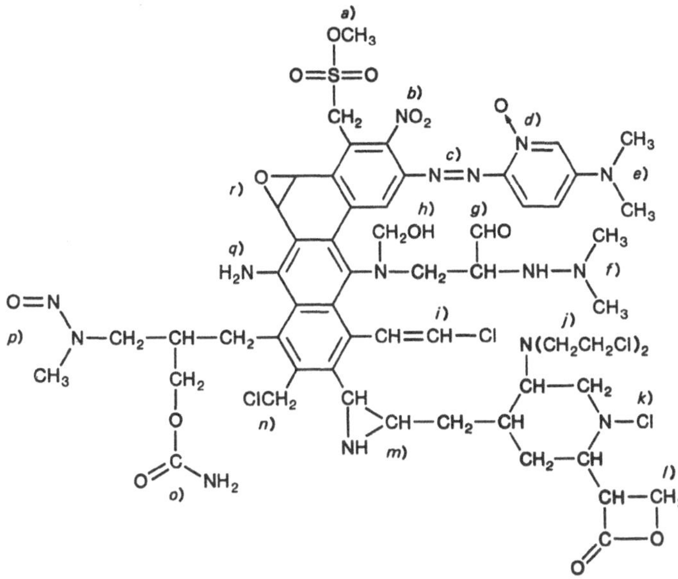

Figure 1. Hypothetical structure containing all known electrophilic centres for a chemical carcinogen (Ashby, Env. Mutag. 7 919-921, 1985). Reproduced by permission of the publisher.

to break down when bulky aromatic carcinogens are considered. Dipple has examined a number of aromatic carcinogens (9) and demonstrated that the ratio of N_2 to 0^6 reaction with guanine varies depending on the electrophilicity of the activated species. The potent carcinogen aflatoxin B_1 (AFB$_1$), however, fails to fit this hypothesis since it reacts with the N^7-position of guanine. Clearly chemical reactivity is not the sole criteria for the position of DNA substitution. Other factors, such as stereochemistry, lipophilicity, goodness of fit, energetic considerations and so forth, could all play a role in determining the site of reaction. In addition, the carcinogenic effects cannot solely be related to chemical reaction with DNA, for example, since DNA repair processes can occur which will remove the 'offending' lesion. There is evidence that these repair processes are more effective in removing carcinogen-adducts in genes which are being actively expressed (10).

Structural Elucidation of Carcinogen-Nucleic Acid Adducts

The first reports that carcinogens could react with nucleic acids were those of Irving and Veazey (11) and Marroquin and Farber (12) for 2-actyl-aminofluorene and the elegant studies by Brooke and Lawley (4), demonstrating that carcinogenic potency of a series of polycyclic aromatic hydrocarbons correlated with their DNA reactivity in mouse skin. Structural characterisation of carcinogen DNA adducts in vivo requires a thorough knowledge of the metabolism of the carcinogen. It is usually necessary to have radiolabelled material available and to have a good synthetic chemistry capability.

One route of adduct identification is set out in Figure 2, in which liver mono-oxygenase enzymes, usually from the rat, are used to metabolise the carcinogen in the presence of DNA. The DNA is recovered, purified and digested to its constituent bases, or deoxynucleosides, before being subjected to chromatography. Evidence for the presence of base- or deoxynucleoside-carcinogen adducts is sought by means of UV or fluorescence absorption, or radioactive peaks. Alternatively, the presumptive reactive metobolite is chemically synthesised, reacted with DNA and the carcinogen-base adduct characterised. Examples of this procedure using liver mono-oxygenases for activation are the aflatoxins (13), certain of the polycyclic aromatic hydrocar-

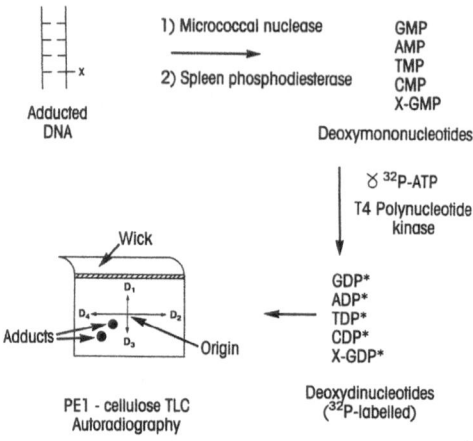

Figure 2. Protocol for ^{32}P post-labelling of DNA adducts. Schematic diagram of method for binding activated carcinogen to DNA _in vitro_

bons (14-16), some aromatic amines (17-19) and halogenated hydrocarbons (20). Synthesis of the presumptive reactive metabolite has been shown to be applicable to 2-acetylaminofluorene (21), benzidene (22), benzo(a)pyrene (23) and many others.

Standard physico-chemical procedures including NMR, mass spectrometry, UV and fluorescence spectroscopy have been used to characterise these adducts. Confirmation that the carcinogen-base adducts characterised are identical to those formed _in vivo_ uses chromatographic procedures and radio-labelled carcinogen. Thus in the case of AFB$_1$, it was established that the _in vivo_ adduct co-chromatographed with the adduct generated _in vitro_, after either liver mono-oxygenase or chemical oxidation (see Figure 3).

Whilst co-chromatography is not the ideal method for demonstrating absolute structure, insufficient material is generated _in vivo_ to chemically characterise. It should always be remembered that chemical reaction with DNA does not always yield stable adducts. This is particularly so for carcinogens which react at the N^7-position of guanine. Here reaction can lead to the opening of the imidazole-ring of guanine, to depurination or to loss of the carcinogen residue itself, yielding the intact base. The situation becomes even more complicated for difunctional alkylating agents, which can form mono- or difunctional adducts; one arm of the alkylating agent can retain its reactive group which can subsequently react with some other nucleophile, such as OH$^-$ or PO$^-_4$. In our laboratories we have characterised some of the _in vivo_ adducts of aflatoxin B$_1$ and G$_1$, the adduct of 1- nitro-pyrene and benzidine, the adduct of methylene orthochlorobisaniline (MOCA) and cyclophosphamide. All of these adducts differ in their site of reaction on DNA and so no generalised statement can be made regarding the mechanism of reaction of their activated metabolites (see Figure 4).

Analysis of Carcinogen-Macromolecular Adducts in Man

The finding that carcinogens react with macromolecules and that such reactions are important in the carcinogenic process, has led to a search for methods that could be used to analyse such interactions in man. Animal studies have indicated that there is a linear relationship between carcinogen dose and DNA binding level, indicating that no binding threshold exists. In addition, multiple dosing with a carcinogen leads to an accumulation of

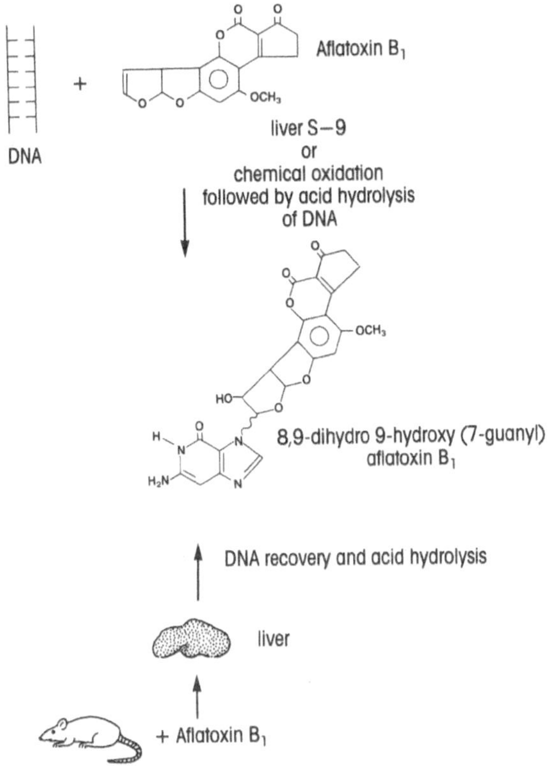

Figure 3. Generation of aflatoxin B_1-guanine in vitro or in vivo

adducts until a steady-state level is reached. These findings mean that, if
a suitable analytical method can be found, it should be possible to measure
human chronic exposure to carcinogens.

Physico-Chemical Methods

Mass spectrometry is the only absolute physico-chemical method of
analysis which is sensitive enough to allow quantitation under real exposure
conditions. The method is expensive in terms of labour and instrumentation,
often requires derivatisation methods to allow sample application and long
work-up procedures. The method has been most widely used in the analysis of
human exposure to ethylene oxide (24). This reactive oxide forms covalent
adducts with histidine and valine in haemoglobin, as well as reacting with
DNA. The chemical is a carcinogen in animals and is suspected of being carcinogenic to man. Ehrenberg and his colleagues in particular, have extensively studied the relationship between the extent of haemoglobin hydroxy-
ethylation and DNA reaction and have concluded that there is a constant ratio
between the two (25). They, on the basis of the level of hydroexthylated
haemoglobin adducts measured in man, estimate the number of DNA adducts to be
formed and reach a conclusion on the cancer risk. This approach, which is
still not widely understood, raises the use of chemical rad equivalents to
measure target dose.

Interestingly, whilst these investigations were being pursued, it was
noted that persons who had not been exposed to ethylene oxide still had
measurable levels of hydroxyethyl groups on their haemoglobin (26). Ehrenberg
has proposed that such adducts arise through exposure to ethene, a ubiquitous
chemical in the atmosphere, which is epoxidised in the body. He goes further

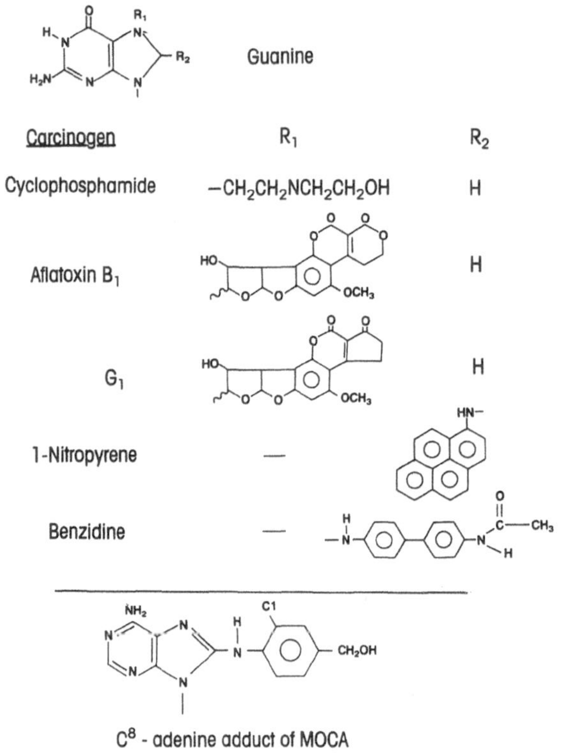

Figure 4. Structure of some adducts studied in our laboratories

on to propose the proportion of human cancers that arise through ethene exposure.

The methods of mass spectrometric analysis of alkylated haemoglobin requires digestion of globin to its constituent amino acids followed by derivatisation for gas chromatography-mass spectrometry.

A recent extension of the mass spectrometry approach is that reported by Tannenbaum and his group (27). They have established a method to analyse adducts of 4-aminobiphenyl, an aromatic amine carcinogen. The method involves recovery of adducted haemoglobin and acid hydrolysis. Any 4-amino-biphenyl (or other aromatic amine) released by this process can be then quantitated by mass spectrometry. A number of aromatic amines are oxidised by haemoglobin, by virtue of a redox cycle between the N-hydroxy and the nitroso-derivative, resulting in significant quantities of the aromatic amine becoming bound. Neumann and his colleagues have examined a whole series of aromatic amine compounds and related the extent of their DNA to haemoglobin binding (28). Whereas it was possible to compare species differences in carcinogenic potency for a particular aromatic amine, it was not possible to make potency comparisons between different aromatic amines structures. Studies on 4-aminobiphenyl adducts in cigarette smokers and non-smokers, have demonstrated the former group to have higher levels of binding (29).

Fluorescence Spectroscopy

There are few physico-chemical techniques which can achieve the sensitivity of mass spectrometry. One such is fluorescence spectroscopy, particularly when computer enhancement is available. The method clearly is only applicable to carcinogen-adducts which have a characteristic fluorescence spectrum. Groups of carcinogens where fluoresence spectroscopy has been used include polycyclic aromatic hydrocarbons, such as benzo(a)pyrene (30,31) and aflatoxin B_1. In the case of benzo(a)pyrene, adducts have been measured in white blood cells of smokers and non-smokers (32), as well as in certain industrial situations, such as coke-oven workers (33). In the case of the major benzo(a)pyrene-diol epoxide (BPDE)-adduct, which is substituted at the extra amino group of guanine, acid hydrolysis of this adduct will release the tetrol, which is intensely fluorescent. Sensitivity of synchromous fluorescence spectroscopy has been estimated as one adduct per 10^7 nucleotides.

In the case of aflatoxin B_1, it has been demonstrated that the major DNA adduct of this potent liver carcinogen, AFB_1-guanine, is present in the the urine of exposed rats (24) as well as of exposed persons (31). Approximate correlations have been found between aflatoxin levels in ingested foods and the amounts of AFB_1-gua excreted in urine.

^{32}P-postlabelling

Randerath and his co-workers reported in 1980 the use of an enzymic method to incorporate ^{32}P ATP into carcinogen-mononucleotides and mononucleotides, with subsequent separation of the carcinogen modified dinucleotides from the non-reacted bases. The method is outlined in Figure 5 and has been reviewed by Randerath (35) and by Watson (36). The limits of sensitivity are determined by the specific activity of the [^{32}P]ATP available. We have used this method in our own laboratories and find it a practical technique for application to human samples. Essentially, the method relies on the ability to separate carcinogen-dinucleotides from dinucleotides using multi-dimensional TLC. If the adduct is indistinguishable chromatographically from the normal dinucleotides, then the procedure cannot be used. Futhermore, certain carcinogens, when reacted with DNA, such as AFB_1, render the DNA resistant to enzymic hydrolysis. Nevertheless, the technique has proven to be particularly useful for PAH adducts. It has the advantage of being non-specific in its applications; in other words, the investigator does not need to know the structure of the adduct he is examining. A footprint adduct pattern could be typical for a particular carcinogen exposure.

In our laboratories (in collaboration with Dr D Phillips, Chester Beatty Labs, Institute of Cancer Research, London) we have analysed DNA samples from human lung removed at thoracic surgery. The results of our study have been reported. A clear linear dose-response relationship was seen between the number of cigarettes smoked per day and DNA adducts (correlation co-efficient = 0.724 p $<$0.001) suggesting that the level of adducts reflects the degree of cigarette smoke exposure. Interestingly, 6 of the patients studied described themselves as non-smokers. Three of these had adduct levels similiar to non-smokers. On closer examination, it turned out that three of the 'former' smokers had given up smoking only a few months previously, whilst the other three had given up at least five years ago. These latter had the low adduct levels. Clearly there is a loss of DNA adducts with time after the cessation of smoking. Could this be the explanation why the incidence of lung cancer declines with increasing time after ceasing smoking?

There are various methods which have been used to enhance the sensitivity of post-labelling, including nuclease-P_1 digestion and butanol extraction. With enhancement, the technique can measure 1 adduct per 10^{10} bases, a higher sensitivity than any other method presently available. Disadvantages of the method are:-

i)   possible incomplete digestion of the DNA sample under investigation to mononucleotides.

ii) failure of the polynucleotide kinase to quantitatively convert the mononucleotide to the dinucleotide.

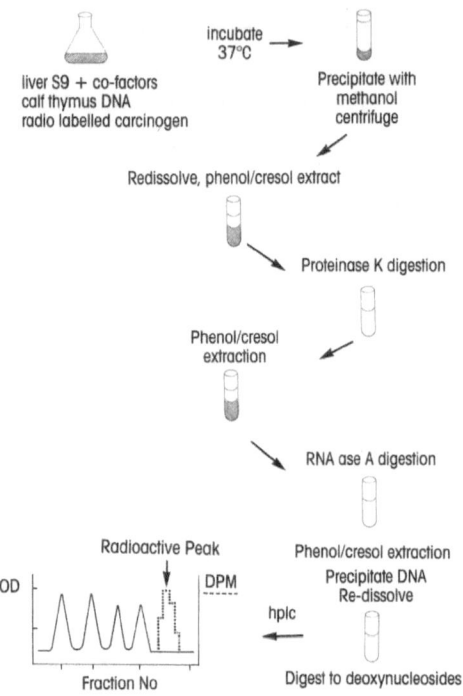

Figure 5. Protocol for ^{32}P post-labelling of DNA adducts

iii) no difference in chromatographic mobility of the carcinogen-adduct
compared with unreacted bases.

iv) failure of an enhancement technique such as nuclease P_1 digestion to
selectively cleave unadducted bases.

With these provisos the ^{32}P-postlabelling technique, or similiar methodology,
offers the most promising method for future adduct monitoring.

Immunological Methods

The generation of antibodies against DNA adducts was first reported over
20 years ago (38). However, it is only during the last decade that antibodies have been used to provide the basis for sensitive assay methods for
adduct measurement. Immunoassays have been long used in the field of
clinical chemistry but their acceptance for population-based DNA adduct
analysis is relatively recent.

The procedures for antibody generation are diverse and include both polyclonal and monoclonal antibodies. It is essential that a good immunogen is
used for antibody generation. Antigens that have been used include carcino-
gen-reacted DNA (cisplatin, aflatoxin B_1, benzo(a)pyrene, 2-acetylamino-
fluorene, dimerised DNA), adduct-protein conjugates (O^6-methyl-, O^6-ethyl-,
O^6-butylguanine, cisplatin, benzo(a)pyrene-diol epoxide-deoxyguanosine).
Antibodies that can be used include both mono- and polyclonal (reviewed in
reference 39). In addition to the use of antibodies in immunoassay, they can
also be used for immunoaffinity concentration. It might be thought that
antibodies are highly specific against the antigen to which they were raised.
Evidence is accumulating that this need not be the case.

We have generated polyclonal antibodies against BPDE-DNA in rabbits and
used these for both immunoassay and immunoconcentration (40). In an experi-

ment in which [3H]dimethylbenzo(a)anthracene was painted on a mouse skin and the mouse skin DNA recovered, it was found that at least one adduct of DMBA was bound to the immunoaffinity column which had been prepared using anti-BPDE-DNA antibodies. Clearly there was some cross-reactivity with at least one DMBA adduct. Presumably there was a group of antibodies in the population which had cross-reactivity with the aromatic region of PAHs. Although this situation is less likely to arise with a monoclonal antibody, there is evidence that cross-reaction can occur with a completely unrelated antigen to that which the antibody was generated against, The word 'mistope' has been coined for an antigenic determinant that is unrelated to the original epitope.

It is my view that the greatest value of antibodies should lie in their use as sample clean-up procedures prior to some other method of analysis, such as ^{32}-postlabelling or mass spectrometry.

CONCLUSION

Procedures have been described for elucidation and characterisation of carcinogen-macromolecular adducts. Quantification methods enabling analysis of human samples in real exposure situations are now available. With these methods it should be possible to relate the levels of DNA adducts to the biological response using animal models as a guide.

This science has now reached a stage where the epidemiologist can combine with the laboratory scientist to carry out population-based studies at the molecular level with a view to determining risk in real exposure situations.

ACKNOWLEDGEMENTS

The author is grateful to the Health and Safety Executive for supporting parts of the work of the Cancer Research Unit, York.

REFERENCES

1   Vainio, H (1987) J. Cancer Res. Clin. Oncol., 113, 403-412.
2   Doll, R. (1978) Cancer Res., 38, 3573-3583.
3   Miller, J.A. (1970) Cancer Res.,30, 559-576.
4   Brookes, P. & Lawley, P.D. (1964) Nature, 202, 781-784.
5   Lawley, P.P. (1984) Chemical Carcinogens vol.1 ed. C. E. Searle American Chemical Society Monograph 182, American Chemical Society, pp 325-484.
6   Moschel, R.C., Hudgins, W.R. & Dipple, A. (1979) J. Org. Chem, 44 3324-3328.
7   Frei, J.V., Swenson, D.H., Warren, W. & Lawley, P.D. (1978) Biochem. J., 174 1031-1045.
8   Loveless, A. (1969) Nature 223, 206-207.
9   Dipple, A., Lawley, P.D. & Brookes, P (1968) Eur. J. Cancer 4, 493- 506.
10 Vos, J.M.H. & Hanawalt, P.C. (1987) Cell,50, 789-799.
11 Irving, C.C. & Veazey, R.A. (1969) Cancer Res., 29, 1799-1804.
12 Marroquin, F & Farber, E. (1965) Cancer Res., 25, 1262- 1269.
13 Garner, R.C. (1973) Chem-Biol. Interact., 6, 125-129.
14 Grover, P.L. & Sims, P. (1969) Biochem. J. 110, 159-160.
15 Gelboin, H.V. (1969) Cancer Res., 29, 1272-1276.
16 Thompson, M.H., Osborne, M.R., King, H.W.S. & Brookes, P. (1976) Chem-Biol. Interact., 14, 13-19.
17 Thorgeirrson, S.S., Jollow, D.J., Sasame, H.A., Green, I. & Mitchell, J.R. (1973) Mol. Pharmacol.,9, 398-404.
18 King, C.M. (1974) Cancer Res., 34, 1503-1515.

19 Nemoto, N., Kusumi, S., Takayama, S., Nagao, M & Sugimura, T. (1979) Chem-Biol. Interact., 27, 191-198.

20 Barbin, A., Bresil, H., Croisy, A., Jacquignon, P., Malaveille, C., Montesano, R. & Bartsch, H. (1975) Biochem. Biophys. Res. Commun., 67, 596-603.

21 Miller, J.A. & Miller, E.C. (1967) Prog. Exp. Tumour Res., 11, 273-301.

22 Martin, C.N., Beland, F.A.,Roth, R.W. & Kadlubar, F.F.(1982) Cancer Res, 42,2678-2696.

23 Sims, P., Grover, P.L., Swaisland, A., Pal, K. & Hewer. A. (1974) Nature, 252, 326-327.

24 Tornqvist, M., Osterman-Golkar, S., Kautiainen. A., Jenson, S., Farmer, P.B. & Ehrenberg, L. (1968) Carcinogenesis, 7, 1519-1521.

25 Calleman, C.J., Ehrenberg, L., Jansson, B., Osterman- Golker, S., Segerback, D., Svensson, K. & Wachtmeister, C.A.(1978) J. Environ. Pathol. Toxicol., 2, 427-442.

26 Farmer, P.B., Bailey, E., Gorf, S.M., Tornqvist, M., Osterman-Golkar, S., Kautianen, A. & Lewis-Enright, D.P.(1968) Carcinogenesis, 7, 637-640.

27 Green, L.C., Skipper, P.L., Turesky, R.J., Bryant, M.S. & Tannenbaum, S.R. (1984) Cancer Res., 44, 4254-4259.

28 Neumann. H-G (1983) in Developments in the science and practice of toxicology eds. A.W. Hayes, R.C. Schnell & T.S. Miya, Elsevier, Amsterdam, pp 135-144.

29 Bryant, M.S., Skipper, P.L., Tannerbaum, S.R. & Maclure, M. (1987) Cancer Res 47, 602-608.

30 Vahakangas, K., Haugen, A.A. & Harris, C.C. (1985) Carcinogenesis, 6, 1109-1116.

31 Autrup, H., Bradley, K.A., Shamsuddin, A.K.M., Wakhisi, J. & Wasunna, A. (1983) Carcinogenesis, 4, 1193-1195.

32 Haugen, A.A., Becher, G., Benestad, C., Vahakangas, K., Trivers, G.E., Newman, M. J. & Harris, C.C. (1986) Cancer Res., 46, 4178-4183.

33 Harris, C.C., Vahakangas, K., Newman, M.J., Trivers, G.E., Shamsuddin, A., Sinopoli, N., Mann, D.L. & Wright, W.E. (1985) Proc. Natl. Acad. Sci. (USA), 82, 6672-6676.

34 Bennet, R.A., Essigmann, J.M. & Wogan, G.N. (1981) Cancer Res., 41, 650-654.

35 Reddy, M.V., Gupta, R.C., Randerath, R. & Randerath, K. (1984) Carcinogenesis, 5, 231-243.

36 Watson, W.P. (1987) Mutagenesis, 2, 319-331.

37 Gupta, R.C. (1985) Cancer Res., 45, 5656-5662.

38 Levine, L., Seaman, E., Hammerschlag, E. & Van Vunakis, H, (1966) Science, 153, 1666-1667.

39 Perera, F.P. (1987) J. Natl. Cancer Inst., 78, 887-898.

40 Tierney, B., Benson. A. & Garner, R.C. (1986) J. Natl. Cancer Inst., 77, 261-267.

THE INDUCTION OF ALKALI LABILE SITES IN DNA OF CHO CELLS BY METHYL METHAN-
SULPHONATE AND DIMETHYLSULFATE AND ITS RELATIONSHIP TO THE INHIBITION OF
OF NASCENT DNA CHAIN ELONGATION

R. Štětina

Department of Drug Carcinogenesis
Institute of Experimental Biopharmacy
Czechoslovak Academy of Sciences
Olesnice v Orlickych horach
Czechoslovakia

SUMMARY

The induction and repair of alkali labile sites (ALS) with methyl methane
sulphonate (MMS) and dimethyl sulphate (DMS) in the DNA of Chinese hamster
ovary (CHO) cells is biphasic. During the treatment of cell and early after-
wards ALS are induced in DNA; they are removed from DNA by a repair mechanism
with a half-life of about 3 hours. Between the 3rd and 6th hour after treatment, secondary ALS developed the repair of which had a half-life of 9 to 16
hours. On treatment of cells, most probably both ALS and potential ALS
(PALS) develop; between the 3rd and 6th hour after treatment they are either
spontaneously of enzymatically converted to secondary ALS. With smaller
doses of MMS (2mM) the secondary ALS do not occur. It seems that as long as
the total amount of induced PALS is small, CHO cells possess a sufficient
capacity to remove them from DNA already prior to their conversion to ALS. A
substantial part of primary and secondary ALS is removed from DNA by a long
type of repair sensitive to araC and novobiocin (NB). Secondarily developed
ALS are effective blockers of elongation of the nascent DNA chains.

INTRODUCTION

Alkylating agents react with various nucleophilic sites in DNA thus
giving rise to a number of chemically induced lesions of DNA (1). Much
effort has been devoted in elucidating the part played by the individual
alkylated bases in the cytotoxicity, mutagenicity and inhibition of DNA
synthesis (for a review see reference 2). Which alkylated bases are formed
depends on the degree of affinity (expressed by Swain-Scott constant) of the
alkylating agent with nucleophilic centres of DNA. Agents with a high constant react in the DNA molecule preferentially with nitrogen atoms which are
more nucleophilic than oxygen atoms. These drugs are generally more toxic
and relatively less mutagenic than those agents with a low Swain-Scott constant which show higher mutagenicity and lower toxicity (3).

It is not clear which of the DNA lesions induced by alkylating agents
cause inhibition of DNA replication. Whereas the replication of the DNA
phage o X174 in vitro is blocked by apurinic sites but not ethyl-adducts on
DNA (4), in hamster and murine cells the alkali labile sites are not the only

cause of the inhibition of replication (5). DNA replication in the rat liver is not blocked by the presence of N^7-methylguanine (N^7-MeG) or O^6-methyl-guanine (O^6-MeG) (6). According to some authors, the blockade of DNA replication in the chinese hamster ovary (CHO) cells is caused by protein-synthesis inhibition (7).

In the present paper an attempt was made to examine the relationship between the repair of alkali labile sites in DNA induced in CHO cells by the action of methyl methane sulphonate (MMS) and dimethyl sulphate (DMS) and the inhibition of elongation of newly synthesized DNA.

Materials and Methods

MMS (Fluka) and DMS (VEB Laborchemie Apolda, GDR) were diluted in ice-cold phosphate-buffered salin (PBS) immediately before use.

The CHO cells were cultivated in MEM supplemented with 3% v/v calf and 3% v/v foetal calf serum and passaged as previously described (8).

Measurement of ALS and the degree of elongation of nascent DNA. These were always measured simultaneously in parallel cultures in one experiment. To determine ALS and elongation, the method of alkaline unwinding of DNA, originally elaborated by Erixon and Ahnstrom (9) and modified by Squires et al. (10), was employed as previously reported in greater detail (10). For ALS determination, cultures labelled with ^{3}H-thymidine (^{3}H-TdR) and culti-vated for 12 hours prior to treatment with the mutagen without thymidine were used. The measurement of DNA elongation was carried out in parallel cultures labelled with ^{3}H-TdR for 10 minutes immediately after treatment with the mutagen and further cultivated without ^{3}H-TdR. At various intervals after treatment with the mutagen the cultures were lysed with an alkaline lysing solution, neutralised and percentage of the double-strand (DS) DNA was determined in the samples by means of chromatography on hydroxyapatite (HA). The decrease in the DS DNA found in cells labelled with the ^{3}H-TdR before treatment with the mutagen reflects the presence of breaks (or ALS) (see 8,9,10). On the other hand, in the cultures labelled after treatment with the mutagen the increase in the DS DNA is the measure of the rate of elongation of newly synthesized DNA strands.

The effect of repair inhibitors on the induction of DNA single strand breaks (alkali labile sites). To the cultures prelabelled with ^{3}H-TdR, hydroxyurea ($HU-10^{-2}M$), 1-β-D-arabinofuranosylcytosine ($araC-10^{-4}M$) or novo-biocin ($NB-10^{-3}M$), (Sigma) were added. To one half of the cultures, the inhibitors were added 15 minutes prior to MMS and these cultures were lysed immediately after treatment. In the second half of the cultures, inhibitors were added immediately after removing MMS and cultures were lysed after a one-hour incubation in the presence of inhibitors. The lysed cultures were processed for the determination of ALS by chromatography on HA as described above.

RESULTS

Induction of ALS in DNA and the elongation of DNA in MMS treated cells

As shown in Figure 1, MMS in high doses of 10 and 20 mM induced within 30 minutes approximately 8 and 11 ALS/10^9 daltons . During the first hour of incubation after treatment a decrease from 8.5 to 6.4 of ALS and from 11 to 6.8 of ALS occurred. The dose of 20 mM was strongly toxic and due to the de-tachment of cells from the bottom of the culture vial it was not possible to determine ALS further. In the dose of 10 mM, an increase in the number of ALS to 10.5 occured between the 3rd and 6th hour after treatment; between the

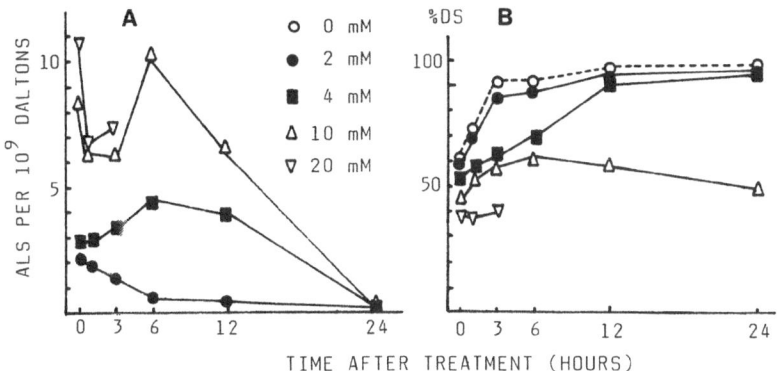

Figure 1. The induction of ALS (A) and the rate of nascent DNA chain elongation (B) after treatment of CHO cells with various concentrations (see graph) of MMS, For details see methods.

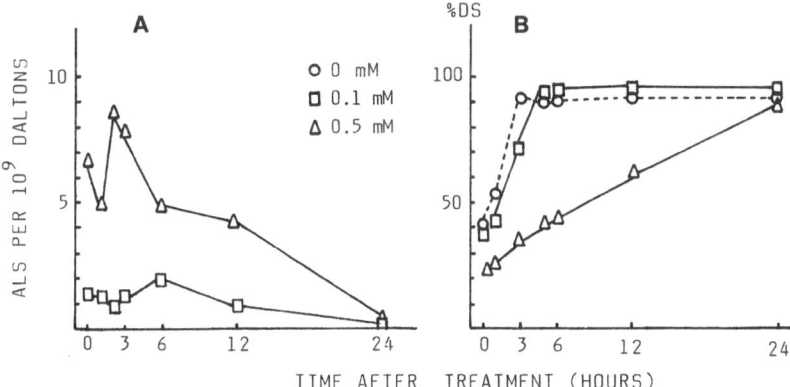

Figure 2. The induction of ALS (A) and the rate of nascent DNA chain elongation (B) after treatment of CHO cells with various concentrations (see graph) of DMS. For details see methods.

6th and 24th hour this was decreased to nearly zero. After a dose of 2 mM MMS the number of ALS gradually decreased (from 2.4 to 0.7) until the 6th hour (half-life of about 3 hours); between the 6th to 24th hour this decrease was very slow. After a dose of 4 mM, however, an increase in the number of ALS (from 3 to 4.5) took place between the 3rd and 6th hour, but this slowly decreased again between the 6th and 24th hour. The process of the repair of these ALS seems to be biphasic. In the first six hours a repair of ALS and their increase simultaneously occur (between the 3rd and 6th hour). This increase is proportional to the dose of MMS; it does not occur at lower doses (2 mM). It seems that the cell is able to repair a certain amount of potential ALS sooner than their conversion to ALS takes place in later periods (between the 3rd and 6th hour).

Elongation of DNA is not severely inhibited with the 2 mM dose; after termination of the first phase of ALS repair in the period of 0 - 6 hours the value of the percentage of %DS DNA approaches that of the control, which it reaches within 12 hours. At a dose of 4 mM, however, the elongation of DNA is markedly decelerated commencing with at a time of 1 hour after treatment at which there was an increase in secondary ALS which reached the value of

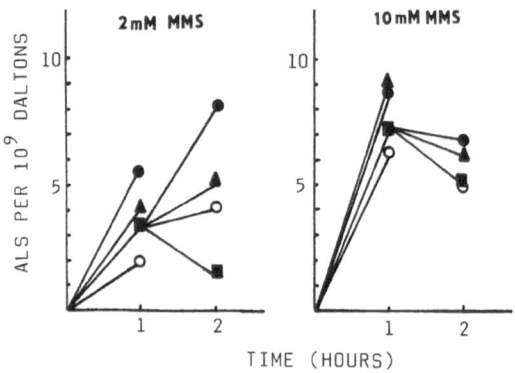

Figure 3. The effect of DNA-repair inhibitors on the acculmulation and repair
of ALS in DNA of CHO cells treated with MMS. MMS was present in the medium
from 0 to 1 hour. Inhibitors were added 15 minutes before MMS or after
removal of MMS for 1 hour. (▲) araC+HU; (o) NB; (●) araC+HU+NB; (■)no
inhibitors.

the control as long as 24 hours after treatment. The dose of 10 mM induced a
considerable deceleration of elongation which began after 6 hours, when the
maximal increase in secondary ALS was found.

The induction of ALS by the action of DMS (see Figure 2), which induced
similar amounts of ALS in concentrations 20-fold lower than those of MMS, was
of a similar character.

Effect of repair inhibitors on ALS induction

Fig. 3 shows that the presence of NB in the medium during the treatment of
the cells with MMS results in a considerable decrease in the number of induced ALS. In the presence of araC+HU, on the other hand an enhanced accumulation occurred. It was interesting to find that NB did not decrease the
accumualtion caused by araC+HU; on the other hand, in its presence together
with araC and HU a further increase in the number of ALS was seen. After the
removal of MMS from the medium, ALS were repaired with kinetics similar to
those shown in Figure 1. When at this time inhibitors were present in the
medium, NB in contrast to the first case (when it was present in the medium
during treatment) blocked the ALS repair. The repair was blocked even more
by the combination of araC+HU; the most effective blockade of ALS repair was,
however, observed when NB was also present.

DISCUSSION

The method of alkaline unwinding of DNA used in the present paper
provides the possibility of observing the course of induction and repair of
DNA breaks simultaneously with the elongation of newly synthesized DNA, or
the effect of various combinations of DNA repair inhibitors on the inducation
or repair of ALS of DNA in one experiment in a rapid time succession. With
the use of the method of centrifugation of DNA in alkaline sucrose gradient
or the method of alkaline filter elution the processing of so many samples in
one experiment of this type is not, on the rule, possible due to practical
reasons.

The amount of ALS of DNA induced in the cells influenced by MMS was similar to other published papers (12). DMS induced approximately the same amount of ALS in concentrations 20-fold lower than MMS.

From the conclusions of a number of authors, alkylated DNA contains alkali labile sites of differing types. As has been discussed (12), alkali labile apurinic sites of phosphotriesters can appear in DNA by the action of glycosylases or spontaneously. Apurinic sites can be repaired by excision and resynthesis of the strand of DNA. In addition, as reported by Abondandollo (13), different apurinic sites can have different sensitivity to high pH, so in the determination of ALS in the present paper, where the alkaline conditions used were relatively mild, short-term acting and of low temperature, probably not all of alkali labile sites have been determined.

The results of the present paper show that ALS formation after treating CHO cells with MMS is biphasic. MMS rapidly induces ALS which are repaired by a process the half-life of which is approximately 3 hours. At high doses of MMS, however, the decrease in ALS after removing MMS from the medium is apparently distorted by further increase in ALS, which probably in later periods of time after treatment is caused by a conversion of potential ALS (PALS) to ALS. On the basis of the results presented here it is not possible to say whether the conversion of PALS to ALS is an enzymatic or spontaneous process. Probably due to this conversion during the period 3 - 6 hours after treating the cells with high doses of MMS an increase in ALS of about 40% occurs. In cultures treated with small doses (2 mM) this increase does not occur, which could suggest that the cell can remove a substantial part of PALS by repair sooner than their conversion to ALS takes place. These secondarily developed ALS are removed from DNA with a longer half-life, approximately 10 - 16 hours. These secondary ALS seem to be a more effective block to DNA elongation. Their quantity (dependent on the doses of MMS) probably decides how strong the inhibition of elongation will be. This can be seen in Figure 1, where after the dose of 4 mM their increase and strong inhibition of elongation occur, whereas after the dose of 2 mM, where no increase in ALS was observed, DNA elongation after an initial deceleration soon reaches the values close to that of the control. These findings are in agreement with the conclusions of Painter (14) and others (15), who found in human cells treated with comparable doses of MMS that the first effect of MMS on DNA replication early after treatment, was the inhibition of initiation of replicons. On the other hand, elongation was most strongly blocked as late as 2 - 2.5 hours after treatment which leads to the authors' conclusion that elongation was inhibited by "a slowly formed product". This chronologically corresponds to our findings of decelerated elongation of DNA in the period of 3 to 6 hours after treating the cultures with higher doses of both MMS and DMS. What the nature of these secondarily developed ALS is cannot be decided on the basis of the results presented here. Apparently not primarily methylated bases, but the products of their enzymatic or spontaneous conversions are involved. Apurinic sites, which in contrast to alkylated bases (6) block the elongation of DNA strands in vitro (4) could be such products. If these ALS develop enzymatically, this process in CHO cells is probably similar to that in human cells (14,15).

It can be judged on the basis of the present study that in the DNA of cells treated with MMS both alkali stable and alkali labile lesions are induced and that a substantial part of both these lesions is removed from DNA by a long type of repair inhibited by both araC+HU and NB. A part of the alkali stable lesions are apparently changed in time (spontaneously or enzymatically) into alkali labile lesions, which can reflect the observed increase in ALS in the period of 3 - 6 hours after treatment. These results are in agreement with similar studies carried out in human cells (12), in which after treatment with MMS two phases of repair of MMS-induced lesions were found; in both phases "the long type of repair" resulting in the accumulation of the DNA breaks in the presence of araC+HU was active.

As it has already been discussed by the present authors (12), the first phase of repair after treating the cells with MMS could represent the repair of 3-methyladenine (the half-life of its repair being 2 - 3 hours) and the second phase could represent the repair of N-7-methylguanine (the half-life of repair in the rodent cells being about 20 hours) (for review see 16). The half-lives of repair in the first and second phases found in the present study approximately correspond to these data (though here the estimated half-life of the second phase is just 9-16).

REFERENCES

1   Singer,B. (1979) J.Natl. Cancer Inst., 62, 1329-1336
2   Saffhill,R., Margison,G.P. and O'Connor, P.J. (1985) Biochem. Biophys. Acta, 823, 111-145
3   Natarajan,A.T., Simons,J.W.I.M., Vogel,E.V. and van Zeeland, A.A. (1984) Mutat. Res., 128, 31-40
4   Lockhart,M.L., Deutch,J.F., Oamanura,I., Cavalieri,L.F., Rosenberg,B.H. (1982) Chem. Biol. Interactions, 42, 85-95
5   Peterson,A.R. (1980) Cancer Res., 40, 682-688
6   Columbano,A., Ledda Columbano,G.M., Rao,P.M., Rajalakshmi,S., Sarmi, D.S.R (1985) Biochemical Archives, 1, 121-130
7   Clarkson,J.M. and Mitchell,D.L. (1979) Mutat. Res., 61, 333-342
8   Stetina,R., Votava,M. (1986) Folia Biol., 32, 128-144
9   Erixon,K. and Ahnstrom,G. (1979) Mutat. Res., 59, 257-271
10 Squires,S., Johnson,R.T., Collins,A.R.S. (1982) Mutat. Res., 95, 389-404
11 Park,S.D., Seong,N.H., Morgan,W.F., Cleaver,J.E. (1985) Chem. Biol. Interactions, 52, 255-263
12 Snyder,R.D. and Regan,J.D. (1982) Carcinogenesis, 3, 7-14
13 Abbondandolo.A., Dogliotti,E., Lohman,P.H.M. and Berends,F. (1982) Mutat. Res., 92, 361-377
14 Painter,S.N. and Regan,J.D. (1973) Mutat. Res., 18, 191-197
15 Buhl,S.N. and Regan,J.D. (1973) Mutat. Res., 18, 191-197
16 Veleminsky,J., Gichner,T. (1982) Biologicke listy, 17, 33-58

FUNCTIONAL SIGNIFICANCE OF RIBONUCLEIC ACIDS MODIFIED BY CHEMICAL

CARCINOGENS: A REVIEW

Jan Hradec

Department of Molecular Biology,
Research Institute for Tuberculosis and Respiratory Diseases
180 71 Prague
Czechoslovakia

ABSTRACT

Interaction of chemical carcinogens with DNA is apparently the most important event in the mechanism of chemical carcinogenesis since it alters the
cellular genome and may thus lead to malignant transformation. Besides DNA,
carcinogens are known to react also with proteins and RNA. It has been
stated that RNA has even a greater propensity for the covalent reaction with
carcinogens than DNA. Of various species of RNA, tRNA seems to be most
frequently modified by chemical carcinogens. Results are reviewed on the
covalent interaction of carcinogens with different tRNAs. Polycylic aromatic
hydrocarbons, amides and alkylating carcinogens seem to modify preferentially
guanosine residues in the molecule of tRNA. Guanosine may be a common target
molecule for carcinogens, in particular for their interaction with tRNA in
vitro. A prior conversion of procarcinogens into active metabolites (ultimate carcinogens) is apparently essential for their covalent interaction with
tRNA. Modification of tRNA by carcinogens is followed by changes in its
acceptance for amino acids and in codon recognition. Chemical carcinogens do
apparently in a specific way react with initiator tRNA in vitro and enhance
its acceptance for cognate L-methionine. A short-term initiator tRNA
acceptance assay for testing carcinogens is briefly described and results
with several carcinogens and non-carcinogenic compounds are reported.
Modification of initiator tRN by carcinogens is probably followed by conformational changes which make this molecule more accessible for the reaction
with cognate aminoacyl-tRNA synthetase. Initiator tRNA modified by some carcinogens is more efficiently utilized by the protein-synthesis factor eIF-2
for the formation of its ternary complex but its modification does not affect
any additional steps of cell-free protein synthesis. Conformational changes
in certain tRNAs induced by a covalent binding of carcinogen residues may
lead to distortions in cellular regulatory processes and contribute thus to
the mechanism of malignant transformation. However, because of a lack of
sufficient data the probable involvement of tRNA modified by carcinogens in
the mechanism of chemical carcinogenesis cannot be clearly defined at the
moment.

INTRODUCTION

It seems to be now generally accepted that interaction of chemical

carcinogens with DNA is of a key importance in the sequence of events involved in the induction of malignant tumours by chemical carcinogens. Alteration of the genetic material resulting from a covalent reaction of carcinogens with DNA changes the cell genome and induces in this way the malignant
transformation of the cell (1).

It has been known for a long time that chemical carcinogens besides reacting with DNA also react with RNA and proteins (2,3). Weinstein and
Grunberger have reviewed the possibility that the nucleic acids (DNA or RNA)
rather than protein, constitute the critical target because of the widespread disturbances in gene expression that accompanies the transformation
process as well as the usual stability of the transformed state once it has
been established (4). Many carcinogens bind in vivo to cellular RNA to an
extent that is equal to or even greater than their binding to DNA (5). Many
studies suggest that RNA may be a critical target during chemical carcinogenesis (6,7). Several years ago, Magee and Farber demonstrated that tRNA
has a greater propensity for methylation by dimethyl-nitrosoamine than DNA
(8).

Evidence has been accumulated suggesting that the tRNA population in
tumor cells may differ qualitatively from that of normal mammalian cells (9).
Every malignant tumor contains a few tRNAs which are different in structure
from the tRNAs in the normal tissue counterpart and there seems to be no
exception (10). These tumor-specific tRNAs contain usually more minor bases
including methylate bases (11). Such modified nucleosides are excreted in
the urine of cancer patients and their utilization as tumor markers has been
suggested (12).

The possibility that cancer represents an abberation in differentiation
rather than somatic mutation suggests that the reaction with RNA may be as
critical (or even worse) as the reaction with DNA (4).

COVALENT INTERACTIONS OF CARCINOGENS WITH tRNA

In a series of papers Weinstein's group provided evidence for the
covalent interaction of aromatic amides with tRNA and presented fundamental
findings on various aspects of this reaction. N-Acetoxy-2-acetylamino-
fluorene was chosen as model carcinogen in their experiments. This compound
represents the final metabolite of a well-known carcinogenic amide, N-
acetylaminocfluorene (13). Fink et al. presented evidence that this compound reacts in vitro with tRNA from E. coli and that covalent attachment of
the acetylaminofluorene residues to unfractionated tRNA is followed by
changes of its chromatographic behaviour on DEAE-Sephadex and benzoylated
DEAE-cellulose columns (13). Administration of procarcinogenic ^{14}C-labeled
N-2-acetylaminofluorene to rats resulted in the binding of radioctivity to
RNA in their liver and the specific activity of tRNA was 2-3 times higher
than the specific radioactivity of 5S, 18S and 28S ribonucleic acids. This
preferential binding of the radioactive carcinogen to RNA was attained by 12
hours after intraperitoneal injection (14). Partial seperation of tRNA
revealed that the carcinogen had reacted with several types of tRNA. However,
feeding of rats with N-2- aceltylaminofluorene-containing diets produced no
gross changes in the profile of the newly synthesized liver tRNAs as revealed
by pulse-labeling with orotic acid (14). Hydrolysis of liver RNA
obtained from rats given acetylaminofluorene in vivo or reacted with N-acet-
oxyacetylaminofluorene in vitro indicated that the major nucleoside target
was guanosine and 8-(N-2-fluorentylacetamido) guanosine was identified as the
product of this interaction (15). Polynucleotides modified by N-acetoxy-2-
acetylaminofluorene showed different template activities when compared with
polymers not modified by the carcinogen (16). Covalent binding of acetyl-
N-2-aminofluorene to guanosine residues resulted in major changes in conform-

ational properties of the modified obligonucleotides (17). These changes
included rotation of the guanine base about the glycosidic linkage and intra-
molecular stacking of the fluorene residue with the adjacent base. This
structure has been designated as the "base displacement model" (4).

Dibenzo(a,e)fluoranthene reacted with tRNA only in a subcellular microsomal system indicating that a prior metabolic conversion of this hydrocarbon
to active metabolites was required for this reaction. Guanosine has been
identified as the target nucleoside (18). Similarly, only metabolically
activated 2-amino-4-(5-nitro-2-phenyl)thiazole reacted covalently with tRNA
(19). N-Acetoxy-2-acetylaminofluorene was demonstrated to modify initiator
tRNA from E.coli (20) as well as yeast tRNA^{Phe} (21). Guanosine residues in
the molecule of tRNA were targets for the interaction of metabolic products
of N-hydroxy-2-acetylaminofluorine with this species of RNA (22). Benzo-
(a)pyrene became covalently bound to tRNA^{Met} only if microsomal enzymes and
NADPH were present in incubation mixtures, indicating that a conversion of
this hydrocarbon into ultimate carcinogens must precede the binding (23).

Alkylating carcinogens have been repeatedly shown to be capable of an
efficient methylation or ethylation of nucleoside residues in tRNA not only
in vivo but also in vitro. Guanosine was the target molecule after treating
tRNA^{Phe} and tRNA^{Val} with N-ethylnitrosourea (24,25). 7-Methylguanosine was
the only methylated nucleoside resulting from the treatment of mammalian
cells with another alkylating agent, 1,2-dimethylhydrazine(26). Radioactivity was found to be covalently associated with initiator tRNA from rat
liver if it was incubated with tritiated N-methyl-N-nitrosourea in the

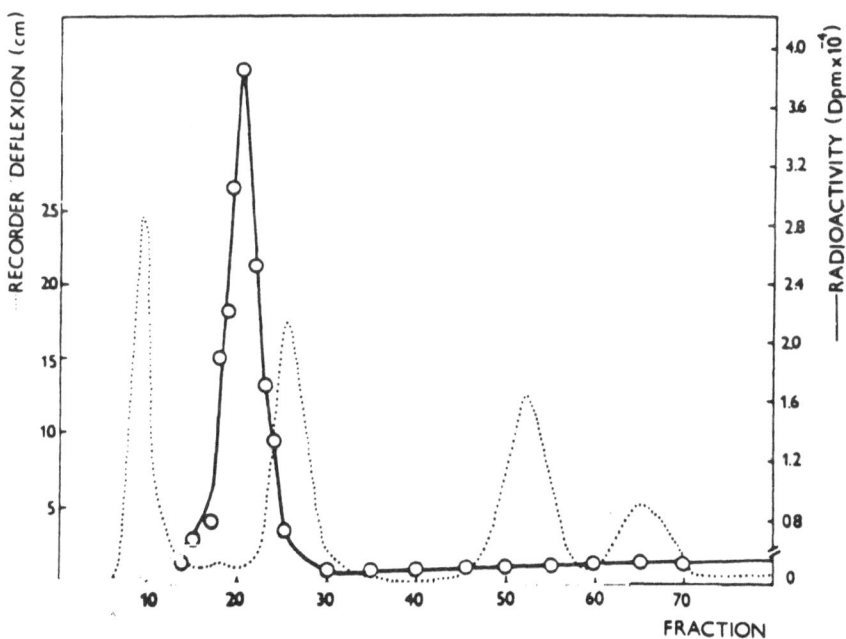

Figure 1. Separation of an enzymatic hydrolysate of initiator tRNA pretreated
with N-(^{3}H)methyl-N-nitrosourea by chromatography on Sephadex G10. A portion
of the hydrolysate corresponding to 5 mg of modified tRNA was chromatographed
on a 15 x 950 mm column, fractions were eluted, absorption ot 260 nm was
continuously recorded and radioactivity in the fraction was assayed.
Position of peaks: 1, uridine, 2, 7-methylguanosine, 3, cytidine, 4, guanosine, 5, adenosine. From Hradec and Macelova (27).

absence of any additional subcellular preparations. 7-Methylguanosine was identified as the only product of alkylation after tRNA^{Met} was pretreated with this carcinogen and subjected to enzymatic hydrolysis followed by a chromatographic separation of nucleosides (27). Initiator tRNA became alkylated by 3,3-dimethyl-1-phenyltriazene and its ring chlorinated derivatives only if microsomal enzymes and NADPH were present in incubation mixtures during the pretreatment of tRNA with these carcinogens. This indicated that a prrior metabolic conversion was required for the modification of tRNA with these compounds, unlike that with the ultimate carcinogen N-methyl-N-nitrosourea. The only product of alkylation was, similarly as with N-methyl-N-nitrosourea, again 7-methylguanosine (28).

It does seem that guanosine is the principal target in the molecule of tRNA for carcinogens irrespective of their chemical structure, at least for the modification of tRNA by these compounds in vitro.

FUNCTIONAL CONSEQUENCES OF THE MODIFICATION tRNA BY CARCINOGENS

Modification of tRNA by N-acetoxy-2-acetylaminofluorene was demonstrated to change the normal function of tRNA in both amino acid carcinogen inhibited the acceptance to tRNA^{Lys} but actually enhanced the acceptance capacity of tRNA^{Val} from bacterial cells (13). An enhancement of the acceptance of certain tRNAs has also been described as a consequence of 5-fluorouracil incorporation (29) or as a result of in vitro methylation with dimethylsulphate (30). On the other hand, treatment of tRNA from rat liver with relative high doses of N-methyl-N-nitro-N-nitrosoguanine was followed by a decrease acceptance capacity for most amino acids tested (31).

Fink et al. provide three possible explanations for the enhanced acceptance capacity to tRNA modified by carcinogens. (1) Such a modification permits other tRNAs to become 'mischarged' with a particular amino acid, (2) residues of the carcinogen present in the molecule of tRNA enhance affinity of a certain tRNA for its corresponding aminoacyl-tRNA synthetases, or, (3) the carcinogen can convert a portion of a particular tRNA from a denatured into a renatured conformation (13).

Modification of polynucleotides by N-acetoxy-2-acetylaminofluorene changes also the template activity of polynucleotides. Evidence was obtained that polymers containing modified guanosine residues are bound to ribosomes but that polypeptide chain growth is blocked when translation encounters these modified residues (16). Polynucleotides containing guanosine residues modified by N-2-acetylaminofluorene affected significantly the function of such oligonucleotides in the binding of aminoacyl-tRNA to ribosomes. If the guanosine residue to which the compound was covalently bound was a part of codon, then the oligonucleotide was completely inactive in the ribosomal binding assay. Modification of guanosine residues next to an adenosine inhibited their ribosomal binding capacity. The stacking interaction with adjacent adenosine, but not with adjacent uridine residues, probably accounted for these effects (32).

SPECIFIC ENHANCEMENT OF THE ACCEPTANCE CAPACITY OF INITIATOR tRNA BY CHEMICAL CARCINOGENS

In earlier experiments performed in our laboratory, post-mitochondrial supernatants of rat liver were incubated with carcinogenic and non-carcinogenic polycyclic hydrocarbons, tRNA was re-isolated and charged with L-methionine in the presenceof aminoacyl-tRNA synthetases from E. coli which are known to charge specifically only initiator tRNA^{Met} (33). A stimulation of the charging was found only after the pretreatment of tRNA with

Table 1. Charging of tRNA^{Met} petreated with DAB and its derivatives with
L-(^{35}S)methionine. From (35)

No	Compound	Carcinogenity	Met-tRNA^{Met}
1	DAB	Strong 250.6	
2	3'-CH$_3$-DAB	Strong	301.2
3	3'-COOH-DAB	Weak	108.8
4	3'-Cl-DAB	Intermediate	141.0
5	3'-OCH$_3$-DAB	Strong	257.3
6	3'-NO$_2$-DAB	Intermediate	148.8
7	3'-Br-DAB	Non-carcinogenic	101.0
8	4'-CH$_3$-DAB	Non-carcinogenic	100.2
9	4'-OH-DAB	Weak	159.7
10	2'-OH-DAB	Weak	161.1
11	DA-1-N	Non-carcinogenic	91.5
12	DA-2-N	Strong	200.0

All values are per cent of controls (postmitochondrial supernatant fractions
pretreated with the solvent only) obtained with saturating amounts of unfractionated tRNA isolated from the S-30 fraction preincubated with 30 pmoles
of compounds tested/ml postmitochondrial supernatant fraction. Charging was
performed using aminoacyl-tRNA synthases from E.coli as described in
Materials and Methods. In control mixtures tRNA^{Met} was charged with 0.717
pmole methionine/nmole tRNA. Abbreviations: DAB, 4-dimethylaminoazobenzene,
DA-1-N, dimethylaminobenzene-1-azo-1'-naphthalene, DA-2-N, dimethylamino-
benzene-1-azo-2'-naphthalene.

carcinogens but not with chemically closely related non-carcinogenic compounds (34) (Table 1). Similarly, only carcinogenic azo dyes stimulated the
acceptance of initiator tRNA for L-methionine whereas non-carcinogenic compounds of this class were ineffective. The charging of unfractionated tRNA
pretreated with p-dimethylaminoazabenzene was also enhanced with 7 out of 8
different amino acids tested but there was no relation between this stimulation and the carcinogenic activity of the tested compounds (35). Only
carcinogenic but not non-carcinogenic alkyltriazenes and imidazoles enhanced
the acceptance capacity of initiator tRNA. Charging of unfractionated tRNA
with most of 12 additional different amino acids tested was also stimulated
by compounds of this class but, similarly as with azo dyes, there was no
difference between carcinogens and compounds without carcinogenic activity
(36) (Table 1 and 2).

Additional evidence that the effect on the acceptance activity of initiator tRNA is specific for carcinogens was provided by further results of
this laboratory. Procarcinogenic benzo(a)pyrene and 3,3-dimethyl-1-phenyl-
triazene exhibited a stimulating effect on initiator tRNA charging only if
incubated with it in the presence of microsomal enzymes and NADPH. These
results indicate that procarcinogens must be converted into active ultimate
carcinogenic metabolites by microsomal enzymes in presence of NADPH (which is
known to be absolutely required for this activation) to be able to react with
tRNA and modify its activity (37). A close relation does exist between the
extent of initiator tRNA modification and the stimulation of its charging.
With both pro-carcinogenic 3,3-dimethyl-1-phenyltriazene and its ring
chlorinated derivatives and the ultimately carcinogenic N-methyl-N-nitrosourea a direct correlation was demonstrated between the extent of alkylation
induced by these compounds in the molecule of initiator tRNA and its enhanced
acceptance for L-methionine (27,28) (Fig 2).

On the basis of these results a procedure was devised for short-term

Table 2. Effect of Triazene and Imidazole Pretreatment of Initiator tRNA on its Acceptance for L-Methionine. From (36)

No.	Compound	tRNA pretreated in S-30 fraction	tRNA alone
		Per cent of controls	
1	5-Aminoimidazole-4-carboxamide (AIC)	290	90
2	5-(3,3-Dimethyl-1-triazeno) imidazole-4-carboxamide (DTIC)	213	99
3	5-Diazoimidazole-4-carboxamide (DZIC)	203	96
4	3,3-Dimethyl-1-phenyltriazene (DMPT)	245	106
5	4-(3,3-Dimethyl-1-triazeno)benzoic acid (DMTBA)	378	104
6	4-(3,3-Dimethyl-1-triazeno)sulphonic acid sodium salt (DMISA Na salt)	142	107
7	1-(4-Chlorophenyl)-3,3-dimethyltriazene (4-Cl-PDMT)	194	119
8	1-(4-Bromophenyl)-3,3-dimethyltriazene (4-Br-PDMT)	161	136
9	1-(2,4,6-Trichlorophenyl)-3,3-dimethyltriazene (2,4,6-Cl_3-PDMT)	120	97
10	3-Methyl-1-phenyltriazene (MPT)	174	232
11	1-(4-Chlorophenyl)-3-methyltriazene (4-Cl-PMT)	198	170
12	1-(4-Bromophenyl)-3-methyltriazene (4-Br-PMT)	245	236
13	1-(2,4,6-Trichlorophenyl)-3-methyltriazene (2,4,6-Cl_3-PMT)	133	182

Unfractionated tRNA isolated from postmitochondrial supernatants and either preincubated with compounds tested, or pretreated alone with these compounds, was charged with aminoacyl-tRNA synthetases from E.Coli. Controls were preincubated with the solvent (ethanol) alone. Charging of tRNA in the control mixtures was 1,24 pmol of L-methionine/nmol of unfractionated tRNA. Test compounds were preincubated at 10^{-4} mg/ml for 60 min at 37° in both experimental systems used.

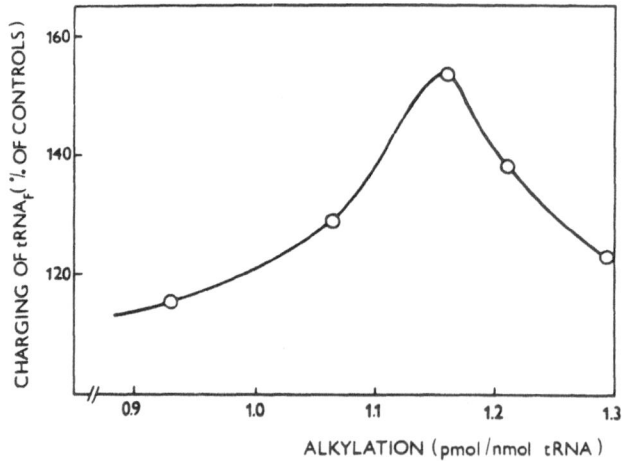

Figure 2. Dependence of the charging of initiator tRNA with L-methionine on the degree of its alkylation by N-(3H)methyl-N-nitrosourea. From Hradec and Macelova (27).

Table 3. Effect of Triazenes and Imidazoles on the Charging of tRNA with Various Amino Acids. From (36)

No	Ala	Arg	Asp	Gly	His	Ile	Leu	Lys	Phe	Thr	Tyr	Val
						Per cent of controls						
1	76.4	542.4	120.5	77.5	87.1	188.2	28.1	505.5	131.1	185.4	108.7	99.2
2	65.6	61.5	69.6	121.4	110.6	107.8	22.6	80.4	134.1	107.8	122.8	323.1
3	100.3	186.9	143.2	209.5	270.7	233.4	18.5	208.5	103.8	60.7	198.2	313.0
4	189.5	201.2	-	209.5	-	188.2	36.9	277.9	101.6	-	319.2	260.8
5	167.9	106.9	82.6	84.5	216.2	164.0	19.1	264.9	124.1	94.8	103.5	123.1
6	135.1	184.9	98.4	221.1	109.4	123.1	25.1	237.5	130.8	110.3	104.3	188.4
7	123.3	93.6	141.4	200.0	93.8	213.2	18.7	223.3	148.8	86.4	244.7	197.1
8	86.8	73.5	184.1	142.2	75.3	165.6	25.6	377.5	118.6	85.4	125.4	139.8
9	99.3	157.2	143.2	123.9	43.9	131.2	22.9	162.8	95.2	109.8	93.8	157.9
10	97.7	110.1	108.7	231.0	-	96.8	13.4	266.3	136.9	68.5	233.3	153.6
11	79.2	232.7	137.9	187.3	69.1	109.3	27.4	252.0	113.8	55.0	97.3	131.8
12	181.6	73.1	103.5	180.2	131.5	103.1	30.3	208.3	98.8	108.3	169.2	137.6
13	198.2	93.4	202.5	142.2	97.5	156.2	27.4	587.4	108.6	173.5	158.7	62.3

Unfractionated tRNA was isolated from postmitochondrial supernatants pre-incubated with compounds tested and charged by aminoacyl-tRNA synthetases from rat liver. Controls were preincubated under the same conditions with the solvent (ethanol) only. The following chargings were obtained in control mixtures (pmol amino acid/nmol tRNA): Ala-3.99, Arg-1.02, Asp-0.66, Gly-0.45, His-0.84, Ile-1.45, Leu-0.13, Lys-1.21, Phe-0.50, Tyr-0.16, Thr-1.04, Val-0.67. Post-mitochondrial supernatants were incubated for 60 min at 37° with 10^{-4} mg/ml ofthe compound tested. See legends for Table 2 for a list of compounds.

testing of carcinogens. In the first step of this method tRNA is prertreated with the compound to be tested and tRNA is re-isolated after the incubation (Fig.3,4). In the second step, this modified tRNA is charged with L-methionine by aminoacyl-tRNA synthetases from E.coli and the extent of charging of $tRNA_F^{Met}$ is compared with that of a control tRNA preparation pretreated with the solvent only (38). This initiator tRNA acceptance assay showed a good reproducibility when used independently in two laboratories (39). In assays of 69 carcinogenic and non-carcinogenic N-nitroso compounds (40) very favourable predictive criteria (sensitivity, specificity, accuracy and predictive value (41) were found. Further results with mycotoxins, aromatic hydrocarbons and amines, as well as cytostatics also seem to be promising (J.Hradec, to be published).

Since this procedure is not based on an interaction of carcinogens with DNA it may complement the usual genotoxic assays for carcinogens. Moreover, it cannot be excluded that it could detect non-genotoxic (epigenetic) carcinogens which cannot be found by genotoxic assays.

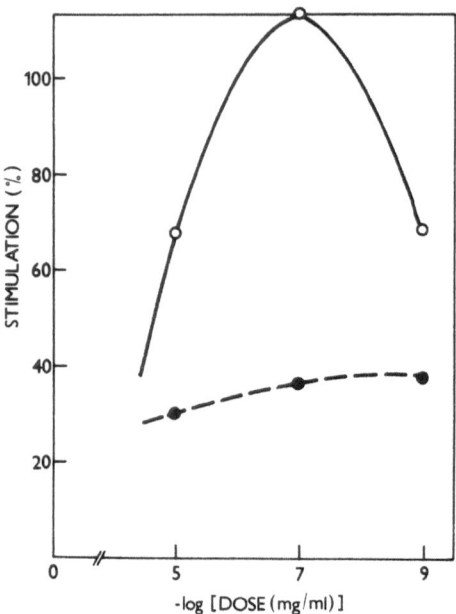

Figure 3. The dose-dependance of different doses of N-methyl-N-nitro-N-nitrosoguanidine ( ) and aflatoxin G1 ( ) added during the pretreatment of tRNA on their stimulatory effect on the charging of initiator tRNA with L-methionine. From Hradec (38).

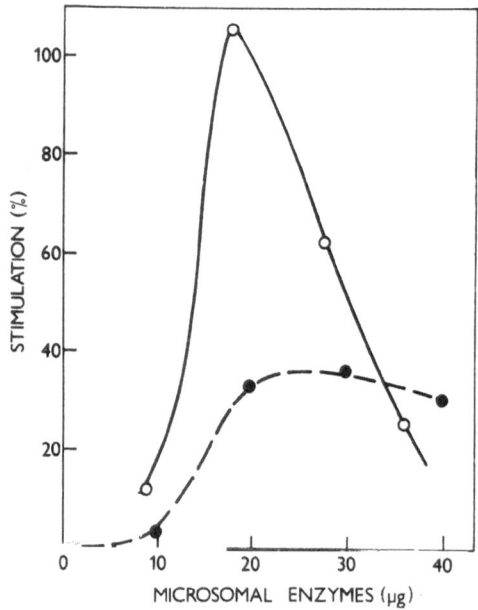

Figure 4. Effect of increasing concentrations of microsomal enzymes present during the pretreatment of tRNA with N-methyl-N-nitro N-nitrosoguanidine ( ) and aflatoxin G1 ( ) on their stimulating effect on the charging of initiator tRNA with L-methionine. From Hradec (38).

THE POSSIBLE ROLE OF tRNA MODIFIED BY CARCINOGENS IN CHEMICAL CARCINOGENESIS

Recent results of our laboratory provided evidence that initiator tRNA modified by carcinogens N-methyl-N-nitrosourea and cholesteryl 14-methylhexadecanoate (42) are significantly more efficiently utilized for cell-free protein synthesis than corresponding control preparations. The modified tRNA apparently specifically interferes with the function of the protein synthesis factor eIF-2. The formation of the ternary complex of this factor with initiator tRNA and GTP is significantly stimulated by modified tRNA. In agreement with this also the formation of the ribosomal initiation complex as well as the over-all protein synthesis in a subcellular system of rabbit reticulocytes is enhanced. On the other hand, no other steps of gene translation are affected with modified initiator tRNA (43). It is well known that reactions catalyzed by eIF-2 represent fundamental regulatory steps is gene translation (44). Similar results were obtained by others in *in vivo* experiments and it has been shown that the administration of carcinogenic 7,12-dimethyl-(a)anthracene was followed by a significant stimulation of a factor-dependent binding of initiator tRNA to ribosomes (45). Thus modification of initiator tRNA by carcinogens can result in an enhanced over-all rate of protein synthesis.

It seems well understandable that the growth of malignant tumors should be accompanied by a stimulated production of cellular proteins. However, it is far from being clear why such an enhancement should follow immediately after the administration of a carcinogen that precedes the formation of a tumor by a very long time. Furthermore, it cannot be explained at the moment why carcinogens should modify the initiator tRNA from normal rat liver although this tissue is not the target tissue for most compounds tested in our laboratory (39, 40). Further investigations will be apparently required to clarify this and several further points and it would be only speculative at the moment to discuss the possible involvement of the modified tRNA in the mechanism of carcinogenesis.

Grunberger and Weinstein suggested that the conformational changes induced in tRNA by chemical carcinogens interfere not only with normal function of tRNA in both amino acid acceptance and codon recognition *in vivo*. It should be also considered that, besides its function as an adaptor in protein synthesis, tRNA may also play an important role as an intermediate in enzyme repression. Carcinogen-induced conformational changes in certain tRNAs might also lead to distortions in cell regulation and contribute thus to the transformational mechanism (4).

Carcinogenic compounds do react probably not only with tRNA but also with mRNA *in vivo*. Such a modification could block translation of certain mRNAs *in vivo*. This may be of a critical importance in transformation if the blocked mRNA normally plays a role in cell regulation, in particular in the synthesis of key repressor proteins (4).

Only few data are available at the moment on the functional significance of the modification of various species of RNA by chemical carcinogens. For this reason it is not possible at the moment to decide whether and how the modification of RNA is involved in the mechanism of chemical carcinogens.

REFERENCES

1 Hathaway, D.E. (1986) Mechanism of Chemical Carcinogenesis, Butterworth and Co. Ltd., London
2 Miller, J.A. (1970) Cancer Res. 30, 559-576
3 Farber, E. (1968) Cancer Res. 28, 1859-1865
4 Weinstein, I.B. and Grunberger, D. (1972) in World Symposium on Model

Studies in Chemical Carcinogenesis (Ts'o, P.O. and Di Paolo, J.A. eds), M. Dekker, New York, pp. 157-171

5 Weinstein, I.B. (1969) Genetic Concepts and Neoplasia. Twentythird Annual Symposium on Fundamental Cancer Research, Williams and Wilkins, Baltimore, MD. pp.380-408

6 Weinstein, I.B. (1968) Cancer Res. 28, 1871-1876

7 Axel, R., Weinstein, I.B., and Farber, E. (1967) Proc. Natl. Acad. Sci. USA, 58, 1255-1259

8 Magee, P.N., and Farber, E. (1962) Biochem. J. 83, 114-122

9 Randerath, K., Agrawal, H.P., and Randerath, E. (1983) in Modified Nucleosides and Cancer (Nass, G. ed.).Springer Verlag, Berlin, pp.103-120

10 Bore, E., Waalkes, T.P., and Gehrke, C.W. (1983) in Human Tumor Markers: Biological Basis and Clinical Relevance (Nieburgs, H.E., Birkmayer, G.D. and Klavins, J.V., eds.). A.R. Liss Inc., New York, pp. 67-71

11 Kuchino, Y., Borek, E., Grunberger, D., Muchinski, J.F., and Nishimura, S. (1982) Nucleic Acids Res. 10, 6421-6432

12 Waalkes, T.P., Gehrke, C.W., Zumwalt, R.W., Chang, S.Y., Lakings, G.B., Tormey, D.C., Ahmann, D.L., and Moertel, C.G. (1975) Cancer, 36, 390-398

13 Fink, L.M., Nishimura, S., and Weinstein, I.B. (1970) Biochemistry, 9, 496-502

14 Agarwal, M.A., and Weinstein, I.B. (1970) Biochemistry, 9, 503-508

15 Kriek, E., Miller, J.A., Juhl, U., and Miller, E.C. (1967) Biochemistry, 6, 177-183

16 Grunberger, D., and Weinstein, I.B., (1971) J. Biol. Chem. 246. 1123-1128

17 Nelson, J.H., Grunberger, D., Cantor, C.R., and Weinstein, I.B. (1971) J. Mol. Biol. 62, 331-346

18 Prin-Roussel, J., Ekert, B., Zajdela, F., and Jacquinon, P. (1978) Cancer Res. 38, 3499-3504

19 Swaminathan, S., Lower, G.M., and Bryan G.T. (1982) Cancer Res. 42, 4497-4484

20 Fujimura, S., Grunberger, D., Carvajal, G., and Weinstein, I,B. (1972) Biochemistry, 11, 3629-3634

21 Schneider, D., and Cramer, F. (1972) Hoppe-Seylers Z. Physiol. Chem. 353, 1565-1566

22 Gutman, H.R., Smith, B.A., and Springfield, J.R. (1985) Proc. Annu. Meet. Am. Assoc. Cancer Res 26, 115

23 Hradec, J. (1983) Cancer Lett. 18, 199-204

24 Vlasso, V.V. Kern, D., Romby, P., Geige, R., and Ebel, J.P. (1983) Eur. J. Biochem. 132, 537-544

25 Barsuczeqski, J., Romby, P., Ebel, E.P., and Giege, R. (1982) FEBS Lett. 150, 459-464

26 Trewyn, R.W. (1981) National Technical Information Service, Springfield, VA, AD-A112 358/7

27 Hradec, J., and Macelova, J. (1988) Biochem. Pharmacol, in the press

28 Hradec, J., and Kolar, G.F. (1985) Carcinogenesis, 6, 995-998

29 Giege, R., Heinrich, J., Weil, J.H., and Ebel, J.P. (1969) Biochem. Biophys. Acta, 174, 53-68

30 Pellinger, D.J., Hay, D., and Borek, E. (1969) Fed. Proc. 28 889

31 Bagewadikar, R.S., and Bhattacharya, R.K. (1977) Indian J. Biochem. Biophys. 14, 334-336

32 Grunberger, D., Blobstein, S.H., and Weinstein, I.B. (1974) J. Mol. Biol. 82, 459-468

33 Gupta, N.K., Chaterjee, N.K., Bose, K.K., Bhaduri, S., and Chung, A. (1970) J. Mol. Biol. 54, 145-154

34 Hradec, J., Dusek, Z., and Bahana, L. (1979) Biochem. Pharmocol. 28, 1157-1161

35 Stiborova, M., Matrka, M., and Hradec, J. (1980) Biochem. Pharmacol. 29, 2301-2305

36 Stiborova, M., Kolar, G.F. (1984) Cancer Lett. 23, 115-120

37   Hradec, J., and Kolar, G.F. (1984) Cancer Lett. 23, 115-120
38   Hradec, J. (1988) Carcinogenesis, 9
39   Hradec, J., Spiegelhalder, B., and Pruessman, R. (1988) Carcinogenesis, 9
40   Hradec, J., Speigilhalder, B., and Preussmann, R. (1988) Carcinogenesis, 9
41   Cooper, J.A., Saracci, R., and Coole, P. (1979) Brit. J. Cancer, 39, 87-89
42   Shabad, L.M., Kolesnichenko, T.S. and Savluchinskaya, L.A. (1973) Neoplasma, 20, 347-348
43   Hradec, J., and pohlreich, P. (1988) This Volume
44   Ochoa, S., and de Haro., C. (1979) Annu. Rev. Biochem. 48, 549-580
45   Chan, I. P., Lendecki, W., and Nichols, D.M. (1977) Cancer Res. 37, 4220-4227

EVALUATION OF CARCINOGENIC AND NON-CARCINOGENIC METAL COMPOUNDS BY THEIR
INTERFERENCE IN THE RENATURATION OF HIGHLY POLYMERIZED DNA AND THEIR
ACTIVATION BY MICROSOMES

H. M. Dani, Sneh Ahuja and Ravinder Mohan

Department of Biochemistry, Panjab University
Chandigarh-160, India

ABSTRACT

Carcinogenic metal compounds specifically inhibit the renaturation of
highly polymerized DNA whereas the non-carcinogenic and physiological metals
either fail to do so or even assist in its renaturation. The carcinogenic
metals tend to keep the DNA unravelled when its two strands are annealed from
around 80° to 2°C as studied by U.V. spectrophotometry. Studies on the
interactions of carcinogenic metals with nucleic acid based have further
revealed that there are major changes in the U.V. spectra of adenine and
cytosine both in terms of shifts in absorption maxima as well as the degree
of absorbance at various wavelengths. Activated metal compounds with microsomal hydroxylase system have been found to detach the ribosomes from rat
liver microsomes more efficiently in comparison to non-activated compounds
but activation was not found to be necessary for their interactions with DNA
or nucleic acid bases. However, activated metal compounds more potently
denatured and interfered with the renaturation of DNA. These simple short-
term techniques can be usefully employed for the prediction of the carcinogenicity of metals/metalloids.

INTRODUCTION

Although metal carcinogens are known to interfere with the fidelity of
RNA transcription (1), no efforts have been made yet to use this observation
for developing a short-term technique for predicting their carcinogenicity.
Development of such a technique was thought desirable considering the failure
of a well established microbial technique in detecting the carcinogenicity of
metals (2). Moreover, it was also not understood whether the microsomal
hydroxylase system was needed for activating the metal compounds/metals in
order to enable them to interact with the nucleophiles in the living cells,
as is required for most of the organic carcinogens. We therefore performed a
number of experiments to study the interactions of highly polymerized DNA
with different concentrations of various metal compounds and have concluded
that only carcinogenic metals promoted the denaturation of DNA and also
interfere with its renaturation while it is cooled from 80° to 2°C. The non-
carcinogenic metals failed to do so, while the physiological metals supported
its reassociation. We have further demonstrated that the pre-incubation of
metal compounds with the microsomal hydroxylase system boosts their
efficiency to denature and interfere with the renaturation of DNA as well as

for the detachment of ribosomes from microsomal preparations. Metals interact with nucleic acid bases as studied by recording the shifts in their U.V.
spectra and absorbance. Our conclusions are based on the results obtained
from 46 compounds/metals studied, but data from only typical experiments
could be included in this paper due to lack of space.

MATERIALS AND METHODS

Highly polymerized calf thymus DNA (Sigma) was dissolved in buffered
saline (0.15 M sodium chloride and 0.015 M sodium citrate of AR grade, pH

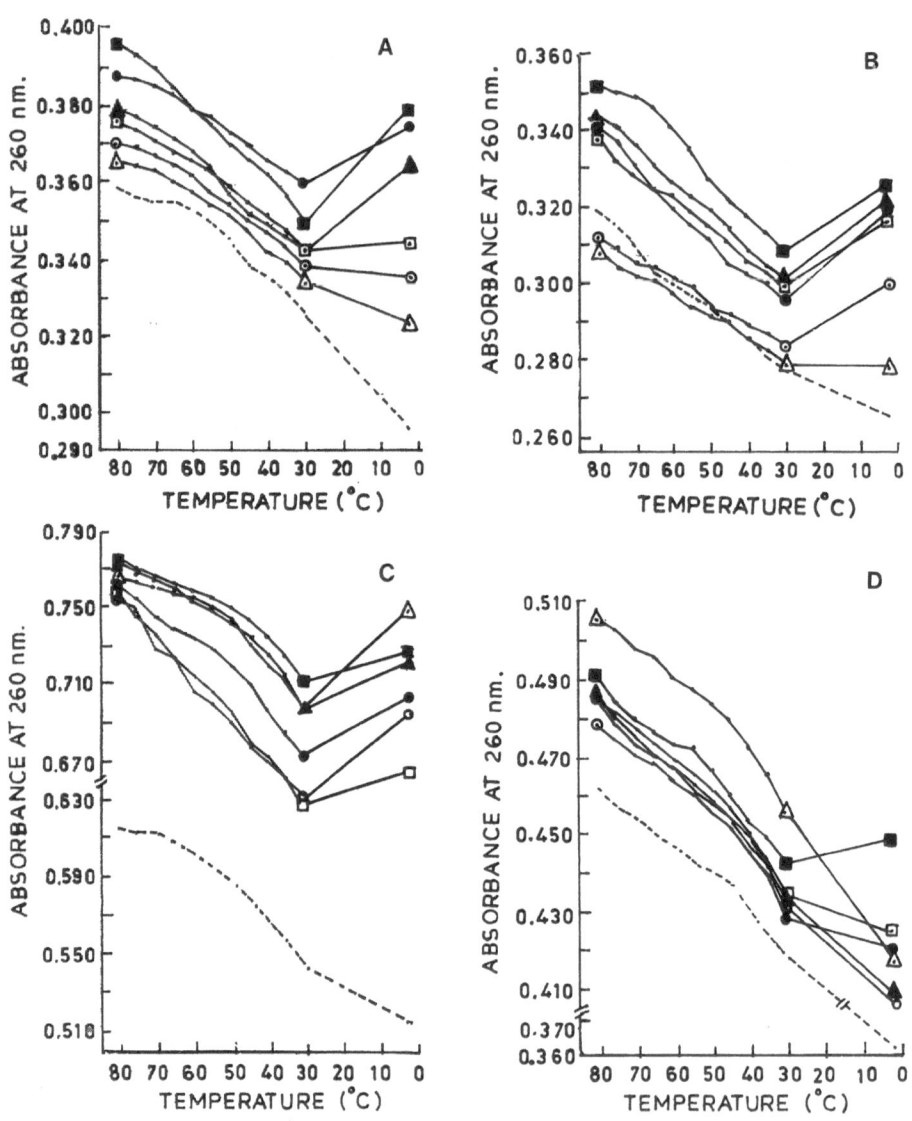

Figure 1. U.V absorbance (260 nm) patterns of denatured DNA during its
renaturation whilst being cooled from 80°C to 2°C (...) alongwith 3 mM
(△); 6 mM (○); 9 mM (□); 12 mM (●); 15 mM (▲) and 18 mM (■) chromium sulphate
(A), nickel chloride (B), beryllium nitrate (C) and sodium hydrogen arsenate.

7.0) at a concentration of 20 µg/ml. The metal compounds were dissolved in double distilled water and added to the DNA solution so as to have their final concentrations as 3, 6, 9, 12, 15 and 18 mM equivalents of the metal. The above solutions were heated in a boiling water bath for 15 minutes. The absorbance values were plotted against temperature to obtain the renaturation curves.

For studying the effects of activation of metals by the microsomal hydroxylase system, to 1 ml of 100 mM equivalent of a metal salt solution in buffered saline, 0.2 ml of microsomal suspension (8-10 mg protein/ml) prepared by the procedure adopted by Gupta and Dani (3) alongwith 1mM NADPH were added and the mixture was incubated at 25°C for 2 h in a constantly shaken water bath. The incubates were then centrifuged at around 5,000 'g' to sediment the aggregated microsomes. Aliquots of clear supernatants were added to highly polymerized DNA (20 µg/ml) solutions to obtain 1mM equivalent of the final concentration of the metals. The U.V. absorbance of these solutions was studied as detailed above. Randomly selected metal compounds were also studied for determining their carcinogenicity by the microsomal degranulation technique (3) at five different concentrations ranging from 10 µg to 80 µg/ml.

Interactions of nucleic acid bases with some metal compounds were also studied by adding 50 µg/ml equivalent of metal to 2 µg/ml of each base and incubating at 37°C for 2 h. These incubations were then scanned from 200 nm to 300 nm using a Spectronic 2,000 spectrophotometer.

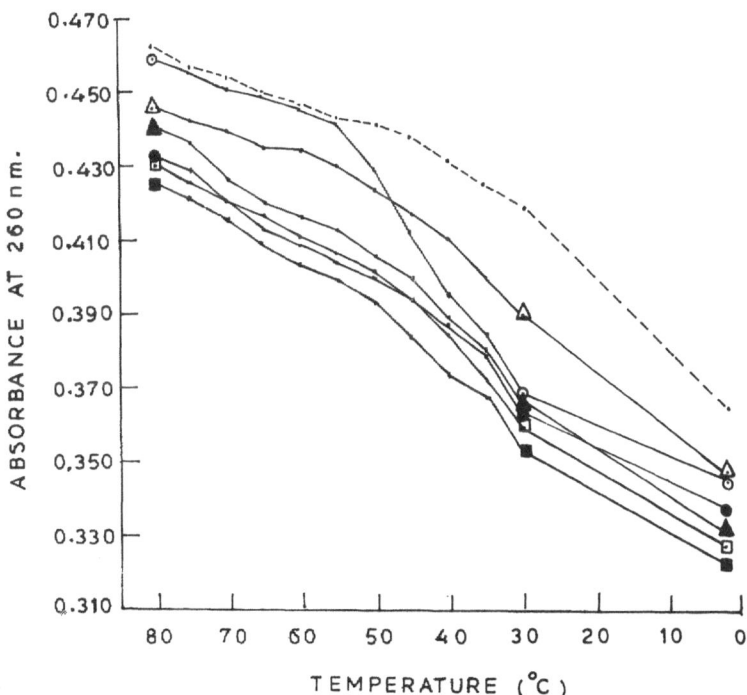

Figure 2. U.V. absorbance (260 nm) patterns of denatured DNA during its re-naturation whilst being cooled from 80° to 2° alone (...) or alongwith 3mM (Δ); 6 mM (o); 9 mM (□); 12 mM (●); 15 mM (▲); 18 mM (■) equivalent concentration of magnessium as magnessium nitrate.

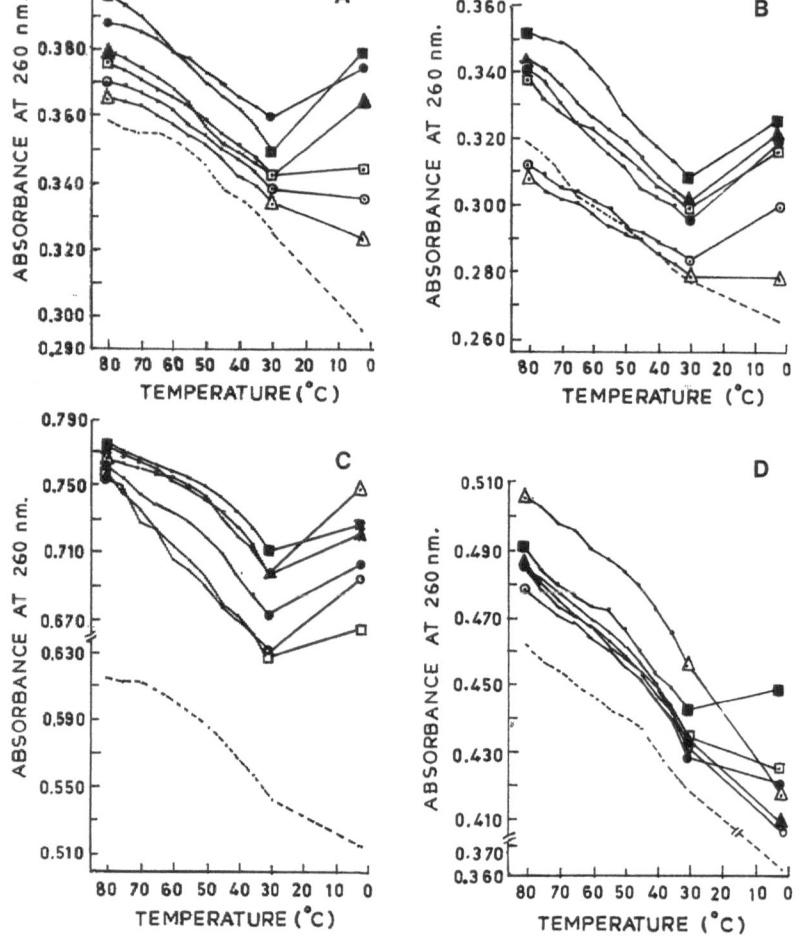

Figure 3. U.V. absorbance (260 nm) patterns of denatured DNA during its renaturation whilst being cooled from 70°C to 2°C in the presence of unactivated (A) and activated (B) chromium trioxide (△); chromic chloride (O);
chromium sulphate (□) and unactivated (C) and activated (D) nickel sulphate
(△); nickel chloride (■); nickel nitrate (●); nickel carbonate (□) and nickel
formate (▲). DNA alone (...).

RESULTS AND DISCUSSIONS

Metal carcinogens failed to denature highly polymerized DNA when
incubated at 37°C. Our preliminary experiments (4) clearly demonstrated that
most of the metal carcinogens (10 mM concentration) enhanced the denaturation
and some of them probably even the depolymerization of DNA when heated in a
boiling water bath. They further interfered with the renaturation of DNA
when cooled from 80°C to 2°C. Non-carcinogenic metals failed to show these
effects and the physiological metal compounds rather supported the
renaturation of DNA.

In order to consolidate our observations, more experiments were performed
employing metal concentrations varying from 3 to 18 mM. Figure 1 (A to D)
illustrates that chromium (A), nickel (B), beryllium (C) and arsenic (D)

Table 1. Effects of different concentrations (10 to 80 µg.ml incubation mixture) of various metal salts/metal powder on the detachment of riobosomes from microsomes as calculated on the basis of RNA/protein ratios according to the method of Gupta and Dani (3).

Compounds/metals studied	Per cent degranulation				
	10 µg	20 µg	40 µg	60 µg	80 µg
Nickel Chloride	0.00	3.84	10.25	14.10	15.51
Nickel sulphate	0.00	3.84	20.00	0.00	0.00
Nickel nitrate	8.97	3.84	30.76	16.66	14.10
Nickel formate	3.77	5.66	14.15	8.96	0.00
Nickel carbonate	2.76	8.91	13.52	12.98	12.32
Nickel powder	12.26	13.68	20.75	21.70	25.47
Beryllium nitrate	26.88	31.60	38.20	36.32	12.73
Beryllium sulphate	16.98	20.28	27.83	16.50	0.00
Chromic chloride	23.28	23.81	34.39	29.62	32.27

enhance the denaturation of DNA and in many cases the denaturation/depolymerization of renaturing DNA suddenly shoots up from around $30^{\circ}C$ to $2^{\circ}C$. These reassociation curves further show that the optimum concentrations depicting the above effects vary from one metal compound to the other but the overall trends are the same. The non-carcinogenic metals failed to show the above effects. Various concentrations of magnesium nitrate, presented here as a typical physiological metal (Figure 2), failed to support the denaturation of DNA and remarkably assisted in its renaturation. On the basis of 46 randomly selected carcinogenic, non-carcinogenic and physiological metal compounds (data for all these could not be included due to constraints on space) studied showing the conclusions drawn from the typical metals presented above, it can be advocated that this short-term technique can be used to predict the carcinogenicity of metal compounds. The metal carcinogens might be disrupting DNA by interfering with the stacking of bases due to the interactions of the metal ions through covalent bonds to purines and pyrimidines (5).

In order to determine the effects of microsomal hydroxylase system on the metal carcinogens for their subsequent interactions with highly polymerized DNA, some more experiments demonstrated that the activation of metals or metal salts with the microsomal hydroxylase system in the presence of NADPH increases their efficiency for denaturing DNA on heating as well as for inhibiting its renaturation (Figure 3). Data for chromium (Figure 3, A and B) and nickel (Figure 3 C and D) have only been presented here while similar results were obtained in case of arsenic, beryllium and cadmium salts. To corroborate this observation, some more experiments were performed to study the effects of activated metal carcinogens at different concentrations on microsomal degranulation (3,6). The activated metal carcinogens have shown a higher degree of microsomal degranulation in comparison to non-activated metal compounds. In most of the cases maximum degranulation was

Table 2. U.V. Absorption maximum and absorbance values of adenine, guanine and cytosine (2 µg/ml) after their incubation with different metal salts (50 µg/ml) at 37°C for 2 h.

Compounds studied	Adenine		Cytosine		Guanine	
	Absorption maxima	Absorbance	Absorption maxima	Absorbance	Absorption maxima	Absorbance
Control	258.8	2.100	264.7	1.140	275.1	1.098
(only base)	223.5	0.438	244.1	0.81	246.3	1.395
	207.0	2.781	200.4	2.731	223.2	0.675
					203.0	2.268
Magnesium	258.8	2.612	265.7	1.021	268.8	1.068
chloride +	223.2	0.462	247.0	0.738	243.7	1.562
base	206.1	2.995	200.1	2.599	221.7	0.495
					200.1	2.261
Sodium	253.3	3.135	263.5	2.310	275.4	1.07
hydrogen	221.5	0.807	227.2	1.685	243.1	1.55
arsenate			200.1	2.158	221.8	0.605
+ base					210.1	3.394
Chromic	250.3	4.386	270.4	2.701	268.7	0.851
chloride	225.3	1.232	239.5	1.100	235.7	1.325
+ base			203.3	3.308	223.8	0.55
					200.1	2.160
Nickel	247.6	3.676	272.1	3.601	269.3	0.851
chloride	222.5	1.333	237.4	0.501	246.6	1.351
+ base			212.6	2.926	223.4	0.401
					200.5	1.994
Beryllium	258.6	2.114	267.1	1.316	275.8	1.178
nitrate +	223.5	0.42	252.9	0.675	254.6	1.868
base	211.2	0.624	220.0	1.210	220.3	0.603
					203.1	2.425

observed at 40 µg/ml of metal concentration in microsomal suspensions (Table 1). Non-activated metals were therefore studied only at 40 µg metal per ml of microsomal suspension and were found to give no microsomal degranulation at all. Therefore the per cent degranulation values as studied by the method of Gupta and Dani (3) have been given only as obtained by using the activated metal compounds/metal powder as shown in Table 1. A decrease in per cent degranulation at carcinogen concentrations higher that 40 µg/ml has also been reported earlier (3,6) as is found in the case of a number of metal carcinogens (Table 1).

To understand the mechanism of interaction of metal carcinogens with DNA bases, some experiments were done by incubating them with adenine, guanine and cytosine and then studying their U.V. absorption maxima (Table 2). As is clear from these data, due to the interactions between nucleic acid bases and metal salts, there are a number of shifts in absorption maxima and the degree of absorbance which varied from metal to metal and from base to base for the same metal. It was quite intriguing to find that the incubation of chromium, nickel and arsenic compounds resulted in the disappearance of an absorption maximum at 207 nm. It can thus be inferred that metal carcinogens might interact with DNA bases and enhance its denaturation and interfere with its renaturation.

Acknowledgement

The authors are grateful to the Indian Council of Medical Research for financial assistance to Sneh and Ravinder.

REFERENCES

1 Costa, M. (1980) metal Carcinogenesis Testing, The Humana Press Inc., New Jersey.
2 Ames, B. N. (1979) Science, 204, 587-589.
3 Gupta, M. M. and Dani, H. M. (1979) Indian J. Exp. Biol., 17, 114-1146
4 Sneh Ahuja. (1987) Ph.D. Thesis, Panjab University, Chadigarh (India)
5 Fuwa, K., Waren, W. E., Druyan, R., Bartholomay, A. and Vallee, B. L. (1960) Proc. Natl. Acad. Sci., 46, 559-563.
6 Purchase, I. F. H., Longstaff, E., Ashby, J., Styles, J. A., Anderson, D., Lefevre, P. A. and Westwood, F. R. (1978) Br. J. Cancer, 37, 873-959.

CHEMICAL CARCINOGENS AND ONCOGENES

D. K. Biswas

Department of Pharmacology
Harvard School of Dental Medicine
Boston
Mass. 02115
USA

ABSTRACT

Association of environmental chemicals as causative agents for cancer has been reported as early as 1761. The geographical pattern of incidence of cancer among human populations and its prevalence in individuals with specific occupations and habits has provided important clues on the etiology of the disease. Studies on the metabolism of potent chemical carcinogens in different cell types have identified active derivatives of many of these compounds and characterized the nature of their direct interactions with the cellular macromolecules. A strong correlation between the mutagenicity and carcinogenicity of large series of compounds suggested that DNA is the ultimate target of these carcinogens. More direct evidence for this concept originated from the classical DNA-mediated transfection studies identifying tumorigenic DNA sequences in chemically transformed cells which, when transferred to nontumorigenic cells resulted in the malignant transformation of recipient cells. Since these early reports we now know more regarding these oncogenic DNA sequences ("oncogenes") and on the molecular aspects of chemical carcinogenesis and accompanying altered gene structure and gene expressions (increase, decrease, inactivation or novel expression). Numerous studies with end stage tumors or with tumor cells in cultures or with in vitro, transformed cells have identified several such ocogenic DNA sequences the cellular unmodified homologue of which are referred to as "proto-oncogenes". It has been postulated that more than one of these cellular protooncogenes are involved in a stage specific fashion in the multistep carcinogenesis process. The cooperative interaction of different cellular protooncogenes activated at different stages of the DMBA-induced in vivo carcinogenesis in hamster buccal pouch epithelium will be elaborated.

The majority of human cancers, estimated between 70-80%, are thought to be caused by chemicals in the environment (1). Some of the chemical exposures can also be related to specific occupations and food and chewing habits. Evidence that chemicals could cause cancer in humans originated about two centuries back when Hill in 1761 (2) and Pott in 1776 (3) reported a high incidence of nasal and scrotal cancer in humans exposed to tobacco snuff and chimney soot respectively. Without going into many details, I will briefly describe the chronology (Table 1) of these studies and will try to

Table 1. SUMMARY OF KEY EVENTS IN EXPERIMENTAL CHEMICAL CARCINOGENESIS
RESEARCH*

1761   John Hill observed the association of nasal cancers and excessive
use of tobacco snuff in men (1).

1775   Percivall Pott associated soot as the cause for the high incidence
of scrotal cancers among chimney sweeps in London (2).

1914   Boveri proposed that neoplastic cells originated from normal cells
as a consequence of chromosomal changes (3).

1915   Yamagiwa and Ichikawa demonstrated the production of cancer in
rabbits' ears by painting with coal tar. This formed the foundation
of experimental chemical carcinogenesis research (4).

1924   Kennaway produced carcinogenic tar by pyrolysis in a hydrogen
atmosphere of several organic compounds and proposed the carcinogenic
agents in coal tar to be polycyclic aromatic hydrocarbons (PAHs) (5).

1930   Kennaway and Hieger demonstrated the carcinogenic activity of
dibenz(a,h)anthracene, the first synthetic PAH (6).

1932   Lacassagne observed mammary cancer development in male mice
treated with estrone. This marked the beginning of hormone-
induced tumors for experimental research. (7).

1933   Cook, et al synthesized and isolated benzo(a)pyrene (BaP) and showed
it to be highly carcinogenic (8).

1938   Bachmann and Chemerda synthesized 7,12-dimethylbenz(a)anthracene
(DMBA) (9).

1937-   Uses of PAHs in the treatment of human tumors (10-13).
1939

1946   Gardner and Heslington experimentally induced osteosarcomas in
rabbits using zinc beryllium silicate and beryllium oxide, the
first experimental demonstration of the carcinogenicity of certain
inorganic metals (14).

1947   Miller and Miller demonstrated the covalent binding of hepatocarcinogenic aminoazo dyes to cellular proteins (15).

1951   Miller reported that BaP covalently bound to proteins of mouse
skin when treated with hydrocarbons (16).

1957   Conney, et al demonstrated that treatment of rats with PAHs markedly
increased the levels of enzymes that metabolize them (17).

1961   Heidelberger and Davenport demonstrated the covalent binding of
carcinogenic hydrocarbons to DNA in mouse skin (18).

1964   Brookes and Lawley demonstrated a correlation between the
carcinogenic potency of six PAHs and the extent to which they became
covalently bound to mouse skin DNA, but not RNA or protein (19).

1968-   Grover and Sims (20) and Gelboin (21) independently showed that PAHs
1970   became bound to DNA in vitro only in the presence of active metabolizing systems.

1973   BaP and other PAHs tested positive in the Ames test (23).

1974   Sims, et al demonstrated that a 7,8-diol-9,10-expoxide is the
"ultimate" carcinogenic form of BaP (24).

---

* Taken from the thesis of Dr. D. T. W. Wong of Postdoctoral Program of
Harvard School of Dental Medicine, 1985, Sponsor: Dr. Debajit K. Biswas.

focus mainly on the subject matter of today's talk, i.e. Chemical Carcinogens and Oncogenes, emphasis being on the oncogenic DNA sequences and on the postulated mode of their action.

Subsequent to the observation made by Hill and Pott, a series of experiments were performed early this century which were mostly directed towards the identification of the carcinogenic compounds in some of these environmental sources.

In doing so, Yamagiwa and Ichikawa (4) and Kennaway and Hieger (5) identified polycyclic aromatic hydrocarbons (PAH) in coal tar and established the carcinogenic activity of dibenz(a,h)anthracene.

Synthetic derivatives of PAH were subsequently isolated and used in 1933. The next phase of the study involved experiments on the interactions of the potent carcinogens with cellular macromolecules (6). In this process Heidelberger and Devenport (7) demonstrated the covalent binding of carcinogenic hydrocarbons to DNA of mouse skin. Several investigators (8-10) subsequently established the metabolic pathway of these chemicals and identified the active carcinogenic derivatives. Chemical carcinogens are subjected to extensive metabolism in animal cells and are converted into electrophilic reactants which exert their biologic effects by covalent interaction with cellular DNA. The majority of chemicals undergo metabolic activation in vivo with the exceptions to this are the alkylating and acetylating agents per se.

The extensive investigations on the chemistry and metabolism of these chemicals as well as on their interactions with cellular DNA have provided a very strong foundation for the further studies on the molecular basis of chemical carcinogenesis. Chemical carcinogens and their electrophilic derivatives induce chromosomal lesions via covalent interactions and also cause further damage during the faulty repair of the original lesion. The question which remains unclarified is: which of these chromosomal lesions are actually responsible for the transformation of cells? The specific affected DNA sequences determine the carcinogenic potential of the chemicals.

A strong correlation between mutagenicity and carcinogenicity suggests that DNA is the ultimate target of these carcinogenic agents (11-13). More convincing evidence for this concept originated from the experiments with retrovirus and from the classical DNA transfer studies, identifying the carcinogenic DNA sequences in cancer cells, transfer of which resulted in the manifestation of malignant phenotypes in non-tumorigenic cells (14-16). Since these earlier reports, a large number of such oncogenic DNA sequences (commonly referred to as oncogenes) have been identified. The cellular inactive form of these sequences are known as Protooncogenes. Phenotypic expression and morphological changes associated with carcinogenesis can be correlated to the concurrent activation of the cellular protooncogenes.

The concept of oncogenes and their role in the carcinogenesis does not need to be redefined to this audience. From the extensive amount of investigations and the voluminous literature the following criteria may be assigned to these oncogenic DNA sequences.

i) These specific DNA sequences are highly conserved.

ii) These sequences are either inactive or exhibit very low level of activity in "normal" cells.

iii) In cancer cells some of these gene are overexpressed.

iv) In most cases the overexpression of the genes can be correlated to the somatic mutation in the "normal" genetic loci.

127

v) These mutated and overexpressed genes can be transferrred from cancer cells into "normal" cells and transformation phenotypes can be manifested in the recipient cells.

On the basis of the biochemical properties and their functions, the oncogenes can be categorised into several classes (17,18). The gene product of one class of oncogenes, identified in cancer cells and also found in animal retrovirus, shows distinct protein kinase activity. The src, abl, erbB, etc are a few examples of this class of oncogenes. The p21 protein product of the ras family of oncogenes identified in chemically-induced mammary gland, bladder, and skin tumors show guanine nucleotide binding property and GTPase activity. The protein product of Class III oncogenes are localized in the nuclear matrix and are believed to be involved in the regulation of transcription. The Class IV oncogenes are associated with growth factor related function and overexpression of this class of oncogenes is responsible for the unregulated growth and proliferation of cancer cells. At present about 55 cellular oncogenic sequences are identified. The exact role and the mechanism of their action in the carcinogenesis process is not yet clearly defined in molecular terms.

Several examples of concurrent activation of cellular protooncogenes in chemically transformed cultured cells and in chemically-induced tumor in vivo systems have been reported. The mechanism of activation of these genes has been established in many cases. These results demonstrate that the over-expression of the genes can be correlated to different chromosomal lesions. These chemicals induce gene translocation, insertion and deletion mutation and occasionally gene amplification. A partial list of these examples is listed below in Table 2. It is evident from these results conducted in different laboratories with a variety of systems that activation of ras family of gene in chemically-induced tumor cells is a consistent event.

There is compelling evidence to believe that carcinogenesis is a multi-step process. If growth rate, cellular differentiation, and function of "normal" cells are operated through an orchestra of events mediated via expression of some genes and suppression of others, the carcinogenesis process may be viewed as a discord in this cellular symphony. Such discord can be brought about by genetic damage such as chromosomal translocation, deletion or insertion mutation gene amplification, etc. The chromosomal lesions, in specific sequences whose role is to remain silent in the orchestra of cellular regulatory process, are randomly caused by genotoxic chemicals. The affected genes start making noise by being overexpressed. One such event may then initiate a series of subsequent reactions which will ultimately result in complete disruption of the cellular growth control process leading to the neoplastic transformation of cells. Activation of several such affected cellular protooncogenes thus can establish another orchestra at a different level than the one in the "normal" cells. To delineate the molecular mechanism of the transition from one level of regulation to another is the key issue in current cancer research.

In the multistep carcinogenesis process, the role of an individual event can be defined only when it is studied during development of the process and not in the end stage tumor or tumor cells. It is very difficult to establish the significance of activation of some genes observed in the end stage tumor cells regarding their role in the transition from one state of growth regulation to another and thereby in the initiation and progression of the carcinogenesis process. However, by DNA-mediated transfer studies, transformation potential of some of these activated genes are established. But their exact role in the process and their mode of actions are yet to be established in molecular terms. In a limited number of in vivo carcinogenesis systems, such as DMBA-induced mouse-skin papilloma and carcinoma system (19,20); rabbit ear keratoacanthoma system (21); and DMBA-induced

Table 2. CHEMICAL CARCINOGENS & ACTIVATION OF CELLULAR ONCOGENES*

Chemical Carcinogen	Oncogene Activated	Mechanism	System
BaP and MCA	LTR of Moloney Murine Sarcoma Virus	Increased Transcription	Mouse C3H 10T1/2 Fibroblasts
Pristine	c-mos	Rearrangement	Mouse Balb/c
DMBA	cHa-ras	Amplification	Mouse skin
DMBA	cHa-ras	Point mutation	Hamster Cheek Pouch
	c-erb B	Amplification	Hamster Cheek Pouch
DMBA	cHa-ras	Point mutation	Rabbit ear Keratoacanthomas
MCA	cKi-ras		Mouse fibro-sarcoma
N-nitroso-N-methyl urea	cHa ras	Point mutation	Rats
Mineral Oil	c-mos	Rearrangement, Hypomethylation Increased Transcription	Mouse myeloma MYOPC 21 cell line
MCA	cKi-ras		Mouse C3H 10T1/2 Fibroblast cell line
BaP and DMBA	c-myc	Loss of cell cycle control	Balb/c 3T3 (A31)cells
N-methyl-N'nitro-N-nitroso-guanidine (MNNG)	k-ras N-ras	Point mutation	Human Osteo-sarcoma cells, Renal cells
Diethyl-nitrosamine (DEN)	cHa-ras cKi-ras N-ras	Increased Transcription	Rat
N-nitroso-methyl urea	N-ras	Increased Transcription	Mice
TPA and Teleocidin	cHa-ras		Mouse C3H 10T1/2 Fibroblasts
BPDE	cHa-ras	Point Mutation	NIH3T3 Mouse Fibroblast cell line
BaP, MCA, MNNG, and DEN	cHa-ras		Guinea pig cell lines

* Taken from the thesis of Dr. D.T.W. Wong in Postdoctoral Program of Harvard School of Dental Medicine, 1985, Sponsor: Dr. Debajit K. Biswas.

hamster buccal pouch epithelial carcinoma system (22,23) attempts have been
made to define the role of the stage specific activation and cooperative
interactions of the cellular protooncogenes in the carcinogenesis process.
In many instances, some of these cellular genes which are found to be overexpressed in tumor cells also play an essential role in the regulation of
growth, proliferation, and differentiation in normal cells. Examples of this
class of oncogene are erb, sis, myc, etc. The protein product of the
recently identified three oncogenes is structurally similar to fibroblast
growth factor (FGF). The FGFs (acidic and basic) are angiogenic agents and
help in the growth of new blood vessels. The product of the oncogene hst
(human stomach cancer) has 206 aminoacids and shows 45% sequence homology
with basic FGF. The int 2 gene originally detected in MMTV also belongs to
the FGF family. The third FGF-like oncogene protein is derived from bladder
cancer. The FGF, like oncogene products, have been suggested to have a role
during embryogenesis. The int 2 gene for example is not active in adult
mice; however, the gene is to be turned on in mouse embryos as 7.5 days of
gestation and is turned off by day 8.5 (24). In the process of solid tumor
growth, new blood vessel formation is necessary, and presence of these
substances may be essential for this purpose.

Oncogenes are an altered form of normal cellular genes which are
modulated to produce either excess amounts or structurally altered products.
The regulatory mechanism of expression of these genes is disrupted and
thereby causes aberrant expression of the gene. The phenotypic expression
and morphological changes associated with transformation process can be
correlated with the concurrent activation of the inactive gene. Alternatively, it is also frequently observed that cellular genes related to the
cell growth and proliferation may be stimulated without undergoing further
genetic modulation, simply by increased transcription probably mediated via
signals from already activated other cellular protooncogenes. Thus the
former class of genes which cooperate in accelerating the cell growth process
may be referred to as cooperator genes; whereas the later class of genes,
activationof which initiates the signal for the overexpression of the cooperator genes, may be designated as initiator gene for the carcinogenesis
process.

A strong correlation exists between activation of ras family of oncogenes
and certain kinds of chemically-induced tumors. The activation of the ras
family of oncogenes detected mostly in the earlier stages of tumor development may play the role of "Initiator" gene in the chemically induced mouse
skin papillomas and carcinomas, rabbit keratoacanthomas, and hamster cheek
pouch carcinomas. The overexpression of ras genes in benign growth or in
growth which ultimately self-regresses suggests that activation of ras alone
may not be sufficent for the fully-tansformed cells. It is proposed that
subsequent activation of other cellular protooncogenes, specifically those
related to cell growth stimulation (such as erb. sis, myc, etc.) may accele-
rate the process and drive the ras-initiated cells to the fully transformed
state. Examples of cooperative interaction of ras and cmyc (25), ras and N-
myc (26) have been reported. Lee et.al.(27) suggested that the ras gene even
with structurally unmodified cmyc is sufficent for the transformation of
mouse embryo fibroblast cells. Along with in vitro transformation studies,
in a limited number of in vivo carcinogenesis systems the cooperative interaction of ras and other cellular protoncogenes has been implicated (28).
Brown et.al.(19) reported that infection if mouse skin with Harvey murine
sarcoma virus (haMSV) did not show neoplastic activity until skin surface was
treated with 12-0-tetradecanoyl-phorbol-acetate (TPA). Interaction of TPA-
induced TGF-beta with ras gene product is implicated in this system (29).

We have proposed cooperative interaction of ras gene product with that of
erb gene in the DMBA-induced hamster cheek pouch carcinoma system. These
results demonstrate stage specific activation of at least two cellular proto-

oncogenes during DMBA-induced hamster buccal pouch epithelial cell (HBPE) carcinoma. cHa-ras gene is activated at an early stage, at a time when no significant pathological changes besides chronic inflammation and hyperkeratosis of the treated tissue is observed (Fig 1). The stimulated level of expression of ras gene persists throughout the rest of the tumorigenesis process. The expression of erb-B, on the other hand, can be detected at a much later stage, and this corresponds with the stage of extensive proliferation and subsequent invasion of the HBPE cells into the underlying connective tissue. Further analysis of the system reveals that over-expression of cHa-ras alone is not sufficient to manifest all the transformation phenotypes (e.g. [i] in vivo tumor formation, [ii] anchorage independent growth on soft agar, [iii] transformation of NIH/3T3 cells, and [iv] finally tumor induction in athymic mice) to the fullest extent. Whereas HBPE cells, by expressing both ras and erb-B genes, are positive by these criteria and exhibit fully transformed phenotypes.

On the basis of the results obtained with the in vitro and in vivo carcinogenesis systems, the following tentative model may be proposed (Fig.2.)

Chromosomal lesions caused by chemicals in the inactive initiator gene (e.g ras) leads to the activation of the gene by somatic mutation in the normal genetic loci. The initiator gene product may then initiate the next event, e.g. by activating (via as yet unestablished mechanisms, three of which are suggested in Fig. 2) the cell proliferation-related genes. This process leads to the production of excess amount of growth factor related substances. The product of the cooperator gene can then act as an autocrine growth factor for the continuous growth phenotype of the transformed cells. This model postulates that the ultimate goal of the altered cellular regulatory mechanism is to produce tumor cells specific autocrine growth factor.

Another example of autocrine growth factor property of the tumor-specific product has recently been demonstrated in our laboratory. This is illustrated in Fig 3. The human lung tumor cells (ChaGo), derived from a bronchogenic carcinoma of a heavy smoker, inappropriately synthesize and secrete the glycoprotein hormone, human chorionic gonadotropin (HCG). These cells, when deprived of their own product HCG either i) by addition of anti-HCG specific antibody or ii) by introduction into the cells of aHCG (alpha subunit) specific antisense RNA-producing expression vector with aHCG cDNA in reverse orientation with respect to the reading direction of the promoter (RSV), lose the characteristic transformation phenotypes. These results provide an autocrine growth factor role for HCG, the product of the inappropriately expressed gene. Our results also reveal that HCG acts as a progression factor in the cells' growth and this fits into the role of cooperator gene product. The question remains unanswered as to which is the initiator gene in this system. This is currently under investigation. It may be postulated again that chromosomal lesions induced by the carcinogenic agent (probably by a chemical) have triggered the activation of otherwise inactive aHCG gene in lung cells. The overexpression of HCG gene, along with another activated cellular protooncogene, selects continuously growing transformed cells.

Before a unified model for the molecular mechanism of chemical carcinogenesis is established, the following as yet unresolved questions have to be resolved: What is (i) the molecular mechanism of the mode of action of the activated genes, (ii) the sequence of activation of these genes in the carcinogenesis process, (iii) the correlation of activation of one gene with the other, and does activation of one gene depend on that of the other or are these independent events?

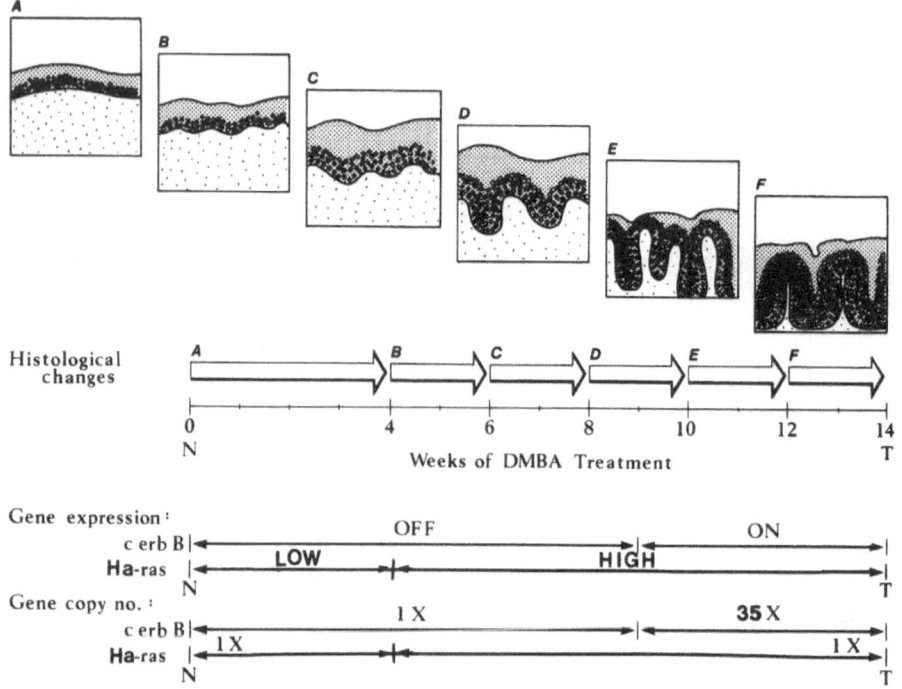

Fig.1. Stage specific activation of cHa-ras and c erb protooncogenes in
relation to histopathological changes during dimethylbenzanthracene-induced
hamster buccal pouch epithelial carcinogenesis.

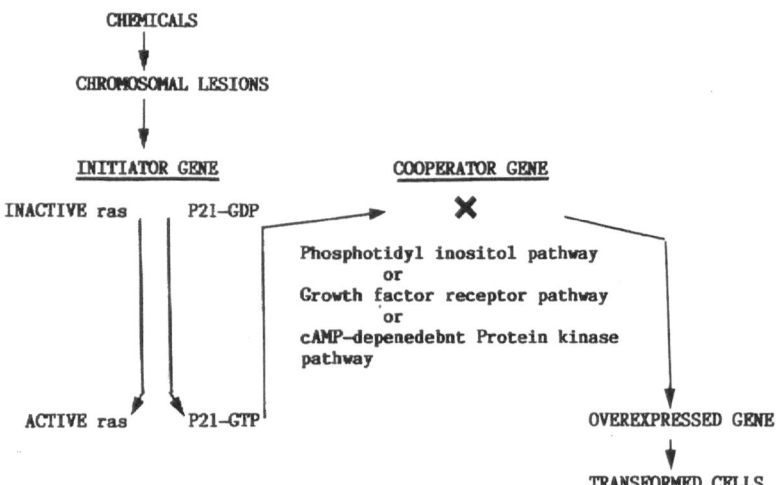

Fig.2. A tentative model for mechanism of chemical carcinogenesis. Activated
ras, gene product (p21) continuously generate signal via one of the indicated
mechanisms to activate the cooperator gene (X) which stimulates cell growth.
Thus (X) may be a growth stimulatory gene of "normal" cells.

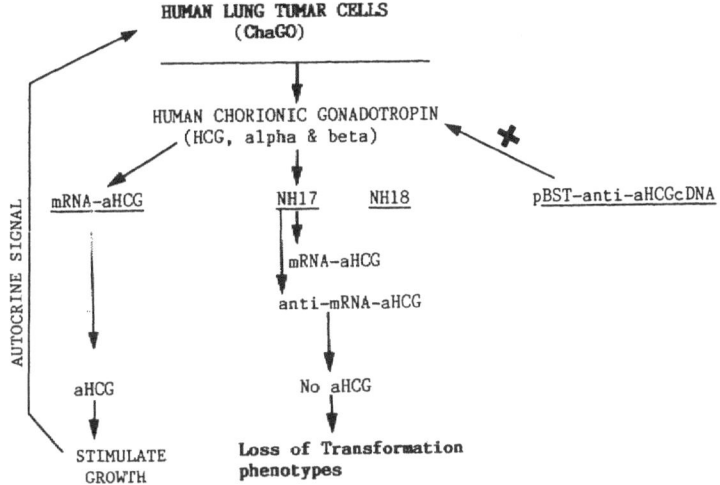

Fig.3. Autocrine function of tumor cell specific product (HCG). ChaGo cells are HCG-producing human lung tumor cells in culture. pBST-anti-aHCG-cDNA is a plasmid expression vector with aHCG-cDNA in reverse orientation with the reading direction of promoter (RSV). Introduced into (+) ChaGo cells by DNA mediated transfection. Both sense and the antisense RNS(aHCG) species identified. Transformation phenotypes assayed.

REFERENCES

1   World Health Organisation (1966) Prevention of Cancer. Tech.
    Rep.Ser.No 276, Geneva
2   Redmond,J.R.,Jr.(1970) New Eng.J.Med. 282, 18-23
3   Pott,P. (1775) in Chirurgical Observations Relative to the Cancer of the
    Scrotum. Reprinted in Natl.Cancer Inst.Monogr. (1963) 10, 7-14
4   Yamagiwa,K. and Ichikawa,K. (1918) J.Cancer Res. 3, 1-297
5   Kennaway,E.L. and Hieger,I. (1930) Br.Med.J.ii,1044-1046
6   Miller,E.C. and Miller,J.A. (1947) Cancer Res. 7, 468-480
7   Heidelberger,C., and Davenport,G.R. (1961) Acta Unio.Int.
    Contra Cancrum, 17,55-63
8   Miller,J.A. (1970) Cancer Res. 30, 559-576
9   Heidelberger,C. (1975) Ann.Rev.Biochem. 4, 79-121
10 Selkirk,J.K. (1980) in Carcinogenesis, Modifiers of Chemical Carcinogen-
    esis, ed.Slaga,T.J., vol.5, pp.1-31, Raven Press, New York
11 McCann,J., Choi,E., Yamasaki,E. and Ames,B.N. (1975) Proc. Natl.Acad.Sci.
    USA 72, 5135-5139
12 Macann, J. and Ames, B.N. (1976) Proc.Natl.Acad.Sci.USA 73 950-954
13 Bonek, N. and diMayorca,G. (1978) Nature 264, 722-727
14 Shih,I.S., Shilo,B., Goldfarb,M., Dannenberg,A.and Weinberg,R.A. (1979)
    Proc.Natl.Acad.Sci.USA 76, 5714-5718
15 Cooper,G.M. (1982) Science, 217, 801-805
16 Weinberg, R.A. (1982)) Adv.Cancer Res. 36, 149
17 Bishop,J.M. (1983) Ann.Rev.Genet. 52, 301-354
18 Varmus,H.E. (1984) Ann.Rev.Gennet. 18, 553-612
19 Brown,K., Quintanilla,M., Ramsden,M.,Kerr,F.B., Young,S. and Balmain,A.
    (1986) Cell, 46, 447-456
20 Quintanilla,M., Brown,K., Ramsden,M. and Balmain,A. (1986) Nature, 300,
    149-152

21 Leon,J.,Kamino,H., Stenberg,J.J. and Pellica,A. (1988) Mol.Cell.Biol. 8, 786-793

22 Wong,D.T.W. and Biswas,D.K. (1987) Oncogene, 2, 67-72

23 Solt,D.B., Polverini,P.J., Ray,S., Fei,Y. and Biswas,D.K. (1988) Carcinogenesis, in press

24 Marx,J.L. (1987) Science, 237, 602-603

25 Land,H., Chen,A.C., Morgenstern,J.P., Parada,L.F. and Weinberg,R.A. (1986) Mol.Cell.Biol. 6, 1917-1925

26 Yancanpoulos,G.D., Nisen,P.D., Tesfaye,A., Kohl,N.E., Goldfarb, M.P. and Alt,F.W. (1985) Proc.Natl,Acad.Sci.USA 82, 5455-5459

27 Lee,W.F., Schwab,M., Westaway,D. and Varmus,H. (1985) Mol. Cell.Biol. 5, 3345-3356

28 Barbacid,M. (1987) Ann.Rev.Biochem. 56, 779-827

29 Akhurst,R.J., Frances,F. and Balmain,A. (1988) Nature, 331, 363-365

RAS P21 LEVEL AND TYROSINE KINASE ACTIVITY DURING CARCINOGENESIS OF

HUMAN COLON TUMORS

O. Csuka and J. Sugár

National Institute of Oncology
Budapest
Hungary

ABSTRACT

The relationship between ras oncogene expression, clinical staging and fraction of S-phase cells has been studied in 25 colon adenomas and 42 colon carcinomas by quantitative immunohisto chemistry. The level of tyrosine kinase activity in 27 colon carcinomas has also been evaluated. Elevated level of ras p21 could be detected in the hyperplastic and dysplastic lesions, in 32% of colon adenomas and 52% of colon carcinomas. A strong correlation could be found between the high rate of cell proliferation and the increased level of ras p21 both in adenomas and carcinomas (r_1 = 0.8418; r2 = 0.9007). There was no correlation between clinical stages and p21 content of colon carcinomas (p = 0.49). A high level of tyrosine kinase activity was found in 26% of colon carcinomas and a close correlation between the level of tyrosine kinase activity and the high ration of S-phase cells has been demonstrated. Our results have shown that the mitotic signal transduction in 52% of colon carcinomas is mediated by ras p21 protein. Elevated level of tyrosine kinase activity in 26% of colon carcinomas suggests the participation of growth factor receptors and/or oncogenes homologous with these receptors in the signal transduction pathway.

INTRODUCTION

Cellular ras proto-oncogenes may be activated by point mutations at high frequencies in rodent tumors induced by chemical carcinogens. These mutations involve the change of a single amino acid at 12, 13, 59, 61 codons of the ras genes (1).

A point mutation at any of these positions is sufficient to elicit the transforming activity of the ras genes. The type and frequency of mutation has been found to depend on the chemical nature of the carcinogen (2). H-ras oncogenes were found to be activated in rat mammary carcinomas induced by nitroso methylurea (MNU) and dimethyl benzanthracene (DMBA) (3,4). Activated H-ras genes could also be identified in DMBA initiated skin carcinomas of mice (5) and N-butyl-N(4-hydroxybutyl) nitrosamine induced rat bladder tumors (6). On the other hand, dimethylhydrazine induced rat colon carcinoma (7) and aflatoxin B, induced rat liver tumors were found to contain activated Ki-Ki-ras genes (8).

Detection of activated ras genes can be measured using a DNA transfection assay (9), analysis of restriction fragment length polymorphism (10) or by techniques utilizing monoclonal antibodies. The frequent detection of activated ras genes in chemically induced tumors strongly suggest that these oncogenes must play an important role in tumor development. Activated ras oncogenes have also been detected in several human cancers including bladder, breast, colon, kidney, liver, lung, ovary, stomach cancers and leukemias (11). These data suggest that ras oncogenes are present in a significant percentage (5-40%) of human tumors.

The ras gene products are thought to be involved in normal cellular growth and differentiation in eukaryotes (12). By analogy with other known G proteins, the p21 proteins encoded by ras genes may act as a regulatory protein in the transduction of signals that lead to DNA synthesis (13). Recent reports have postulated a common signal transduction pathway for both the ras proteins and growth factors by the activation of phosphatidylinositol turnover (14,15). Determination of the specific mitotic signal transduction pathway of a particular tumor type might be important from both a theoretical and practical point of view. The work reported here investigates the role of p21 and tyrosines kinase in the signal transduction pathway of human colon tumors. Since the mitotic signal transduction could be mediated by growth factor receptors having tyrosine kinase activity, we studied the correlation between the level of tyrosine kinase activity and the rate of cell proliferation in 19 colon tumors. In the present work we also wished to study whether increased expression of ras oncogene is an early or a late event during colon carcinogenesis. Therefore we have surveyed 25 human colon adenomas and 42 human colon carcinomas measuring the ras p21 level by quantitive immuno-histochemistry. We studied the relationship between ras oncogene expression, clinical staging and the fraction of S-phase cells. Our data suggest that the level of ras p21 correlates with the rate of cell proliferation rather than the stage of the disease. Our results support a role for the ras onco-genes in the control of cell proliferation of colon tumors.

Materials and Methods

Materials: 25 colon adenomas and 42 carcinomas were utilized for the immunohistochemical detection of ras oncoprotein. The specimens were fixed in 10% v/v formalin and embedded in paraffin. Imprint smears were taken from the same specimen for the determination of their cellular DNA content.

Monoclonal antibody: Murine Ig G2a MAb RAP5 was a generous gift of Drs A Thor and J Schlom (Bethesda, NIH). This monoclonal antibody was raised against a synthetic peptide corresponding to amino acid positions 10-17 of the human ras gene product p21. This antibody reacts with both point-mutated and proto oncogene forms of ras p21 (16).

Immunohistochemical assay: 5-7 µm thick sections of formalin fixed paraffin-embedded tissue were processed immunocytochemically by the PAP method (17). Endogenous peroxidase was inactivated by incubation with 0.3% v/v H_2O_2 in PBS at room temperature for 30 minutes. Sections were then treated with 3% v/v normal goat serum for 30 minutes followed by the ras p21 specific monoclonal antibody for 30 minutes. In the PAP method, first anti-body was followed by goat anti mouse immunoglobulin and by mouse peroxidase-antiperoxidase complex. Finally, in immunoperoxidase procedures, peroxidas was visualized by DAB (18).

DNA cytophotometry: The DNA content of Feulgen's stained cells was determined with an MPM01 computer-controlled absorption cytophotometer. The diploid reference standard was the mean DNA content of 50 leukocytes present in the smears. The percentage distribution of G_0, $_1$-S-G_2 phase cells was determined by the mathematical analysis of DNA histograms. The mathematical

136

analysis was based on the maximum likelihood method and the Bayesian iteration for mixture decomposition (19).

Cytophotometry of p21 protein: To determine the amount of reaction product of peroxidase and DBA (a polymer of oxidized DAB) on individual cells, the absorbance is directly proportional to the amount of poly-DAB (20). Absorbance measurements were performed with a Karl-Zeiss scanning microscope photometer interfaced with a Hewlett Packard 9825 B computer. For the standardization of the measurement a ratio of extinction measured in adenomas or carcinomas and in the adjacent intact epithelium of the same section was calculated.

Tyrosine kinase assay: Tyrosine kinase activity was measured according to the method of Swarup et al. (21). The reaction volume of 100 µl contained 50mM TRIS-Cl pH = 7.8, 50mM MgCl2, 10uM sodium vanadate, 0.1% Nonidet P-40, 0.5 nmol (^{32}p) ATP, 1mM of Angiotensin II, 60 ul of homogenised tissue (2-15 µg of protein). The assay was initiated by the addition (^{32}p) ATP. After incubation for 10 minutes at 30°C the reaction was stopped by the addition 150 µl of 10% v/v trichloroacetic acid, and subsequently 10µl of 20 mg/ml of bovine serum albumin were added. The precipitated protein was removed by centrifugation (3200 g, 25 min) and two 50 µl aliquots of the supernatant fluid were spotted on phosphocellulose paper (Whatman P81). The phosphocellulose paper squares were washed 6 times in 0.5% v/v phosphoric acid and once in acetone. The dried papers were counted for radioactivity in 5 ml of scintllation fluid. For each sample, appropriate reaction mixture containing no peptide were run as controls.

RESULTS

Level of Ha-ras p21 protein in normal, hyperplastic colonic epethelium and colonic carcinomas

Quantitative immunohistochemical studies were performed on formalin fixed tissue sections of 25 adenomas and 42 colorectal carcinomas using MAB RAP5 (16) and peroxidase-antiperoxidase method. Since this monoclonal antibody reacts with both proto-oncogene and point mutated forms of ras p21, the level of this protein has been determined in both the normal colonic epithelium and adenomas or carcinomas in the same section. The average extinction at 480nm was calculated from data determined in 50 different fields. As shown in Fig 1 the average extinction of normal epithelium ranged from 0.15 to 0.38 while colon carcinomas have extinction values between 0.5 - 0.97.

These data clearly demonstrate that the ras p21 level is significantly higher in carcinomas than in normal colonic epithelium (p=0.001). The expression of ras p21 protein in hyperplastic and dysplastic areas of colon adenomas has been determined. The extinction range of hyperplasias and

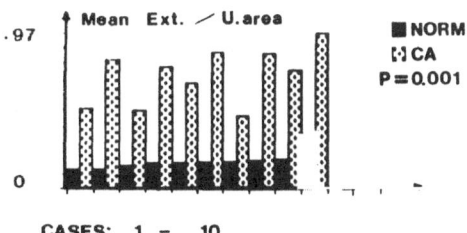

Figure 1. Average extinction values of ras p21 protein in normal colonic epithelium and colon carcinomas.

Table 1. Expression of ras p21 protein in human colon

NUMBER OF CASES	LESIO	MEAN E_{480} nm S D OF p21
67	Intact epithelium	0.160 ± 0.049
25	Hyperplasia	0.354 ± 0.064
25	Grade III dysplasia	0.752 ± 0.083

$$p = 0.001$$

dysplasia is 2–5 times higher than that of the adjacent normal epithelium indicating an elevated ras p21 level in these pathological lesions.

Determination of p21 level and fraction of S-phase cells in colorectal adenomas and carcinomas.

The level of p21 protein has been determined in different types of adenomas and colon carcinomas and in the adjacent normal colonic epithelium by quantitative cytophotometry. The extinction value of the end product of the p21 specific innumohistochemical reaction depends on the thickness of the sections. To correct for variation in thickness our results are expressed as the ratio of the extinction value obtained in adenomas or carcinomas and in the adjacent normal epithelium of the same section. This approach allows the comparison of p21 level in a large series of sections with various thickness. Apart from the estimation of ras expression, the rate of cell proliferation in the same specimen has also been determined. We evaluated the ras p21 expression in 25 colon adenomas with different grades of dysplasia. Eight out of 25 adenomas have 2.5 times higher level of p21 protein visualized by the PAP-diaminobenzidine method are summarized in Table 2.

The higher level of ras p21 expression was accompanied by a higher fraction of S-phase cells. The correlation between the level of ras p21 protein and the proportion of S-phase cells was found to be significant $(r = 0.8414)$.

This group of adenomas could be characterized by a villous structure and with moderate to severe dysplasias. The ras gene expression and the rate of cell proliferation have also been studied in 42 colon carcinomas. Colon tumors can be divided into 3 groups on the basis of their ras p21 content and their proliferative capacity (Table 3).

22 out of 42 carcinomas could be characterized by three times higher level of ras p21 than that of the normal epithelium. In group III a high rate of cell proliferation is accompanied with moderate or high levels of p21 protein. A strong correlation could be found, however between increased ras p21 levels and the fraction of S-phase cells (Figure 2).

Concerning the relationship between clinical stage and ras expression, there was no significant difference in the ras p21 level af colon carcinomas belonging to clinical stage II or stage III. The clinical staging was based on the TNM classification (Figure 3).

Moreover, the rate of cell proliferation of colon carcinomas does not depend on the clinical stage, since the fraction of S-phase cells could either be low or high in the same clinical stages (Table 4).

Table 2. Correlation between the ras p21 level and fraction of S-phase cells in colon adenomas.

NUMBER OF CASES	LEVEL OF p21* (ADENOMA/CONT)	FRACTION OF S-PHASE CELLS %	CORRELATION COEFFICIENT
I GROUP 17/25	0.6 - 2.5	14.7 - 32.8 24.2 ± 5.3	0.4107
II GROUP 8/25	2.5 - 4.9	24.5 - 34.5 29.5 ± 4.02	0.8418

*Relative level of p21 = extinction values of adenoma/extinction values of intact epithelium at 480 nm

Table 3. Correlation between the p21 level and fraction of S-phase cells

NUMBER OF CASES	p21 LEVELS* (TU/CONTR)	FRACTION OF S-PHASE CELLS %
I GROUP 13/42	2.6 ± 0.27	17.0 - 23.0 19.6 ± 3.88
II GROUP 8/42	3.5 ± 0.29	23.0 - 28.0 24.2 ± 2.37
III GROUP 14/42 7/42	5.8 ± 0.64 2.1 ± 0.19	28.0 - 49.0 32.0 ± 6.92

* Relative level of p21 = extinction value of carcinoma/extinction value of intact epithelium at 480 nm

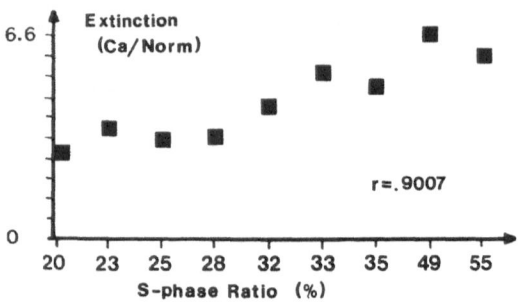

Figure 2. Relationship between _ras_ oncogene expression and the fraction of S-phase cells

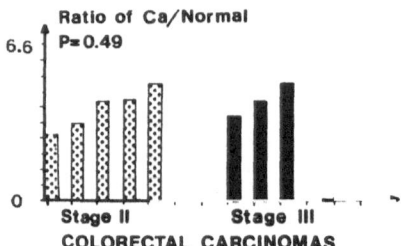

Figure 3. Relationship between stages and _ras_ p21 level of colorectal carcinomas

We have also evaluated the level of tyrosine kinase in colon carcinoma. Seven out of 27 colon carcinomas had a very high tyrosine kinase activity compared with normal colonic mucosa. The high level of tyrosine kinase activity was accompanied by a high rate of cell proliferation (Table 5).

DISCUSSION

Previous immunohistochemical studies have shown that in breast carcinoma (22) stomach (23) and colon carcinomas (16, 24) higher levels of _ras_ p21 positive cells was determined, but the actual level of p21 expression was not quantitated. To our knowledge the present work is the first report dealing with the quantitative cytochemical determination of p21 _ras_ protein does not distinguish the mutant and protooncogene forms of p21 (16), therefore it is suitable for the estimation of both forms. Our results have shown that p21 protein is detectable in normal colonic epithelium but its level is sig- nificantly lower than in colon adenomas or carcinomas. Elevated expression of _ras_ protein could be observed not only in severe dysplasias but in colonic hyperplasias, as well. These findings agree with previous data which have also demonstrated an increased _ras_ protein expression in a typical ductal hyperplasias of human breast (22). It could be suggested that the elevated expression of p21 might be related to the well known high proliferative capacity of colonic hyperplasias. We verified our original proposal by studies on the relationship between the level of cell proliferation of colon adenomas and carcinomas. Our studies have shown that the amplification of _ras_ gene product strongly correlates with the high rate of cell proliferation both in colon adenomas and carcinomas. Our finding is compatible with pre- vious data demonstrating that the presence of _ras_ protein was required for

140

Table 4. Cell cycle parameters of Colorectal tumors in various clinical stages

NO OF CASES	CELL CYCLE PARAMETERS %			CLINICAL STAGES
	G1	S	G2	
1	64.4	22.7	12.9	STAGE II
2	70.4	24.5	5.1	STAGE II
3	58.5	28.4	13.1	STAGE II
4	63.2	32.5	4.3	STAGE II
5	37.3	52.4	19.3	STAGE II
6	63.3	20.1	16.6	STAGE III
7	64.5	26.0	9.5	STAGE III
8	56.6	31.8	11.6	STAGE III
9	50.9	34.6	14.5	STAGE III
10	43.9	49.0	7.1	STAGE III

Table 5. Correlation between the level of Tyrosine Kinase Activity and Fraction of S Phase Cells

NUMBER OF CASES	TYROSINE KINASE ACTIVITY (TU/CONT)	FRACTION OF PHASE CELLS	CORRELATION COEFFICIENT
I GROUP 7/19	2.5 – 19.0	26.3 – 38.7 29.7 ± 3.88	0.8418
II GROUP 12/19	0.0 – 2.5	13.8 – 29.3 24.13± 5.02	0.4107
III GROUP 8	HIGH AUTOPHOSPHORYLATION	19.5 – 38.9 29.42± 6.92	–

the initiation of S-phase in NIH 3T3 cells (25). The fact that high levels of p21 protein could already be detected in hyperplasia of colon adenomas suggests that the ras protein might be involved in the immortalization of these cells. Previous data appear to contradict the immortalizing capacity of ras gene because myc genes were supposed to fulfil this function (26).

Subsequent studies, however showed that full transformation, including immortalization, has been obtained by activated ras gene both in primary rodent cell and human cell cultures (27,29). It has been emphasized,

however, that continuous presence of a threshold level of ras protein is
indispensable not only for the initiation but also for the maintainance of
the transformed phenotype (29).

Since the permanent overexpression of ras genes significantly contributes
to the malignant transformation, adenomas with amplified ras genes might have
a higher risk for cancer development than those having low level of ras
oncogene expression. Our results, which demonstrate that the amplification
of ras oncogenes is an early event in colon carcinogenesis are consistent
with the data of Gallic et al. (30) but disagree with some previous reports
(16,24). 32% of adenomas examined by us could be characterized by an elevated p21 level. This ratio is lower than the average frequency of adenoma –
carcinomas sequence because previous studies reported that at least 60% of
all colorectal carcinomas have their origin in pre-existing adenomas (31). A
follow up study is required to define the usefulness of high level in the
identification of those adenoma bearing individuals who are at high risk of
developing malignancy. We also found that the ras gene amplification is not
restricted to colon adenomas but occurs in the colon carcinomas, as well.
Previous studies have also shown that ras gene are amplified in different
human tumors (32,33).

These previous studies were not designed to test the role of the amplified ras protein in the signal transduction. We found that the increased
level of ras p21 strongly correlates with the high rates of cell proliferation in colon carcinomas, indicating that the mitotic signal transduction is
presumably mediated via p21 protein. Consistent with our results, a higher
growth rate of NIH 3T3 cells with amplified Ha-ras gene has also been reported (34). Moreover, a four fold increase in Ha-ras mRNA has been detected in colon carcinomas (35).

We could not find any correlation between different clinical stages, and
the level of ras gene expression, confirming previous observations (36).
These results might indicate that neither the high rate of cell proliferation
nor increased ras gene expression are accompanied with tumor progression.

The immunohistochemical detection of p21 protein may provide an additional diagnostic and prognostic tool in tumor pathology. The diagnostic value
of p21 detection is limited because an increased ras expression both in benign and malignant lesions of the human colon could be observed. The prognostic value of the high fraction of S-phase cells has already been proved
(37,38). Studies on ras expression in colon adenomas and carcinomas, however, might have a good prognostic impact because of the close correlation
between elevated level of ras p21 and high rate of cell proliferation.

Apart from p21 oncoproteins, tyrosine kinases also play a very important
role in mitotic signal transduction. They can be either growth factor receptors localized in plasma membrane of products of oncogenes localized in
membrane or cytoplasm interfering with various stages of the signal transduction pathway (39). The aim of this study was to investigate the role of
tyrosine kinases in the signal transduction pathways of human colon tumors.
In order to investigate the possible correlation of tyrosine kinase activity
with cellular growth we measured tyrosine kinase activity and the fraction of
the S-phase cells in human solid tumors. From the 27 cases tested we found a
high level of tyrosine kinase activity in 7 cases (26%). In these cases we
found a close correlation between the level of tyrosine kinase activity and
the high ratio of the S-phase cells (r=0.8418). The high level of tyrosine
kinase activities in these samples might be attributed to the occurrence of a
high number of growth factor receptors or oncogenes with protein products belonging to the tyrosine kinase family.

In 12 cases the solid colon tumor samples had detectable but low tyrosine kinase activities and the correlation with the fraction of the S-phase cells was also low (04107).

High autophosphorylation accompanied with high proliferation rate was detected in 8 cases (33%), which might indicate that the level of some intracellular substrates increased in these samples.

Our results suggest that a significant number of human colon tumors have an active tyrosine kinase signal transduction pathway that might be important to consider both from a diagnostic and from a therapeutic point of view.

REFERENCES

1   Baracid,M. (1987) Am.Rev.Biochem. 56: 779-827
2   Topal,M.D., (1988) Carcinogenesis, 9: 691-696
3   Sukumar,S., Notario,V., Martin-Zanca,D., Barbacid,M. (1983) Nature, 306: 658-661
4   Zarbl,H., Sukumar,S., Arthur, A.V., Martin-Zanca, D.,Barbacid,M. (1985) Nature, 315: 382-385
5   Quintanilla,M., Brown,K., Ramsden,M. and Balmain,A. (1986) Nature, 322: 78-80
6   Fujita,J., Ohuchi,N., Ito,N., Reynolds,S.H., Yoshida,O., Nakayama,H., and Kitamura,Y. (1988) J.Natl.Canc.Inst. 80: 37-43
7   Caignard,A., Kitagawa,Y., Sato,S., and Nagao,M. (1988) Gann, 79: 244-249
8   McMahon,G., Hanson,L., Lee,J., and Wogan,G.N. (1986) Proc.Natl.Acad.Sci. 83: 9418-9422
9   Graham.F.L., and A.J.Van der Eb (1973) Virology 52: 456-462
10 Futija,J., Srivastava,S.K., Kraus,M.H. et al. (1985) Proc.Natl.Acad.Sci. 82: 3849-3853
11 Barbacid,M. (1986) In: Important Advances in Oncology, 1986. Ed.V. DeVita, S.Hellman, S.Rosenberg, pp.3-22. Philadelphia, Lippincott
12 Furth,M.E., Aldrich,T.H., and Cordon-Cardo.C. (1987) Oncogene 1: 47-58
13 Hurley,J.B., Simon,M/I. Teplow,D.B., Rabishaw,J.P., Gilman,A.G. (1984) Science, 226: 860-862
14 Balk,S.D., Riley,T.M., Gunther,H.S., Monisi,A. (1985) Proc.Natl.Acad. Sci. 82: 5781-5785
15 Benjamin,C.W., Tarpley,W.G., and Gorman,R.R. (1987). Proc.Natl.Acad.Sci. Sci. 84: 546-550
16 Horand Hand,P., Thor.A., Wunderlich,D., Muraro,R., Caruso,A., Schlom,J. (1984) Proc.Natl.Acad.Sci. 81: 3227-3231
17 Sternberg,I.A., Hardy,P.H., Cuculis,J., Meyer,H.G. (1970) J.Histochem. Cytochem.18: 315-319.
18 Graham,R.C., Karnovsky,M.J. (1966) J.Histochem.Cytochem. 14: 291-302.
19 Pick,R., Tusnady,G. (1980) Studio Sci. Mathematicarum Hung.15: 31-42
20 Hardine,G. (1982) J.Histochem.Cytocem. 30: 1983-1989
21 Swarup,G., Dasgupta,J.D., and Garbers,D.L. (1984) J.Bio.Chem. 258: 10341 - 10347
22 Ohuchi,N., Thor,A., Page,D.L., Horand Hand,P., Halter, A.S., Schlom,J. (1986) Cancer Res. 46: 2511-2519
23 Noguchi,M., Hirohashi,S., Shinosato,Y., Thor,A., Schlom,J., Tsunokowa, Y., Terada,M., Sugimura,T. (1986) J.Nat.Cancer Inst.77: 379-385
24 Thor,A., Horand Hand,P., Wunderlich,P., Caniso,A., Muraro,R., Shlom,J. (1984) Nature 311: 562-565
25 Mulcahy,L.S., Smith,M.R., Stacey,D.W. (1985) Nature 313: 241-243
26 Land., Parada,L.F., Weinberg,R.A. (1983). nature 304: 596-602
27 Spandidos,D.A., Wilkie,N.M. (1984) Nature 310: 469-475
28 Yoakkum,G.H., Lechner,J.F., Gabrielson,E.W., Kozba,B.E., Malan-Shibley, L., Willey,K., Valerio,M.G., Shammsuddin,A.M., Trump,B.F., Harris,C.C. (1985) Science 227: 1174-1179

29 Winter,E., Yamamoto,F., Almoguera,C., Perucho,M.(1985) Proc.ynat.Acad.
    Sci.82: 7575-7579
30 Gallick,G.E., Kurzrock,R., Kloetzer,W.S., Arlinghaus,R.B., Guttermann,
    I.U.(1985) Proc.Nat.Acad.Sci.82: 1795-1799
31 Eide,T.J. (1982). Cancer: 1866-1983
32 Pulciani,S.E., Santos,L., long,V., Sorrention,L., Barbacid,M. (1985)
    Mol.Cell.Biol.5: 2836-2841
33 Schwab,M., Alitalo,K., Varrmus,H.E., Bishop,J.M., George,D. (1983)
    Nature 303: 497-501
34 Sistonen,L., Keski-Oja,I., Ulmanen,I., Holtta,E., Wikgren,B., Alitalo,
    k., (1987) Exp.Cell.Res.168: 518-539
35 Monnat,M., Tardy,S., Sarage,P., Diggelmann,H., Costa,J. (1987) Int.J.
    Cancer 40: 293-299
36 Kerr,I.B., Spandidos,P.A., Finlay,I.G., Lee,F.P., McArdle,C.S. (1986)
    Brit.J.Cancer 53: 231-235
37 Barlogie,B., Johnston,A.D., Small Wood,L., Raber,H.M., Maddox,A.M.,
    Latreille,J., Swartzendruber,P.E., Drewinko,B.(1982) Cancer.Gen.Cytogen.
    6: 17-28
38 Szentirmay,Z., Csuka,O., Sugar,J. (1985) In: M.I,Filipe and J.R.Jass
    (eds), Current Problems in Tumor Pathology, Gastreic carcinoma pp.68-86,
    Churchill Livingstone, Publ. Ltd. London
39 Slamon.D.J., Dekenion,J.B., Verma,J.M., and Cline,M.J. (1984) Science,
    224: 256

UTILIZATION OF INITIATOR TRANSFER RIBONUCLEIC ACID MODIFIED BY CARCINOGENS
CHOLESTERYL 14-METHYLHEXADECANOATE AND N-METHYL-N-NITROSOUREA FOR CELL-FREE
PROTEIN SYNTHESIS

Jan Hradec and Petr Pohlreich

Department of Molecular Biology, Research Institute for
Tuberculosis and Respiratory Diseases, 180 71 Prague
Czechoslovakia

ABSTRACT

(^{3}H)Cholesteryl 14-methylehexadecanoate becomes bound to initiator tRNA
from rat liver if incubated with it at 37°C. The binding of the ester was
competitively inhibited by N-methyl-N-nitrosourea which has been previously
demonstrated to alkylate guanine. It thus appears that the lipid reacts
with guanosine residues in the molecule of tRNA$_F$. Initiator tRNA preparations pretreated with cholesteryl 14-methylhexadecanoate were charged
significantly better with L-methionine in the presence of aminoacyl-tRNA
synthetases than comparable control preparations. The effect of the ester
was in this respect comparable with that of N-methyl-N-nitrosourea. Incubation of (^{35}S) Met-tRNA$_F$ modified by the lipid or N-methyl-N-nitrosourea with
partially purified eIF-2 resulted in an enhanced formation of the ternary
complex of this protein-synthesis factor. Also the sythesis of the 80S initiation complex was stimulated in the presence of these modified Met-tRNAs.
Protein synthesis in a system composed of rabbit reticulocyte polyribosomes
and partially purified initiation and elongation factors was stimulated if
unfractionated aminoacyl-tRNA from rat liver pretreated with cholesteryl 14-
methylhexadecanoate of N-methyl-N-nitrosourea was used. No changes were
found in poly(U)-dependent peptide elongation when using aminoacyl-tRNA
modified by either of these agents. Evidence was presented previously that
cholesteryl 14-methylhexadecanoate is involved in the control of protein synthesis by affecting the binding sites of protein-synthesis factors and ribosomes for aminoacyl-tRNA, and protein phosphorylation. The present results
suggest that the modification of initiator tRNA by this lipid may represent
another pathway for its modulation of gene translation.

INTRODUCTION

Cholesteryl 14-methylhexadecanoate (CMH) is a naturally occurring lipid
that was isolated and the chemical constitution of which was elucidated in
our laboratory in 1967. Its concentration is significantly increased in
tissues of tumor bearing animals. The synthesis of this lipid is enhanced
in the liver tissue during the growth of experimental tumors in rats in
parallel with the stage of malignant growth; CMH also shows carcinogenic
activity (see (1) for a review).

CMH is involved in protein synthesis. It is a natural constituent of several enzymes and protein factors required for gene translation. This applies for aminoacyl-tRNA synthetases, eIF-2 and both peptide-elongation factors. CMH is also present in both ribosomal subunits and modulates the activity of individual ribosomal binding sites. The removal of this lipid by extraction with organic solvents or treatment of proteins with immobilized cholesterol esterase is followed by a decrease of enzymatic activity which can be fully restored by the addition of pure CMH. All results obtained with these preparations suggest that the ester is involved in the activity of binding sites for aminoacyl-tRNA.

Furthermore, cholesteryl 14-methylhexadecanoate modulates post-translational modifications of protein molecules. It affects significantly the endogenous GTPase activity of eEF-1 and the autocatalytic phosphorylation of this purified protein-synthesis factor (2). This may represent an alternate pathway by which CMH is involved in the control of protein synthesis.

Recent results by our laboratory demonstrated that carcinogens modify in a specific way eukaryotic initiator tRNA (3). Since CMH is a carcinogen, its possible interaction with this species of tRNA was tested in the present experiments. Evidence is presented that CMH becomes bound to initiator tRNA from rat liver and modulates in this way its function in some steps involved in the translation of the genetic message.

MATERIALS AND METHODS

(^{3}H) Cholesteryl 14-methylhexadecanoate labelled in the sterol moiety was synthesized from (^{3}H) cholesterol (Amersham) (50 Ci/mmol) and 14-methylhexadecanoic acid isolated from wool fat (4) as described by Helmich and Hradec (5). The same method was used for the preparation of CMH labelled in its fatty acid moiety from 14-methyl(2,3-^{3}H)hexadecanoic acid (6) (51 Ci/mmol). L-(^{35}S)Methionine was purchased from Amersham and N-(^{3}H)methyl-N-nitrosourea from NEN. tRNA from rat liver was isolated as described by Rogg et al. (7) and initiator tRNA was separated by BD cellulose chromatography (8).

Isolations

Aminoacyl-tRNA synthetases from E.coli B and from rat liver were isolated as described by Stanley (8). Crude peptide-initiation factors were the ribosomal wash fraction (9) and were used for the assay of the 80S initiation complex formation. This fraction was further precipitated by $(NH_4)_2SO_4$ and the fraction precipitated at 40-70% of the saturation was used as crude eIF-2 for assays of the ternary complex formation. The usual technique of our laboratory (10) was used for the isolation of the polyribosomes from rabbit reticulocytes and for the separation of 40S and 60S ribosomal subunits by centrifugation in the zonal rotor. pH 5 enzymes from rat liver were isolated as described by Falvey and Staehelin (11).

Incubations

Incubation procedures as well as methods used for the preparation of samples for assays of radioactivity with a Beckman LS 5801 liquid scintillation spectrometer are described in detail in Legends for Figures for individual experiments.

RESULTS

(^{3}H)Cholesteryl 14-methylhexadecanoate became bound to tRNA$_F$ from rat liver if incubated with it at slightly alkaline pH and at 37^{0}C. The quantity

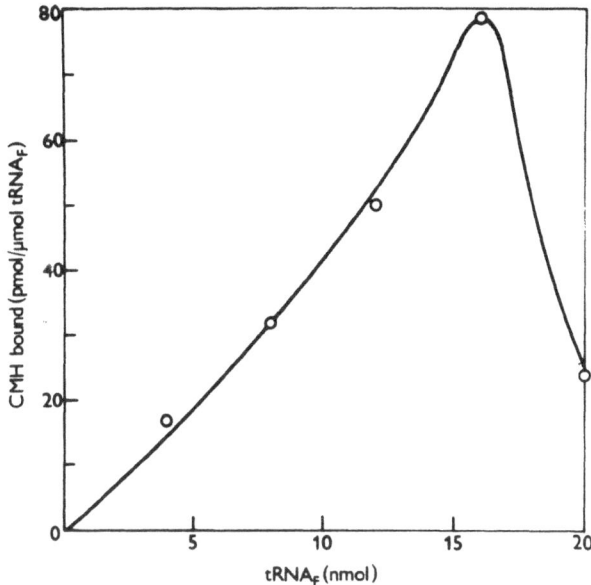

Figure 1. Binding of (3H)cholesteryl 14-methylhexadecanoate to initiator tRNA

of bound CMH was dependent on the amount of tRNA added and at optimum Incubation mixtures contained in a final volume of 1.0 ml: 50 mM-tris/HC1 buffer, pH 7.5, 2.5 mM-MgC1$_2$, 12.9 pmol of (^{3}H)cholesteryl 14-methylhexade-canoate, and quantities of tRNA$_F$ as indicated. They were incubated for 60 min. at 37° C.

After incubation the pH was brought to 5.0 by 2 M-potassium acetate buffer and tRNA was precipitated by 2 volumes of chilled ethanol. Precipitates were centrifuged, washed with ethanol-ether (1:1) (twice) and dried in a stream of N$_2$. Dried precipitates were dissolved in 1M-NaOH. Aliquots of this solution were plated on to discs of Whatman GF/A glass fibre paper. After drying, the radioactivity on filters was counted by liquid scintillation using a toluene-based scintillation mixture.

conditions approximately 80 pmol of CMH was combined with 1 μmol of initiator tRNA. This bond was rather unstable and certainly not covalent since it was almost fully destroyed by the treatment with phenol or trichloroacetic acid. The exact nature of the bond has not yet been elucidated.

    Binding of (^{3}H)cholesteryl 14-methylhexadecanoate to tRNA is significant-ly inhibited by N-methyl-N-nitrosourea (MNU). Previous evidence obtained in our laboratory indicated that MNU reacts with guanosine residues in the mole-cule of tRNA with the resulting formation of 7-methylguanosine (J. Hradec and J. Macelova, Biochem. Pharmacol., submitted). Since the inhibition by MNU is of a competitive nature (results not shown) it seems probable that CMH inter-acts with the same nucleoside residues as MNU.

    If initiator tRNA has been pretreated with CMH or MNU and charged with L-(^{3}L)methionine in the presence of aminoacyl-tRNA synthetases from E.coli B, a significantly higher quantity of the amino acid became combined with tRNA$_F$ than in control mixtures. This effect was strictly dependent on the dose of both compounds used for the pretreatment of tRNA, the optimum dose being in the range of 10^{-7} μmol/incubation mixture.

If initiator tRNAs modified by CMH or MNU were incubated with crude eIF-2 and GTP, an enhanced formation of the ternary complex of this protein-synthesis factor with Met-tRNA$_F$ and GTP was found. Since the same radioactivities of all L-(^{35}S)Met-tRNA$_F$s were added, the possibility can be excluded that this stimulation effect may be due to an enhanced charging of preparation pretreated with both compounds.

Formation of the 80S initiation complex by crude initiation factors in the presence of both ribosomal subunits was stimulated by Met-tRNA$_F$ preparations pretreated with certain doses of CMH or MNU. The same radioactivities rather than quantities of different preparations were again used to compensate for a higher charging of tRNAs modified by both carcinogens.

If unfractionated tRNA from rat liver was pretreated with CMH or MNU, charged with L-(^{35}S)methionine and all essential amino acids (cold), and added to a translation system from rabbit reticulocytes, a stimulation of endogenous globin mRNA synthesis was found in mixtures containing tRNA modified by preincubation with certain doses of CMH or MNU.

DISCUSSION

Recent results of our laboratory provided evidence that MNU alkylates guanosine residues in initiator tRNA (J. Hradec and J. Macelova, in preparation). The same target molecule was also found when using procarcinogenic derivatives of 3,3-dimethyl-1-phenyltriazene (14). In the present experiments the binding of radioactive CMH to initiator tRNA was competitively inhibited by MNU. This indicates that CMH reacts with the same nucleoside residues in tRNA as does MNU. Guanosine seems to be the most frequent target for interaction with various carcinogens (15). However, it is not the only one. Thus a directly acting roden carcinogen -propiolactone was demonstrated to react in vitro with all bases in DNA (16,17).

Weinstein and Grunberger suggested that modification of tRNA molecules by chemical carcinogens may induce conformational changes (18). It seems possible that these changes make tRNA molecules more accessible for the interaction with enzymes and protein factors required for protein synthesis.

Charging with L-methionine of initiator tRNA modified by the binding of CMH proceeded more efficiently than that of corresponding controls. Evidence was provided by our previous results that such an enhanced acceptance is characteristic for tRNA$_F$ modified by chemical carcinogens. This is an agreement with the carcinogenic activity of CMH in animals (20).

Met-tRNA$_F$ modified by the binding of CMH was significantly better utilized by eIF-2 for the formation of the ternary complex of this factor. Also the synthesis of the 80S initiation complex was stimulated in the presence of tRNA$_F$ modified by both carcinogens used. This indicates that eIF-2 may be the target protein-synthesis factor the activity of which can be modulated by its substrate modified by carcinogenic CMH or MNU. This interaction may be significant for the whole protein-synthesis since reactions catalyzed by eIF-2 are apparently the most important regulatory steps modulating the rate of protein synthesis (21).

Earlier experiments in the laboratory showed that gene translation in a complete subcellular system is stimulated by the addition of carcinogenic polycyclic aromatic hydrocarbons but peptide elongation was apparently not affected (22). Although naturally much less sophisticated methods than those available today were used in this study, the results are in good agreement with our present experiments where only the peptide initiation step was affected. This resulted in a stimulation of proteins dynthesis in a complete polyribosomal translation system. On the other hand, the poly(U)-dependent peptide elongation was not affected. These results seem to suggest that the enhancing effect of tRNA modified by carcinogens involves in a specific way the initiation step of translation and that eIF-2 is the most probable, if not the only protein-synthesis factor affected.

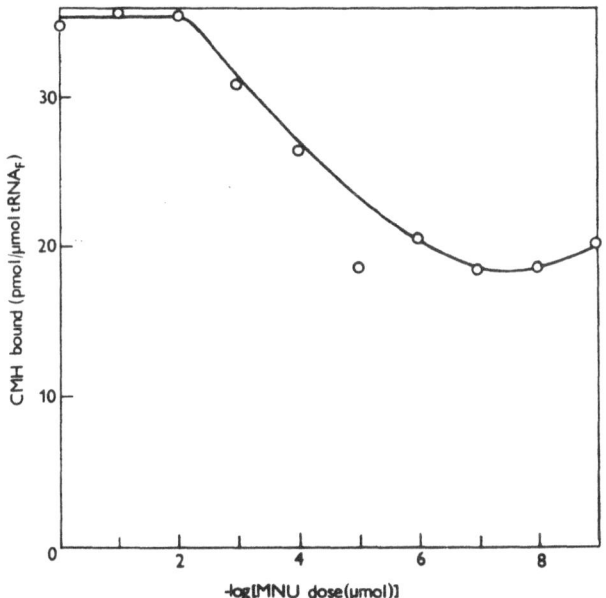

Figure 2.   Inhibition of the binding of CMH to TRNA$_F$ by N-methyl-N-nitro-sourea.
Incubation mixtures composed as described in Legends for Figure 1 contained 10 nmol of tRNA$_F$ and quantities of MNU as indicated.
After the incubation, mixtures were processed as given in Legends for Figure 1.

CMH was previously shown to modulate protein synthesis by interactions with binding sites of some protein-synthesis factors and ribosomes for amino-acyl-tRNA and by affecting protein phosphorylation (23). The modification of tRNA$_F$ described here may represent an additional pathway by which the regulatory role of this lipid in translation is expressed.

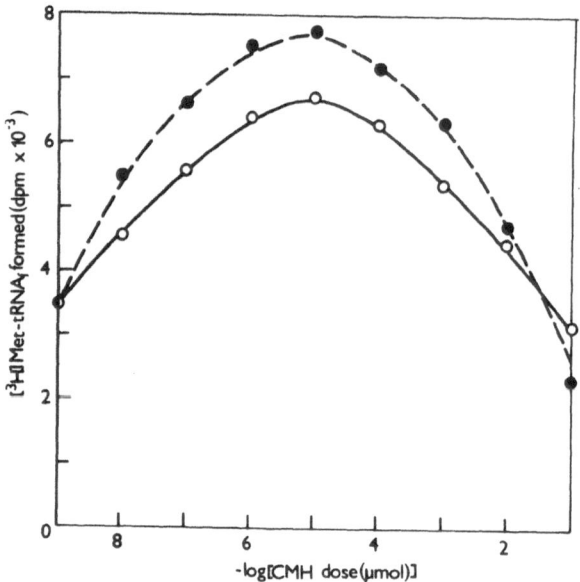

Figure 3. Effect of the pretreatment of tRNA$_F$ with CMH (----) of MNU (- - -) on its charging with L-(^{35}S)methionine

Initiator tRNA was preincubated with CMH or MNU as described in Legends for Figure 1.

Incubation mixtures for the charging with L-methionine contained in a final volume of 0.1 ml: 150mM Tris/HCl buffer, pH 7.4, 50 mM-KCl, 10 mM-MgCl$_2$, 5mM ATP, 1 mM-CTP, 8 mM-phosphoenol pyruvate, 20 ug pyruvate kinase, 7 mM-2-mercaptoethanol, 5 nmol-tRNA, 25 ug of protein of partially purified amino-acyl tRNA synthetases from E.coli B and 4.5 X 10^4 dpm of L-(^{35}S)methionine. Mixtures were incubated for 20 min. at 37°C.

After the incubation 75 ul portions of incubation mixtures were plated on to discs of Whatman GF/A glass fibre paper and filters were processed as described by Hradec et al. (10).

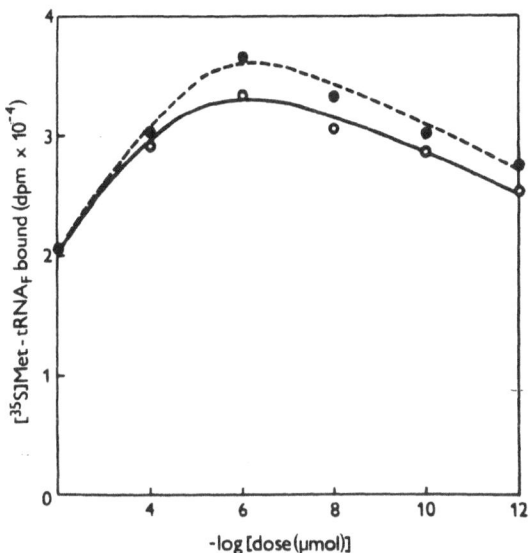

Figure 4. Effect of the pretreatment of initiator tRNA with CMH (————) or
MNU (- - -) on the formation of the formation of the ternary complex of eIF-2
with Met-tRNA and GTP
Incubation mixtures composed as described by Thomas et al. (12) contained in
a final volume of 25 ul: 20 mM-Hepes/KOH buffer, pH 7.6, 1mM-ATP, 0.4 mM-GTP,
mM-creatine phosphate, 1 mM-DTT, 120 mM-potassium acetate, 2.4 mM-magnesium
acetate, 0.05 units of creatine kinase, 5 ug of protein of crude eIF-2, and 5
X 10^4 dpm of L-(^{35}S)Met-tRNA.
Mixtures were incubated for 10 min at 37°C, diluted with buffer (12) and
filtered through cellulose nitrate filters. The retained radioactivity was
counted by liquid scintillation.

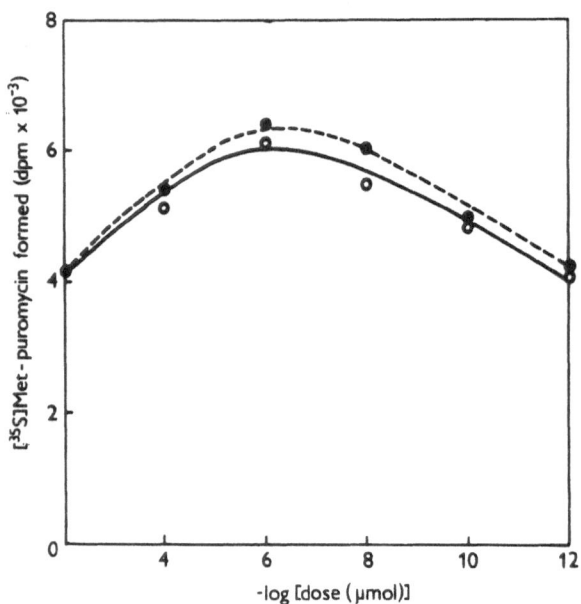

Figure 5. Formation of the 80S initiation complex with tRNA preparations
pretreated with various doses of CMH (----) or MNU (- - -)
Initiator tRNA was preincubated with various doses of CMH or MNU as described
in Legends for Figure 1 and charged with L-(^{35}S)-methionine as given in
Legends for Figure 3.
Incubation mixtures composed as described by Thomas et al. (12) contained in
a final volume of 25 ul: 20 mM-Hepes/KOH buffer, pH 7.6, 120 mM-potassium
acetate, 2 mM magnesium acetate, 1 mM-ATP, 0.4 mM-GTP, 5 mM-creatine
phosphate, 1 mM-DTT, 0.05 units of creatine kinase, 0.1 A_{260} units of 40S
ribosomal subunits, 0.25 A_{260} units of 60S subunits, 0.05 A_{260} units of AUG,
1mM-puromycin, 15 ug of protein of crude peptide initiation factors, and 3 x
10^4 dpm of different L-(^{35}S)Met-tRNA preparations. Mixtures were incubated
for 60 min. at 37°C.
After the incubation, methionyl-puromycin was extracted with ethylacetate as
described by Leder and Bursztyn (13) and assayed for radioactivity by liquid
scintillation.

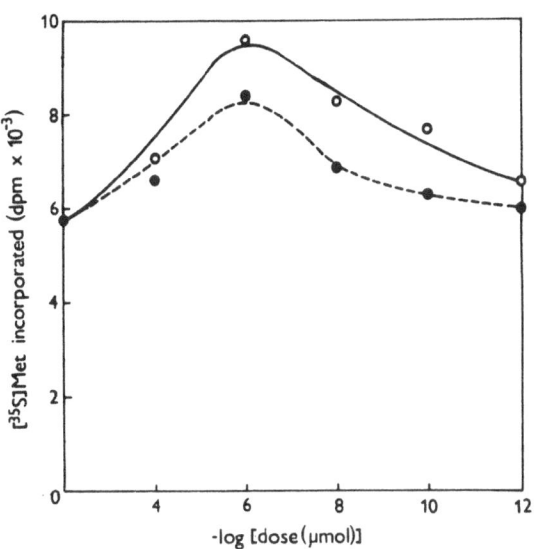

Figure 6. Translation of the endogenous message in a system containing aminoacyl-tRNA modified by CMH (------) or MNU (- - -).
Unfractionated tRNA from rat liver was preincubated with CMH or MNU as described in Legends for Fig.1 and charged with L-(^{35}S)Met and other 19 cold amino acids.
The translation system contained in a final voloume of 25 ul: 20 mM-Hepes/KOH buffer, pH 7.6, 2mM-magnesium acetate, 120 mM-potassium acetate, 1 MM-DTT, creatine kinase, 25 ug of protein of pH 5 enzymes, 30 ug of protein of crude peptide-initiation factors, 0.4 A_{260} units of polyribosomes from rabbit reticulocytes, and 3 x 10^4 dpm of L-(^{35}S)Met-tRNA. Mixtures were incubated for 90 min. at 37° C.
After the incubation, 20 ul portions of incubation mixtures were plated on to discs of Whatman GF/A glass fibre paper and filters were processed as described by Hradec et al. (10). Radioactivity was assayed by liquid scintillation.

REFERENCES

1   Hradec,J. (1975) in Lipids and Tumors. (K.K.Carroll,ed.). S.Karger, Basel, pp. 197-226
2   Tuhackova,Z. and Hradec,J. (1981) Eur.J.Biochem. 146, 365-370
3   Hradec,J. (1988) Carcinogenesis 9,
4   Helmich,O. and Hradec,J. (1981) J.Labelled Comp. Radiopharm. 18, 747-754
5   Helmich,O. and Hradec,J. (1980) J.Chromatogr. 193, 153-156
6   Helmich,O., Streibl,M., Filip,J. and Hradec,J. (1985) J.Labelled Comp. Radiopharm. 22, 917- 923
7   Rogg,H., Wehrli,W. and Staehelin,M. (1969) Biochim.Biophys.Acta, 195, 13-15
8   Stanley,W.M. (1972) Anal.Biochem. 48, 202-216 (1972)
9   Schreier,M.H., Erni,B. and Staehelin,T. (1977) J.Mol.Biol. 116, 727-753

10   Hradec,J., Dusek,Z., Bermek,E. and Matthaei,H. (1971) Biochem.J. 123, 959-966

11   Falvey,A.K. and Staehelin,T. (1970) J.Mol.Biol.53, 1-19 (1970)

12   Thomas,A., Goumans,H., Amesz,H., Benne,R. and Voorma,H. (1979) Eur.J. Biochem. 98, 329-337

13   Leder,P. and Bursztyn,H. (1966) Biochem.Biophys.Res.Commun. 25, 233-238

14   Hradec,J. and Kolar,G.F. (1985) Carcinogenesis, 6, 995-998

15   Hradec,J. (1988) This Volume P105.

16   Segal,A., Solomon,J.J., Dino,J. and Mignano,J. (1981) Cem.Biol. Interact. 35, 349-361

17   Chen,R.F., Miyeyal,J.J. and Goldwait,D.A. (1981) Carcinogenesis, 2, 80 73-80

18   Weinstein,I.B. and Grunberger,D. (1972) in World Symposium on Model Studies in Chemical Carcinogenesis Ts'o,P.O. and Di Paolo,J.A. eds) M.Dekker, New York, pp. 217-235

19   Hradec,J., Spiegelhalder,B. and Preussmann,R. (1988) Carcinogenesis, 9,

20   Shabad,L.M., Kolesnichenko,T.S. and Savluchinskaya,L.A. (1973) Neoplasma, 20, 347-348

21   Ochoa,S. and de Haro,C. (1979) Annu.Rev.Biochem. 48, 549-580

22   Hradec,J. (1967) Biochem.J. 105, 251-259

23   Tuhackova.Z. and Hradec,J. (1985) in The Pharmacological Effect of Lipids II (J.J.Kabara, ed.) Am.Oil Chemists Soc., Champaign, IL, pp. 157-171

EFFECT OF THIRAM ON CYTOCHROME P-450 STUDIED BY THE CYPIA-TEST

Iwonna Rahden-Staron, Maria Szumilo, Teresa Szymczyk

Dept of Biochemistry, Inst of Biopharmacy, Medical School
02-097 Warsaw, ul. Banacha 1, Poland

ABSTRACT

The double CPIA-test (cytochrome P-450 induction assay) as described by
Lesca et al. (7) was used to discriminate between the forms of cytochrome
P-450 [the phenobarbital (PB)-3-methylcholanthrene (3-MC)-type] induced by
thiram, a dithiocarbamate fungicide. The test consisted of studies by the
technique of Ames for the ability of the S-9 preparation obtained from livers
of rats induced by thiram to cause metabolic activation of ethidium bromide
(EtBr) (for 3-MC type) or cyclophosphamide (CPA) (for PB-type of cytochrome
P-450). The mutagenicity of activated EtBr or CPA was checked with TA 98 and
TA 100 strains of S. typhimurium respectively.

It was found that thiram at 100 mg/kg of body weight, while only slightly
increasing the 3-MC-type of cytochrome P-450 (400 his^+ revertants/ug of EtBr
as compared to 2200 his^+ revertants/ug of EtBr for 80 mg of 3-MC/kg of body
weight) caused a noticeable induction of the PB-type (800 his^+ revertants/
800 ug of CPA as compared to 1300 his^+ revertants/800 ug of CPA for 50 mg of
PB/kg of body weight). The direct measurements performed on livers of rats
induced with thiram showed no overall changes in the level of cytochrome P-
450 but a shift in absorption maximum to 449 nm.

INTRODUCTION

Thiram is of interest in genotoxicity studies firstly, because of its
occurence in the environment (it is extensively used as a crop fungicide and
as industrial antioxidant) and secondly, because of its close chemical
resemblence of disulfiram, a drug used in aversion-therapy of chronic
alcholics.

Although many chemicals administered to animals are inactive in their
native forms, they can be activated in a chain of metabolic reactions, the
final stage of which is always linked with one of two specifically induced
forms (PB- or MC- type) of cytochrome P-450. Thus, the ability of a compound
pound to induce PB- or MC-type of cytochrome P-450 serves as a sensitive
measure of its genotoxic effect.

In the present work we used the CYPIA dual assay to distinguish between
the forms of cytochrome P-450 induced in animals upon administration of

thiram. The assay is based on the differential transformation of ethidium bromide (EtBr) and cyclophosphamide (CPA) to mutagenic metabolites by liver preparations containing either MC- or PB-type of cytochrome P-450 (1,2,3).

MATERIALS AND METHODS

Commercial chemicals were of the purest grade available. Thiram (purity 98%) was from Organica-Azot; biotine, L-histidine, 2-aminofluorene, 2-nitro-fluro- rene from Koch-Light Lab. Ltd.; 3-methylcholanthrene, glucose-6-phosphate, NADPH, cyclophosphamide were from Sigma Chemical Co., benzo(a)- pyrene from EGA-Chemie, DMSO, 1-oxide-4-nitroquinoline (NQU) from Fluka AG; Aroclor 1254 from Analabs Inc.; ethidium bromide from Calbiochem. Bacto- trypton and Bactoyeast were from Difco. Other chemicals were from Ciech (Poland).

Treatment of animals

Four male Wistar rats weighing 100g each were pretreated with chemicals as follows. All chemicals, in sunflower oil except PB which was in saline, were administered interperitoneally. Animals receiving a single dose of thiram (80mg/kg body weight) and 3-MC (80 mg/kg body weight) were killed after 48h and those receiving Aro after 5 days. Animals were treated with PB (50 mg/kg body weight) repeatedly for 3 days and were sacrificed 24h after the last injection. Control animals received the vehicle alone.

Liver Preparations

S9 was prepared according to Ames et al.(4) and stored at -80°C. Cyto- chrome P-450 was determined according to Matsubora et al.(5) and protein according to Lowry et al.(6). S9 mix was prepared as described in (4).

Cytochrome P-450 Induction Assay (CYPIA)

A dual CYPIA assay was performed as described by Lesca et al.(7). The Salomella typhimurium strains TA 98 and TA 100, a generous gift from Dr. B. N. Ames (Berkley, Ca., USA) were used for the MC- and PB-type of induction respectively.

Determinations of the mutagenicity of the EtBr (0.25 - 5 ug/plate for induction of 3-MC-type of cytochrome P-450) and CPA (200 - 800 ug/plate for induction of PB-type of cytochrome P-450) were performed in triplicates with 100 and 200 ul of S9/plate respectively.

To check the ability of strains to revert to histidine prototrophs positive controls were performed using 2 ug of 2-nitrofluorene (for TA 98) and 1 ug of 4-nitroquinoline (for TA 100). The amounts of his^+ revertants observed in the presence of these standard mutagenes were in agreement with those reported in the literature(4). Each type of S9 mix fraction was ex- amined in the absence of substrate to rule out the possibility of his^- reversion being caused by the S9 fraction alone. The resulting 50 - 100 revertant colonies were subtracted from the number of revertants counted in mutagenicity tests. Spontaneous reversions (without S9 mix) were between 30 - 50 colonies for TA 98 and 120 - 200 colonies for TA 100.

RESULTS AND DISCUSSION

The results presented in Fig.1A show that thiram administered at 100 mg/ kg body weight caused the induction of 3-MC-type of cytochrome P-450 at a level corresponding to 400 rev. his^+/ug EtBr. Although this induction was relatively low compared to that achieved by administration of 3-MC at 80mg/kg

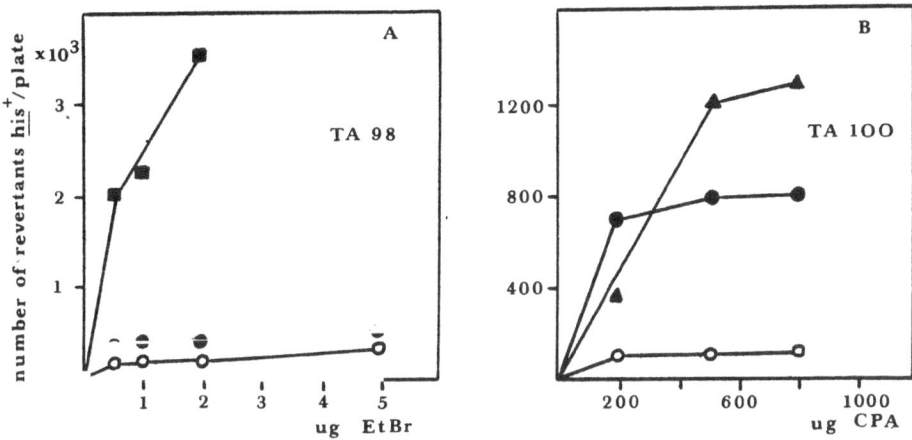

Figure 1.   Dose response plots of mutagenicity of A/. ethidium bromide
            to S. typhimurium TA 98 and B/. cyclophosphamide to S.
            typhimurium TA 100. S9 was isolated from livers of rats treated
            with 3-MC (■) thiram 80 mg/kg (o), thiram 100 mg/kg (●) and PB
            (▲).

body weight (2200 rev. $\underline{his}^+$/µg EtBr), according to Lesca et al. (7), a compound can be classified as an indicator of the 3-MC-type of cytochrome P-450
if the amount of $\underline{his}^+$ revertants/plate is more than 2-fold the amount for the
control (the S9 mix without EtBr), in our case 40-60 revertants/plate.

    The same dose of thiram resulted in the induction of the PB-type of cyto-
chrome P-450 corresponding to 800 rev. $\underline{his}^+$/800 µg CPA. This is comparable
to the induction achieved upon administration of PB at 50mg/kg body weight
(1300 rev. $\underline{his}^+$/800 µg CPA) (Fig. 1B).

    The total level of cytochrome P-450 in S9 fraction isolated from thiram-
treated rats remained unchanged as compared to control (Table 1) whereas for
animals treated with 3-MC and PB it increased similarly to that reported

Table 1.   Ethidium bromide and cyclophosphamide mutagenicity and the level
of cytochrome P-450 in S9 from livers of rats pretreated with thiram.

Chemicals	Dose mg/kg	Revertants $\underline{his}^+$/µg EtBr	$\underline{his}^+$/800 µg CPA	Cyt.P-450 nmole/mg protein	Protein mg/ml
oil control		20	10	0.232	15.6
3-MC	1 x 80	2200		0.546	15.0
PB	3 x 50		1300	0.685	14.1
thiram	1 x 80	160	150	0.163	12.8
thiram	1 x 100	400	800	0.219	11.8

(7,8,9). On the other hand we found that in thiram-treated rats the absorption maximum of cytochrome P-450 was shifted towards lower wavelength (449 nm).

In conclusion, we have shown that the treatment of rats with thiram at 100 mg/kg body weight results in variable induction of two types of cytochrome P-450: a slight one for the 3-MC type and a more distinct one for the PB-type. Higher doses of thiram (200-1000 mg/kg body weight) were hepatotoxic and resulted in the decrease of the level of cytochrome P-450 (10, 11).

ACKNOWLEDGEMENT

This work was supported by a grant from Centrum Onkologii (CPBR 11.5.4).

REFERENCES

1   Lecointe, P., Bichet, N., Fraire, C. and Paoletti, C. (1981) Biochem. Pharmacol.30, 601-609
2   Hales, B.F. and Jain, R. (1980) Biochem. Pharmacol., 29, 256-259
3   Hales, B.F. and Jain, R. (1980) Biochem. Pharmacol.,29, 2031-2037
4   Ames, B.N., McCann, J. and Yamasaki, E. (1975) Mutation Res 31, 347-364
5   Matsubora, T., Koike, M., Touchi, A. Tochino, Y. and Seguno, K. (1976) Anal. Biochem.,75, 596-603
6   Lowry, O.H., Rosenbrough, N.J., Farr, A.L. and Randall, R.J. (1951) J Biol Chem., 193, 265-270
7   Lesca, P. Fournier, A., Lecointe, P. and Cresteil, T. (1984) Mutation Res., 129, 299 - 310
8   Harada, N. and Omura, T. (1981) J. Biol. Chem. 89, 237-248
9   Le Prevost, E.T., Gretail, T., Columelli, S. and Leroux, J.P. (1983) Biochem Pharmacol
10 Dalvi, R.R., Robbins, T.J., Williams, M.K., Deoras, D.P., Donastorg, F. and Banks, C. (1984) J Environ. Sci Health, B19 (8 & 9), 703-712
11 Zemaitis, M.A., and Greene, F.E. (1979) Toxicol. and Applied Pharmacol., 48, 343-350

TRANSLOCATION OF PHOSPHATIDATE PHOSPHOHYDROLASE FROM THE CYTOSOL TO

MICROSOMAL MEMBRANES IS INCREASED IN EXPERIMENTAL LIVER TUMORS

C. Cascales, L. Bosca, A. Martin, D. Velasco, D. N. Brindley and M. Cascales

Instituto de Bioquimica, Facultad de Farmacia, Universidad Complutense, Plaza de Ramon y Cajal s/n 28040 Madrid, Spain

SUMMARY

The effect of oleate, spermine, chlorpromazine and phorbol ester (TPA) on the translocation of phosphatidate phosphohydrolase activity from the cytosol to the microsomal membranes was assayed in liver homogenates from long-term thioacetamide-treated rats (100 mg/Kg body weight, daily for 90 days). The translocation of phosphatidate phosphohydrolase induced by oleate was higher (two fold) in liver homogenates obtained from thioacetamide-treated rats than in those from control rats. These differences between thioacetamide-treated and control livers were noticeably higher (four-fold) in the presence of physiological concentrations of salt (0.15 M KCl). The response to oleate-induced translocation of this enzyme activity in thioacetamide-treated liver homogenates did not change significantly in the presence of 0.15 M KCl. However, in homogenates from control rats there was a lack of response when physiological concentrations of salt were present. The enhanced response to translocate phosphatidate phosphohydrolase activity in liver homogenate from thioacetamide-treated rats was due to an increased binding ability of microsomal membranes. These data indicate that tumor-promoting effect of thioacetamide may play some role in the modulation of phosphatidate phosphohydrolase activity and in glycerolipid synthesis in the liver.

INTRODUCTION

In the course of investigations concerning the translocation of phosphatidate phosphohydrolase (PAP) (EC 3.1.1.4) and its modulation by a series of substances, the effect of long-term thioacetamide (TAM) administration was investigated since some of these substances are also involved in the mechanism of cell proliferation. It is well known that ornithine decarboxylase increases markedly in the liver of TAM-treated rats (1) and spermine, one of the polyamines derived from the action of this enzymes, is one of the modulators of PAP (2).

In rat liver homogenates and isolated hepatocytes it has been demonstrated that oleic acid (3,4), polyamines and chlorpromazine (5), are very active substances that promote or modulate the transfer of PAP from the cytosol to the membranes. The effect of increasing salt concentration (NaCl) on PAP activity in rat lung has been recently reported (6), as well as the fact that chlorpromazine, a Ca^{2+} antagonist agent, which produces a considerable decrease in the hepatic content of Ca^{2+} (7) and in accumulation of Ca^{2+} produced in centrilobular necrosis associated with the administration of thioacetamide (8).

Previous studies in liver of long-term TAM-treated rats concentrated on the enzymes involved in fatty acid and glycerolipid synthesis have shown that lipogenic enzymes activities are lowered, together with an enhancement both in NADPH-generating enymes (9) and in the total PAP activity (1).

The aim of the present investigation was to study in liver of long-term TAM-treated rats, the translocation of PAP from the cytosol to the membranes in a cell-free system in response to oleate, spermine, chlorpromazine and phorbol esters. The properties of some of these substances on promoting or modulating translocation have been demonstrated (2,3,4,5). To determine the effect of physiological ionic strength on the transfer of PAP to microsomal membranes, the addition of 0.15 M KCl to the incubation media was investigated. The results obtained show that in liver homogenates from TAM-treated rats, translocation of PAP induced by promoters or modulators is more active than in non-treated controls. Microsomal membranes obtained from TAM-treated livers show an enhanced ability to translocate PAP activity from soluble fraction.

MATERIAL AND METHODS

Animals and Experimental Design

Male albino Wistar rats (200–250 g body weight) three months old were used in all experiments. They were maintained on standard laboratory diet and water 'ab libitum', at room temperature (+21°C) and constant day/night rhythm. The rats were received daily i.p. injections of freshly prepared solution of TAM in 0.15 M NaCl at a dose of 100mg/kg body weight, during a 90 day period. The control group received a similar volume of 0.15 M NaCl. At the end of the treatment with TAM, rat livers developed hypertrophia with numerous nodules (3–5mm) which is termed hyperplastic noduligeneisis. Rats were decapitated and the liver was immediately chilled in ice-cold buffer, and a biopsy specimen (10 g) was taken for homogenization. Livers were homogenized in 4 volumes of ice-cold 0.25 M sucrose containing 20 mM HEPES, pH 7.4, 0.2 mM dithioerythritol and 2mM EDTA.

Subcellular distribution of PAP

The supernatant obtained from rat liver homogenates after centrifuging for 18,000 xg (r=10.8cm) for 10 min at 4°C in KONTRON centrifuge was incubated at 37°C for 10 min either in absence or presence of 0.15 M KCl with various combinations of 0.4 mM chlorpromazine, 0.5 mM oleate, 150 nM 12-0-tretradecanoyl phorbol-13-acetate (TPA), 1 mM spermine. In the case of oleate plus chlorpromazine there was a previous incubation for 10 min at 37°C with 0.5 mM oleate to cause the translocation of phosphatidate phosphohydrolase onto the microsomal membranes. The concentrations of chlorpromazine, spermine and oleate have previously been reported to be optimal in producing the translocation of PAP from the cytosol onto microsomes with no inhibition of the enzyme (2,5,10). TPA concentration is the same as that used by most research groups (11,12). Protein concentrations were about 24 and 21 mg/ml in control and TAM-treated post-mitochondrial supernatant of rat liver.

The microsomal and cytosolic fractions were collected after centrifugation at 105,000 xg (r=5.99cm) in a KONTRON ultracentrifuge for 45 min at 4°C. Microsomal pellets were resuspended in 0.25 M sucrose containing 0.2 mM dithioerithritol, 20 mM HEPES pH 7.4 and 2 mM EDTA. Protein concentration was about 17 mg/ml and 14 mg/ml (cytosolic fraction) and about 3.4 mg/ml and

2.4 mg/ml (microsomal fraction) of post-mitochondrial supernantant of liver from control and TAM-treated rats, respectively.

Crossing Experiments

Soluble and microsomal fractions were obtained from liver homogenates either from control or TAM-treated rats as described above. Microsomal pellet was washed twice before recombining with the soluble fraction. The recombination was carried out by crossing the fractions as follows: Control-soluble fraction plus TAM-microsomes and TAM-soluble fraction plus control-microsomes. Both mixtures were incubated for 10 min at $37^{\circ}C$ either in the presence or in the absence of 0.5 mM oleate and /or 0.15 M KCl. Samples were then centrifuged (105,000 xg for 45 min at $4^{\circ}C$) and PAP activity was determined either in the supernatant (soluble) or in the resuspended pellet (microsomes).

PAP Activity

[3H] Phosphatidate the substrate was synthesized on endoplasmic reticulum membranes obtained from rat liver as previously described (13,14). The enzyme PAP activity was assayed as previously described (14).

RESULTS

In Table 1 is shown the percentage or relative distribution of PAP activity in the soluble and microsomal fractions obtained after the addition of spermine, chlorpromazine and TPA with or without oleate to the supernatant of liver homogenates from control rats and those treated with TAM. 0.5 M KCl was added to the incubation media, to demonstrate the relevant effects of physiological concentrations of salt on the translocation of PAP activity, since hypotonic media favours unspecific binding of cytosolic proteins to membranes. The oleate-induced translocation of PAP from the cytosol to the membranes was higher (three times) in liver homogenates obtained from TAM-treated rats than those of control (from 20% to 63% and from 16% to 31%, p<0.001, respectively). When 0.15 M KCl was added to the incubation medium, the differences between TAM-treated and control homogenates were still higher (six fold) (from 21% to 65% and from 10% to 17% p<0.001, for TAM-treated and for control, respectively). The comparison of the results obtained with and without KCl, demonstrated that liver homogenates from control rats, when incubated with 0.15 M KCl, lost the capacity of transferring PAP activity from the cytosol to membranes and except for oleate plus TPA, the values obtained were not significantly different. However, in TAM-treated liver homogenates, 0.15 M KCl did not inhibit the response to translocation promotors; on the contrary, the transfer of this enzyme activity was markedly higher. For spermine plus oleate, the activity incorporated to the microsomal fraction was 59%, and when 0.15 M KCl was added the value was 74% (p<0.001). The differences of distribution of this enzyme activity in microsomal fraction was still much higher in the presence of oleate plus spermine (74% and 12% p<0.001), for treated and non-treated rats respectively. The loss of response in control rats due to salt concentration can be explained by an inhibiting effect of the ionic strength in the medium since it is well known that below the physiological concentration of salt, most of the proteins are attached to the membranes (15). The effect of chlorpromazine in reversing the oleate-induced translocation of PAP has been reported (3). In the present investigation similar results were observed either in the presence of absence of KCl. In homogenates obtained from TAM-treated livers, chlorpromazine when added alone did not change the subcellular distribution of the enzyme. However, when the chlorpromazine was incubated after oleate, reverse translocation was especially noteworthy when KCl was absent and PAP activity in microsomes decreased to less than half (63% to 24%, <0.001), for oleate

Table 1. Effect of Long-Term Thioacetamide on the Translocation of PAP from the Cytosol to Microsomal Membranes of Rat Liver Homogenates.

Additions	Control		Thioacetamide	
	−	0.15M KCl	−	0.15M KCl
None	16 ± 2	10 ± 2	20 ± 3	12 ± 3
0.4 mM Chlorpromazine	8 ± 1	9 ± 1	16 ± 2***	18 ± 2***
150 nM TPA	25 ± 3	11 ± 1	21 ± 2	19 ± 2**
1 mM Spermine	19 ± 2	8 ± 2	22 ± 4	23 ± 4***
0.5 mM Oleate	31 ± 9	17 ± 3	63 ± 7**	65 ± 7***
0.5 mM Oleate + 0.4 mM Chlorpromazine	11 ± 2	13 ± 2	24 ± 3***	29 ± 4***
0.5 mM Oleate + 150 nM TPA	46 ± 6	31 ± 4	76 ± 10***	71 ± 7***
0.5 mM Oleate + 1 mM Spermine	51 ± 6	12 ± 2	59 ± 12	74 ± 13***

Results are expressed as relative distribution of activity in microsomes(%) and are the average of four experiments. The significance of the differences between control and TAM-treated rats is indicated by *p 0.01; ***p 0.001. Incubations were carried out at 37°C for 10 min. In case of oleate+chlorpromazine there was a previous incubation with 0.5mM oleate adding chlorpromazine for another 10 min.

and oleate plus chlorpromazine, respectively. The presence of KCl did not affect the reverse translocation of PAP produced by chlorpromazine (65% to 29%, p<0.001).

The increase in the total PAP activity (Table 2) in liver homogenates of both TAM-treated and control rats, when 0.15 M KCl was present, was detectable in nearly all conditions assayed, reaching in some occasions values over 150%. Total enzyme activity was enhanced by the effect of long-term TAM administration in values about 120%.

In order to see if there was any modification of the membranes or the presence of some soluble factor induced by the chronic administration of TAM, experiments were performed by crossing the fractions (microsomal and soluble) obtained from liver homogenates of both the control and TAM-treated rats. Fig. 1 shows the relative distribution of PAP activity in microsomes obtained from the recombined samples. The translocation of the PAP was higher when liver microsomes from TAM-treated rats were combined with the hepatic soluble fraction from control rats, either in the presence or the absence of oleate and/or physiological concentration of KCl. No significant differences could be observed in total PAP activity determined in the recombined samples.

DISCUSSION AND CONCLUSION

The short term administration of TAM is known to produce an increase both in lipogenesis and in the concentration of triacylglycerol in the liver (16). However, in the hyperplastic noduligenesis caused by long-term TAM

Table 2. Effect of Long-Term Thiocetamide on the Total Activity of PAP in Rat Liver Homogenates.

Additions	Control –	Control 0.15M KC1	Thioacetamide –	Thioacetamide 0.15M KC1
None	114 ± 16	159 ± 22	114 ± 14	183 ± 14
0.4 mM Chlorpromazine	108 ± 13	133 ± 11	147 ± 10***	187 ± 11**
150 nM TPA	130 ± 14	160 ± 11	155 ± 11*	197 ± 11***
1 mM Spermine	150 ± 13	167 ± 23	158 ± 17	200 ± 18
0.5 mM Oleate	120 ± 21	159 ± 20	131 ± 11	166 ± 11
0.5 mM Oleate + 0.4 mM Chlorpromazine	107 ± 12	169 ± 12	141 ± 12**	179 ± 15
0.5 mM Oleate + 150 nM TPA	149 ± 12	182 ± 13	152 ± 13	193 ± 10
0.5 mM Oleate + 1 mM Spermine	125 ± 11	161 ± 18	143 ± 21*	182 ± 17

Total enzyme activity is expressed as nmol of diacylglycerol formed per min and are the average of four experiments. The significance of the differences between control and TAM-treated rats is indicated by *p<0.05; **p<0.01: ***p<0.001. Incubations were carried out at 37°C for 10 min. In the case of oleate+chlorpromazine there was a previous incubation with 0.5 mM oleate before adding chlorpromazine.

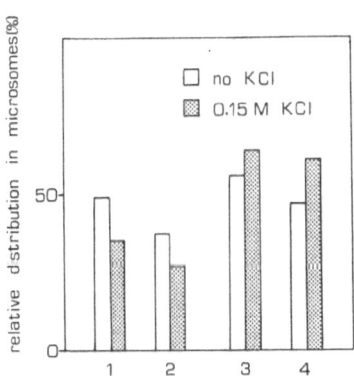

Figure 1. Effect of crossing hepatic subcellular fractions between control and thioactamide-treated rats.
1 & 3: Microsomes from TAM plus Soluble from Control incubated in the absence (1) or in the presence (3) of 0.5 mM Oleate. 2 & 4: Microsomes from Control plus Soluble from TAM incubated in the absence (2) or presence (4) of 0.05 mM Oleate.
Results are the mean of two experimental data. Total PAP activity was determined in sample and the values obtained were close to that observed in Table 1.

administration, lipogenesis as well as the activity of lipogenic enzymes and triacy- lglycerol content, showed a marked decrease, as has been recently demonstra- ted (17) both by the measurement of the lipogenic flux "in vivo" and of enzymatic activities releated to fatty acid synthesis and triacyl-glycerol concentration.

Although there is an apparent paradoxical relationship between the rates of hepatic lipogenesis and the rates of glycerolipid synthesis in the liver of long-term TAM-treated rats, control of glycerolipid synthesis differs from that of lipogenesis because the liver is required to esterify fatty acids, either if they are derived from synthesis "de novo", from peripheral fat or from the diet.

The relative proportion of PAP activity distributed between the cytosolic and the microsomal fractions can be modified by altering the ionic strength of the medium in rat adipose tissue (18). The ability of PAP to attach to the membranes is also affected by hormones (19,20).

The role of polyamines in controlling PAP activity and glycerolipid synthesis and the growth promoting effect of polyamines could have a single mechanism. Diacylglycerol, the main product of PAP activity, is not only a central intermediary metabolite in glycerolipid synthesis but also can be considered as a second messenger in the mechanisms in which stimulus-response are coupled. Generation of diacylglycerol is involved in the activation of Ca^{2+} and phospholipid sensitive protein kinase C. PAP translocation, through diacylglycerol, is releated to protein kinase C activation and therefore can be related to cell proliferation and tumor promotion.

Considering that PAP translocation indicates the changes of this enzyme to an active form (21), the promotion of the translocation in liver homo-genates of TAM-treated rats compared to that of control, by more than six fold, when spermine plus oleate were added in a medium woth 0.15 M KCl may suggest that·the activity of this enzyme is closely connected to·that mechanism of cell proliferation since as well as ornithine decarboxylase, these two enzymes increase their activities when cell proliferation increases as occurs in the case of TAM-induced hyperplastic noduligenesis. These results agree with previous studies on regenerating liver (22).

It is also significant to point out that in liver homogenates of TAM-treated rats, 0.15 M KCl did not influence the promoting translocating effect of oleate. On the contrary, the same salt concentration produced an absolute and negative effect on the oleate induced translocation in non-treated rats.

Phosphatidate and diacylglycerol are important intermediates for phos-pholipid synthesis, and the control of their metabolism is needed for the regulation of bile secretion, membrane proliferation and the growth of new tissues such as tumor formation. Phorbol esters are tumor promoters and long acting analogues of diacylgycerol. They increase PAP activity in adipocytes (23).

Physiological significance

The aim of this study was to investigate in what way the metabolism of TAM, a typical hepatocarcinogenic substance, can influence the responses to effectors or modulators of an enzyme so involved in glycerolipid synthesis as PAP. This enzyme is of interest not only because its metabolic position makes it closely connected to cellular membrane biosynthesis but also because it must be attached to membranes to be active.

The results obtained clearly demonstrated that in our experimental conditions, long-term TAM-treatment increases the hepatic response to oleate

164

of PAP translocation from the cytosol to microsomal membranes. The increased response, versus control, was especially remarkable when physiological concentration of KC1 were present in the incubation media. These results add a further significance because it is well known that increasing ionic strength in the medium produces detachment of the membranes (15), which explains the negative effect of KC1 in the response to translocation of this enzyme in homogenates from control livers. As the translocation of PAP from the cytosol to the membranes means activation of the enzyme, it can be concluded that in the physiological conditions of iconic strength, TAM-treatment promotes the activation of PAP liver in a striking manner. The mechanisms by which this activation occurs is still unknown, but experimental evidences in the present study suggest that the response should be located at the membranes whose ability of binding PAP is increased. However, further investigations are required to determine if this control exerted on the synthesis of glycerpolipid is directed to membrane production. Hepatic hyperplastic noduligenesis in the rat, generated by different xenobiotics, should exhibit several unique biochemical properties and these results provide more insight into the metabolism of these substances.

ACKNOWLEDGEMENTS

We thank Mr Erik Lundin for his help in the preparation of the manuscript. The work was supported by CSIC TD 816.

REFERENCES

1 Cascales, C., Martin-Sanz, P., Pittner, R.A., Hopewell, R., Brindley, D.N. and Cascales, M. (1986) Biochem. Pharmacol. 35, 2655-2661.
2 Martin-Sanz, P., Hopewell, R. and Brindley, D.N. (1985) FEBS Lett. 179 (2) 262-265.
3 Hopewell, R., Martin-Sanz, P., Martin, A., Saxton, J. and Brindley, D.N. (1985) Biochem. J. 232, 485-491.
4 Cascales, C., Mangiapane, E.H. and Brindley, D.N. (1984) Biochem. J. 219, 911-916.
5 Martin, A., Hopewell, R., Martin-Sanz, P., Morgan, J.E. and Brindley, D.N. (1986) Biochem. Biophys. Acta 876, 581-591.
6 Walton, P.A. and Possmayer, F. (1984) Biochim. Biophys. Acta 796, 364-372.
7 Landon, E.J., Naukam, R.J., and Sastry, B.V.R. (1986) Biochem. Pharmacol. 35, 697-705.
8 Anghileri, L.J., Crone-Escanye, M.C., Martin, J.A. and Robert, J. (1986) Int. J. Clin. Pharm. Ther. Toxicol. 24 (5), 270-273.
9 Cerdan, S., Cascales, M. and Santos-Ruiz, A. (1981) Mol. Pharmacol. 19, 451-455.
10 Martin-Sanz, P., Hopewell, P., and Brindley, D.N. (1984) FEBS Lett. 175 (2), 284-288.
11 Hirasawa, K. and Nishizuka, Y. (1985) Ann. Rev. Pharmacol. Toxicol. 25, 147-170.
12 Patel, T.B. (1987) Biochem. J. 241, 549-554.
13 Brindley, D.N. and Bowley, M. (1975) Biochem. J. 148, 461-469.
14 Martin, A., Hales, P. and Brindley, D.N. (1987) Biochem. J. 245, 347-355.
15 Eby, D. and Kirtley, M.E. (1979(Arch. Biochim. Biophys. 198, 608-613.
16 Franke, H., Zimmermann, T. and Dragel, R. (1983) Wirchows. Archiv. (Cell. Pathol.) 44, 99-113.
17 Martin-Sanz, P., Cascales, C., Gomez, A., Brindley, D.N. ans Cascales, M (1978) Carcinogenesis 8 (11) 1685-160.
18 Moller, F. and Hough, M.R. (1982) Biochim. Biophys. Acta 711, 521-531.
19 Pittner, R.A., Fears, R. and Brindley, D.N. (1985) Biochem. J. 230, 525-534.

20 Butterwith, S.C., Martin, A., Cascales, C., Mangiapane, E.H. and
    Brindley, D.N. (1985) Biochem. Soc. Trans. 13, 158-159.
21 Brindley, D.N. (1984) Prof. Lipid. Res. 23, 115-133.
22 Mangiapane, E.H., Lloyd-Davues, K.A. and Brindley, D.N. (1973) Biochem.
    J 134, 102-112.
23 Hall, M., Taylor, S.J. and Saggerson, E.D. (1985) FEBS Lett. 179 (2),
    351-354.

INDUCTION OF HEPATOMA IN SEVEN-DAY-OLD C57BL/6 MICE WITH A SINGLE DOSE OF

AFLATOXIN AND THE INFLUENCE OF DIETHYLDITHIOCARBAMATE

F. Habálek

Institute of Experimental Biopharmacy
Czechoslovak Academy of Sciences
Department of Drug Carcinogenesis
Olešnice v Orlických horách
Czechoslovakia

SUMMARY

The frequency of hepatomas was investigated in mice of both sexes, strain C57BL/6, after a single dose of 10 mg $AFB_1.kg^{-1}$, which was administered i.p. on day 7 after birth. In males, tumors were found beginning in the 12th month in 66% of animals, and in the 20th month in 80% of animals. In females, no tumors were found within 20 months. In the course of the 20 month hepatocarcinogenesis study the activities of microsomal cytochrome P-450 dependent enzymes (cyt P-450 and b_5, NADPH- and NADH- cyt c reductase, aniline hydroxylase and aminopyrine demethylase) decreased in the liver by 25-33%. The hepatic detoxifying enzymes (epoxide hydrase and glutathion S-transferase) had various activities. The activity of epoxide hydrase in the initiating stage of carcinogenesis decreased, whereas in the early stage of the progression of the hepatoma it achieved the fetal levels. The activity of glutathion-S-transferase increased in the preneoplastic stage, i.e. during the promotion of hyperplastic nodules, and then decreased.

In the hepatoma, 1/2 to 1/3 of the normal liver activities of activating enzymes were found; the activities further decreasing with intensifying dedifferentiation of the hepatoma. Glutathion-S-transferase behaved in a similar way, whereas the activity of epoxide hydrase increased in the early stage of the progression of the hepatoma.

Diethyldithiocarbamate administered continually in drinking water commencing with the 9th month of carcinogenesis did not decrease the number of hepatoma, but it produced a slight increase in the activities of cyt P-450 dependent enzymes and the activity of glutathion-S-transferase too. On the other hand, it decreased the activity of epoxide hydrase in the livers of both sexes and in the hepatoma.

Abbreviations

AFB_1 - aflatoxin B_1, DMSO - dimethyl sulphoxide, MFO - mixed function oxygenase system, P-450 - cytochrome P-450, b_5 - cytochrome b_5, NAD/P/H - NAD/P/H cytochrome c reductase, ADM - aminopyrine demethylase, AOH - aniline hydroxylase, EH - epoxide hydrase, GST - glutathion-S-transferase, DTC - diethyldithiocarbamate.

Biochemistry of Chemical Carcinogenesis
Edited by R. Colin Garner and Jan Hradec
Plenum Press, New York, 1990

INTRODUCTION

The hepatocarcinogenicity of aflatoxin B_1 in rats, ducks and rainbow trout has been convincingly demonstrated (1). In contrast, adult mice appear to be resistant to the toxic and carcinogenic effects. Newborn and infant mice are more sensitive to the effects of different hepatocarcinogens (2,3) and also to AFB_1, which induced hepatomas in newborn C57BL/6 x C3H F1 mice (4).

A single dose model of aflatoxin hepatocarcinogenesis was tested in infant C57BL/6 mice. AFB_1 was administered i.p. on day 7 after birth and the following data were investigated:

1) Changes in the activities of microsomal cytochrome P-450 dependant enzymes and conjugating enzymes in the course of the 20-month-long aflatoxin hepatocarcinogenesis study. At intervals of 6,8,12,18 and 20 months after AFB_1 administration, the concentrations of cyt P-450 and b_5 and the activities of NADPH-, NADH-cyt c reductase, AOH, ADM, EH and GST were determined.

2) Effect of continually administered diethyldithiocarbamate on the course of aflatoxin carcinogenesis.

MATERIALS AND METHODS

Aflatoxin B_1 was prepared biosynthetically in our laboratory and was checked for purity chromatographically, spectrophotometrically and by NMR. It was administered to seven-day-old mice i.p. in a dose of 10 $mg.kg^{-1}$, dissolved in DMSO. The administered volume represented 0.5 ug per g of body weight. Diethyldithiocarbamate (DTC) was administered to experimental animals from the 9th month of age until the termination of the experiment in the 20th month as a 0.05% solution in drinking water. The daily dose corresponded to 100 mg of $DTC.kg^{-1}$ of body weight.

Mice C57BL/6 came from the breeding station, Sumice and the standard Larsen diet and water were given ad libitum. At time intervals the animals were killed and the liver, or the tumorous tissue, was removed into a cooled solution of 0.25 M sucrose in 0.1 M sodium phosphate buffer, pH 7.4. Microsomes were isolated from the homogenate by differential centrifugation and the concentrations of cyt P-450 and b_5 (5) and the activities of NADPH- and NADH-cyt c reductases (6), aminopyrine demethylase (7), aniline hydroxylase (8), and epoxide hydrase (9) were determined in them. The activity of glutathion-S-transferase was determined in cytosol (10). The results were calculated as mean and standard deviation.

RESULTS

Induction Of Hepatomas

It follows from Table 1 that hepatomas after one dose of AFB_1 to newborn mice were induced in males between the 8th and 12th months. They were determined histologically as hepatomas composed of large light or basophilic hepatocytes. In the 12th month after AFB_1 administration the incidence of tumors was 66.6%, and in the 20th month, 88.8%. DTC alone as well as DMSO did not influence the survival of animals. In combination with AFB_1 the percentage of surviving animals was the same as after AFB_1 as was the frequency of tumors.

In females, no macroscopic changes were found in the livers at any time.

Table 1. Induction of hepatomas after a single-dose i.p. administration of 10 mg $AFB_1 \cdot kg^{-1}$ to 7-day-old male mice of strain C57BL/6. Diethyldithio-carbamate was administered continually in drinking water as a 0.05% solution commencing with the 9th month, i.e. in a dose of about 100 mg kg^{-1} day^{-1}. DMSO was administered to 7-day-old mice i.p. in a single of 0.5 ug per g of body weight.

Groups	Age at sacrifice (mounts)	Survival No. surviving total No.	No. mice with hepatoma	% mice with hepatoma	No. hepatomas per mice
AFB_1	6	4/5	0	–	–
	8	4/6	0	–	–
	12	6/7	4	66.6	1.25
	18	9/13	8	88.8	1.87
	20	10/18	8	80.–	2.0
AFB_1+DTC	20	12/24	10	83.3	3.9
DMSO	20	6/6	0	–	–

Enzymatic Changes

Changes in the activities of activating and conjugating enzymes in the liver and hepatomas are tabulated in Table 2. In the course of 20-month-long aflatoxin carcinogenesis experiment in males, the concentration of cyt P-450 decreased by 25%, the activity of ADM, AOH, and GST in cytosol by 33 to 75%. On the other hand, the activity of both reductases was slightly increased (by 25 to 35%) and the activity of EH was markedly increased (by 240%). Other quantities under study did not change.

In the female liver, less marked, though similiar changes were found. The concentration of cytochromes P-450 and b_5, activity of both reductases, AOH, ADM and GST decreased more slightly. In the 20th month the activity of EH showed a slight increase (by 39%). (Data not presented).

DTC administration commencing with the 9th month of carcinogenesis did not affect the frequency of hepatomas. DTC administration increased the concentration of cyt b_5 and P-450 both in the male livers (by 40-50%) and in the hepatoma (by 60 to 148%). Also the activity of GST was increased: slightly in the liver (by 24%) and markedly in the hepatoma (by 230%). On the other hand, the activity of EH in the hepatoma decreased by a half.

It has thus been found that in the course of the aflatoxin hepatocarcino-genesis the concentation of cytochromes and activity of GST decrease, whereas the activity of EH increases, and that DTC acts against this change.

DISCUSSION

The frequency of the incidence of hepatomas after one dose of AFB_1, found

by the present author, is in good agreement with the results of
Vesselinovitch (4), who after 6 mg $AFB_1.kg^{-1}$ administered to 7-day-old male
C57BL/6 x C3H F1 mice after 12 and 19 months found hepatomas in 75% or 85% of
animals, respectively.

The latent period in our model of aflatoxin carcinogenesis was 8-12
months. In rats, various dosage regimens can induce up to 100% incidence of
hepatomas in 12 to 20 months in both sexes (11, 15-17).

The strong hepatocarcinogenicity of AFB_1 in infant mice is in contrast
to other hepatocarcinogens. In 7-day-old mice, 0.032 umol of AFB_1 or 0.51
umol of ethynitrosourea, or 16.84 umol of urethane induce 90% incidence of
hepatoma (4).

Differences between sexes in the frequency of hepatomas are usual also
after other carcinogens, e.g. dimethylnitrosamine. They are explained by the
hormonal environment for neoplastic expression. In promotion stage as after
administration of carcinogen there is no difference between the primary impairment of the cell population of male or female hepatocytes (11). It is
obvious, therefore, that the outcome of carcinogenesis must be viewed as the
result of a multifactorial interaction.

The activities of MFO enzymes were found to achieve 40-100% of values
of adult animals as early as the 7th to 10 days of life, no sex differences
being found (Table 3). They most probably do not exist at such an early
stage of development as the MFO activity is secured by the fetal forms of
P-450 (12). To reveal the intersexual differences, the enzymatic activities
specific for each sex ought to be examined.

AFB_1 in mice 48 hours after administration slightly increased the
activity of the MFO system and GST. This is in agreement with our results
obtained in Wistar strain rats, where the inductive action of AFB_1 was measured after 24-48 hours; after 72 to 120 hours inhibition of the MFO system
was seen (Data are not presented). Similar results in rats were obtained by
Dent and Graichen (13). These changes reflect the actute-toxic action by
aflatoxin B_1.

For the carcinogenic effect of AFB_1 its biotransformation into active
metabolite, 8,9-epoxy AFB_1 is necessary. It follows from the acute-toxic
and carcinogenic effect of AFB_1 and our determinations of MFO activities that
as early as day 7 of life sufficient biotransformation of AFB_1 takes place.
Unlike cyt P-450, cyt b_5 is fully developed already in the neonate and its
concentration in the liver did not change during hepatocarcinogenesis. No
differences between the sexes were found, either.

In the hepatoma about half the concentrations of cyt P-450 and b_5 and
half MFO activity were found in comparison with the liver. With the dedifferentiation of the hepatoma the activities further diminished so that
after the 20th month they represented approximately 30% of the normal liver
values. This is essentially consistent with the literature data (22-24).

The activity of GST is not quite developed in young males, so it reaches
the maximal values about the 3rd month of age (18). In the course of aflatoxin hepatocarcinogenesis the activity increased till the 8th month when it
was double the value of control animals. During this promotion stage of
hepatocarcinogenesis, prenoplastic cells such as enzyme-altered foci and
hyperplastic nodules are known to appear and express the deviation patterns
of preneoplastic marker enymes (19,20). The cytosolic GST may be useful as a
marker of hepatic preneoplatia.

The activity of EH is higher in young than in adult mice and with age it

Table 2. Activities of MFO enzymes and detoxication enzymes during the course of a 20 month-long hepatocarcinogenesis experiment in C57BL/6 male mice initiated by a single dose of aflatoxin B_1. AFB_1 was administered to 7-day-old mice i.p. at the dose of 10mg kg^{-1} in DMSO. Control animals received on day 7 only 0.5 ul of DMSO per g. DTC was administered continually beginning with the 9th month in drinking water as a 0.05% solution.

Tissue	Groups	Age at sacrifice (mounts)	No of samples animals	P-450	b_5	NADPH	NADH	AOH	ADM	GST	EH
						Activities of enzymes					
Liver	AFB_1	6	4/4	0.800 ±0.095	0.489 ±0.030	247.6 ±20.2	2135.9 ±53.5	59.3 ±5.8	136.7 ±12.3	8.82 ±0.5	18.3 ±3.6
		8	4/4	0.764 ±0.115	0.461 ±0.038	237.7 ±12	2107.5 ±145.8	59.4 ±2.4	119.8 ±5.5	9.97 ±3.2	52.6 ±11.1
		12	6/6	0.594 ±0.134	0.403 ±0.039	302.4 ±37.1	2232.9 ±289.2	43.2 ±18.4	91.6 ±13.8	4.06 ±1.9	51.7 ±20
		18	9/9	0.606 ±0.245	0.509 ±0.198	306.9 ±122.5	2859.0 ±1176.0	50.3 ±27.2	90.0 ±12.6	2.12 ±0.9	84.9 ±34.4
		20	10/10	0.547 ±0.128	0.473 ±0.099	362.1 ±137.2	2096.2 ±396.6	42.0 ±14.0	99.4 ±28.8	4.12 ±1.3	99.5 ±18.1
	AFB_1+DTC	20	8/8	0.833 ±0.129	0.657 ±0.102	362.8 ±62.8	2514.9 ±442	37.8 ±2.8	110.0 ±16.3	2.66 ±0.2	68.2 ±19.1
	Control	20	6/6	0.735 ±0.128	0.602 ±0.133	351.7 ±82.7	2119.8 ±495.3	50.4 ±4.43	100.5 ±18.5	4.03 ±1.49	40.0 ±13.2
Hepatoma	AFB_1	12	2/3	0.291 ±0.089	0.271 ±0.020	297.2 ±405	2088.5 ±379	6.9 ±4.7	71.4 ±30.5	4.3 ±0.8	62.8 ±25.5
		18	3/8	0.156 ±0.042	0.253 ±0.08	371.4 ±143.0	1069.7 ±700	5.5 ±0.9	34.0 ±3.3	3.2 ±2.4	76.9 ±30
		20	7/10	0.231 ±0.135	0.186 ±0.09	264.6 ±117.9	2569.9 ±1167.1	4.9 ±3.4	21.8 ±5.7	2.6 ±1.9	123.7 ±35.5
	AFB_1DTC	20	7/10	0.372 ±0.157	0.464 ±0.159	296.8 ±48.1	2564.9 ±495.2	3.1 ±1.5	15.1 ±4.4	8.58 ±5.1	37.1 ±12.8

Table 3. Effect of aflatoxin B_1 on the liver mono-oxygenase system and detoxicating enzymes in 7-day-old mice of strain C57BL/6. The withdrawal was carried out 48 hours after administration. AFB_1 was administered i.p. in DMSO in a dose of 10mg kg^{-1}. Control animals received DMSO alone.

Sex of mice	Groups	No. of animals	P-450	b_5	NADPH	NADH	AOH	ADM	GST	EH
male	AFB_1	14	0.397 ±0.047	0.419 ±0.025	349.7 ±31.4	1015.4 ±94.9	35.7 ±8.7	99.7 ±6.1	1.47 ±0.06	106.3 ±19.1
	Control	13	0.318 ±0.104	0.360 ±0.042	404.3 ±93.6	1141.2 ±182.6	27.9 ±6.5	76.4 ±17.5	1.14 ±0.13	112.7 ±14.0
female	AFB_1	13	0.431 ±0.015	0.408 ±0.032	328.1 ±14.1	913.0 ±72.1	39.0 ±9.9	101.4 ±13.9	1.44 ±0.08	109.7 ±18.3
	Control	13	0.360 ±0.049	0.390 ±0.018	492.6 ±126.4	1126.8 ±125.4	31.0 ±2.4	65.1 ±22.1	1.04 ±0.09	113.2 ±32.5

Concentration of cyt P-450 and b_5 expressed in nmoles per mg of microsomal protein. Activities of NADH NADPH cyt c reductase expressed in nmoles of the formed product per mg of microsomal protein per minute. Activity of AOH in nmoles of the formed p-aminophenol per mg of microsomal protein in 20 minutes. Activity of ADM in nmoles of the formed formaldehyde per mg of microsomal protein in an hour. Activity of GST in nmoles of the formed adduct of glutathion with dinitro-chlorobenzene per mg of cytosol protein in a minute. Activity of EH in umoles of the formed NADH per mg of microsomal protein in a minute.

decreases to about one half (21). In the progressive stage of aflatoxin hepatocarcinogenesis (18th to 20th month) it is increased. In males this increase in the EH activity was perceptible especially in hepatoma. Also in females by the 20th month an increased activity of EH in the liver appeared which could mean that there is a delay in malignant transformation in comparison with males but that tumors eventually appear (results not shown). In our earlier work it was found that the activity of EH, e.g. in Ehrlich ascitic carcinoma, was also higher than in the liver (unpublished data). This supports the conclusions that the activity of EH is an important neo-plastic indicator.

In our previous papers agents were examined which increase the activity of GST in the liver, and thus inhibit hepatocarcinogenesis. DTC, which increased the activity of GST twice in 7-day administration, was selected to influence aflatoxin carcinogenesis. A therapeutic model analogous to that used by Novi to inhibit cellular carcinoma in rats by gluthathione (15) was selected. DTC in the given scheme did decrease the number of hepatomas in male mice (Table 1). In the liver it increased the concentration of both the cytochromes and GST activity by 25 to 40%, but mainly it increased the GST activity in the hepatoma. The hepatoma thus kept its capability of being induced. This is in agreement with the results published by Okita (22).

After DTC administration in the course of aflatoxin hepatocarcinogenesis both an increase in the concentrations of cytochromes and GST activities and a decrease in the EH activity in the liver and hepatoma occured. At the same time the number of hepatomas was not diminished. In the future we will attempt to produce changes in the incidence of hepatocellualr carcinoma to different DTC dosing.

REFERENCES

1   Busby, W. F. and Wogan, G.N. (1981) In Mycotoxins and N-nitroso compounds. Environmental Risk. Vol II (Shank, R.C. ed.) pp. 134-144. CRC Press, Boca Raton
2   Liebelt, R.A., Liebelt, A.G. and Lane, M. (1964) Cancer Res., 24, 1869-1876
3   Delclos, K.B., Tarpley, W.G., Miller, E.C. and Miller, J.A. (1984) Cancer Res., 44, 2540-2550
4   Vesselinovitch, S.D., Milhailovich, N., Wogan, G.N., Lombard, L.S. and Rao, K.V.N. (1972) Cancer Res., 32, 2298-2303
5   Omura, T. and Sato, R. (1964) J. Biol. Chem., 239, 2730-2738
6   Jeffery, E., Kotake, A., Azhary, R. and Mannering, G.J. (1977) Mol Pharmacol. 13, 415-425
7   Mazel, P. (1972) in Fundamentals of Drug Metabolism and Drug Disposition (Mandel, H.G., Way, E.L. eds.) pp 527-545. Williams and Wilkins Co., Baltimore
8   Imai, Y., Ito, A., and Sato, R. (1966) J Biochem, 60, 417-428
9   Guengerich, F.P. and Mason, P.S. (1980) Anal. Biochem 104, 445-451
10   Habig, W.H. and Jakoby, W.B. (1981) Methods Enzymol., 77, 388-405
11   Moore, M.R., Drinkwater N.R., Miller, E.C., Miller, J.A. and Pitot, H.C. (1981) Cancer Res., 41, 1558-1593
12   Maeda, K., Kamataki, T., Nagai, T., Kato, R. (1984) Biochem. Pharmacol., 33, 509-512
13   Dent, J.G. and Graichen, M.E. (1982) Carcinogenesis 3/7/, 733-738
14   Busby, W.F., Paglialunga, S., Newberne, P.M. and Wogan, G.N. (1976) Cancer Res., 36, 2013-2018
15   Novi, A.M. (1981) Science, 212, 541-542
16   Lin, W., MacKenzie, J.W., McCoy, J.R. and Clark, I. (1983) Cancer Biochem. Biophys., 7, 61-68
17   Wogan, G.N. and Newberne, P.M. (1967) Cancer Res., 27, 2370-2376

18   Gregus, Z., Varga, F. and Schmelas, A. (1985) Comp. Biochem. Physiol., 80C, 85-90

19   Sato, K., Kitahara, A., Yin, Z., Ebina. T., Sataoh, K., Tsuda, H., Ito, N., and Dempo, K., (1983) Ann New York Acad Sci, 213-223

20   Kitahara, A., Satoh, K., and Sato, K. (1983) Biochem. Biophys. Res. Commun., 112, 20-28

21   Rouet, P., Dansette, P., and Frayssinet, Ch. (1984) Dev. Pharmacol. Ther, 245-258

22   Okita, K., Noda, Y., Fukumoto, Y., and Takemoto, T. (1976) GANN, 67, 899-902

23   Miyake, Y., Gaylor, J.L. and Morris, H.P. (1974) J Biol Chem., 249, 1980-1987

24   Ohmachi, T., Sagami, I., Fujii, H. and Watanabe, M. (1985) Arch. Environ. Cont. Toxicol., 14, 197-202

INHIBITION OF CHEMICAL CARCINOGENESIS

Dietrach Schmähl, Barbara Bertram, Eva Frei, Reinhold Klein,
Beatrice L. Pool, Peter Schmezer, and W. Jens Zeller

German Cancer Research Centre, Im Neuenheimer Feld 280
D-6900 Heidelberg, Federal Republic of Germany

ABSTRACT

The inhibition of chemical carcinogenesis by specific antidotes has to
be considered a long-term objective in cancer prevention. Taking three
examples, the possibilites and limitations of such inhibition are elucidated.

INTRODUCTION

According to the state of present-day knowledge, the possibility of
primary cancer prevention exists in only 35% of all cancer morbidities at the
most (1). This stresses the need to support theoretical approaches to the
inhibition of chemical (or viral) carcinogenesis by administration of
specific antidotes capable of modifying carcinogenesis. Obviously, we are
yet far from this ideal, in particular because the mechanisms of action on
the molecular level are still widely unknown. Below three examples from the
field of chemical carcinogenesis are discussed which cast a light on the
problem of inhibition of carcinogenesis from different vantage points. These
examples were chosen, because they refer to investigations carried out in our
institute.

A. Disulfiram

Disulfiram (DSF)

$$C_2H_5 \diagdown \atop N - C - S - S - C - N \diagup \atop C_2H_5 \qquad \overset{S}{\|} \qquad \overset{S}{\|} \qquad \diagup C_2H_5 \atop \diagdown C_2H_5 \qquad (I)$$

is a thiolic compound which inhibits the toxicity and carcinogenicity of
several chemical carcinogens, such as aromatic hydrocarbons and N-nitroso-
compounds (2,3). Concomitant administration of DSF and N-nitrosodimethylamine or N-nitrosodiethylamine in rats altered the organotropism of the
carcinogenic effects. Whereas the liver was the target organ of carcinogenesis induced by the nitrosamines alone, concomitant administration with
DSF resulted predominantly in cancer of the nasal cavity or the oesophagus.
It was thus possible to protect one organ, but the carcinogenesis was shifted
to other organs.

Biochemistry of Chemical Carcinogenesis
Edited by R. Colin Garner and Jan Hradec
Plenum Press, New York, 1990

The inhibition of oxidizing enzymes seems to be the main mechanism thereof and is probably related to the metal-binding capacity of DSF. Moreover, an influence of DSF on the glutathione/glutathione-S-transferase system and direct chemical reaction between DSF and nitrosamine metabolites were detected. The results of several experiments demonstrating these three effects of DSF are summarized below.

Since nitrosamines belong to those chemical carcinogens which are not carcinogenic per se but need enzymatic activation, it seemed expedient to study the influence of DSF on enzyme activities. It is well known that DSF is a strong inhibitor of several enzymes but it was not known whether this inhibition persisted over the long time period of a carcinogenesis experiment. In fact, we demonstrated that the inhibition is still statistically significant after 28 weeks of treatment. Only a marginal rescue of enzyme activities occurred (4,5).

Only a few facts are known about the nature of nitrosamine-metabolizing enzymes such as, for instance, that different enzyme forms exist depending on the substrate concentration. Nothing is known, however, about the three dimensional structure. Assuming that they might be enzymes needing metal ions as cofactors, we studied the influence of DSF on metal ions in vivo.

After a ten-day period of DSF treatment in rats, the liver concentrations of three elements out of a whole series (K, Fe, Cu, Co, Mn, Se and Zn) were elevated:

Cu: 6 x higher
Co: 3 x higher
Zn: 0.8 x higher

This seems important in view of findings reported by Brown et al. who measured a decrease in Cu and Zn content of the liver after treatment with the carcinogen N-nitrosodiethylamine (6).

The elevation of the metal content was still higher after 32 weeks of DSF treatment:

Cu: 60 x higher
Co: 10 x higher
Zn: 1.5 x higher
Cd: 110 x higher
Se: 1.8 x higher

Mo was the only element reduced in concentration (- 50%). The diagrams of copper, cadmium and zinc content are shown as examples for the influence DSF, NDEA, or DSF + NDEA on the metal content in the liver of rats (Fig 1).

We assume that the elevation of the metal content is caused by formation of unsoluble chelate complexes between metals and mixed disulfides (protein-diethyldithiocarbamate). In this form the metals are no longer available for enzymes, which results in inhibition of enzyme activity.

A further parameter of biochemical interest is the detoxifying system glutathione/glutathione-S-transferase (GSH/GST). GSH is responsible for several detoxifying processes:

(1) GSH reduces cytotoxic peroxides,

(2) GSH scavenges aldehydes formed in the metabolism of foreign compounds,

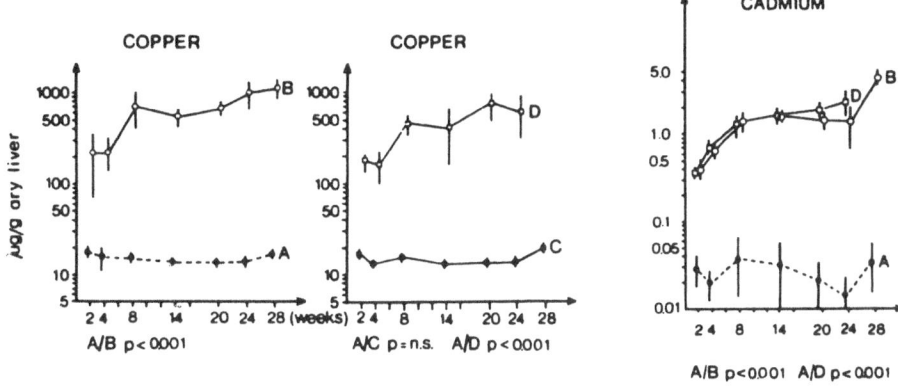

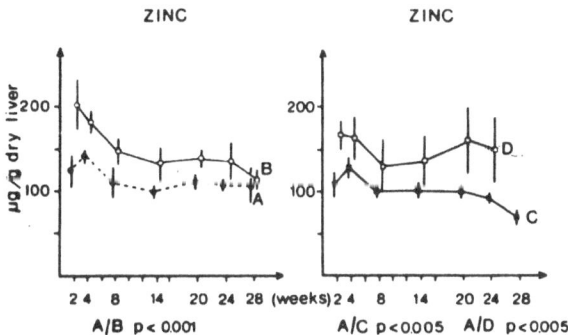

Figure 1. Influence of prolonged treatment with disulfiram (2 g/kg diet) and NDEA (12 mg/kg p. o. 1 x weekly) on the concentration of copper, cadmium and zinc
A, control: B, DSF; C, NDEA; D, DSF+NDEA

(3) GSH leads to drug conjucation of epoxidic metabolites formed in the metabolism of aromatic hydrocarbons. This reaction is catalyzed by glutathione-S-transferase.

In 1982, Sparnins et al. demonstrated that the activity of GST is induced by DSF treatment (7). This finding was confirmed by us in short-term and long-term experiments. Moreover, a marked increase in GSH content in the liver was revealed. The increase reached its maximum after four weeks and thereafter was not as striking as before. We concluded that the observed effects of DSF are related to its anticarcinogenic activity. However, because of its strong enzyme-inhibiting effects it seems unsuitable for human intake over a long time.

B. Mesna

Cyclophosphamide-induced urinary bladder carcinoma is one example of drug-induced cancer. Several years ago when cancer chemotherapy was not very effective and did not result in prolonged survival times, side effects such as carcinogenic effects were not observed. With increasing survival times, however, drug-induced second tumors manifested themselves clinically (8). In a survey on second tumors reported world wide after administration of anti-cancer drugs, we found a definite correlation between the administration of drugs and the occurence of certain tumors (Fig 2).

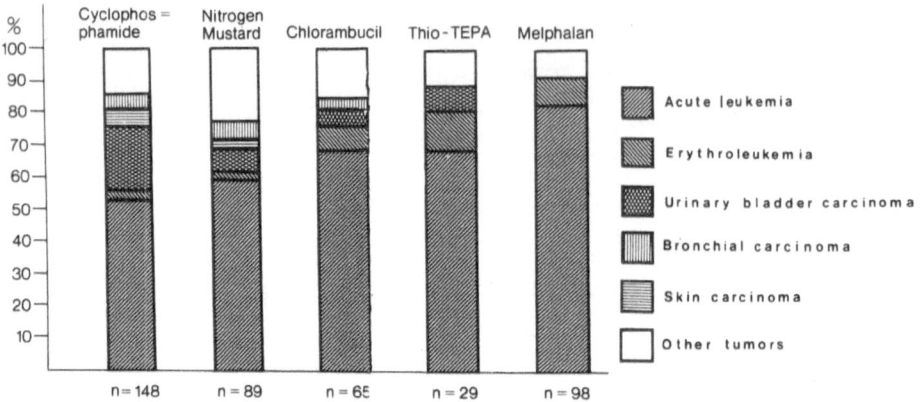

Figure 2. Organotropism of second tumors induced by cytostatic drugs in humans.

Among others, urinary bladder cancer was observed after oxazaphosphorine administration. It is remarkable but not unexpected, that in particular, alkylating cytotoxic agents proved to be potentially carcinogenic (9).

We investigated the occurrence of second tumors induced by cyclophosphamide in rats (10,11). Groups of 40 female and 40 male Sprague-Dawley rats received low doses of cyclophosphamide in drinking water lifelong. Doses of 2.5, 1.25, 0.6. and 0.3 mg/kg b.w. of cyclosphosphamide, respectively, were administered five times a week. In the untreated control, the incidence of malignant tumors was 11% (males) and 15% (females). In treated animals, however, a dose-related increase in tumor incidence was observed. Among cyclophosphamide-treated animals 16 males and one female bore transitional-cell carcinomas of the urinary bladder, while in 18 male and 10 female rats mostly multifocal papillomatosis had formed in the urinary bladder, which regularly resulted in early hematuria. No malignancies were found in the urinary bladders of untreated animals, only one female rat had a papilloma in the bladder.

The model of cyclophosphamide-induced urinary bladder tumors was used to examine the efficacy of uroprotectors in long-term experiments in rats. Sodium 2-mercaptoethanesulfonate (II) (mesna)

$$ SH - CH_2 - CH_2 - \overset{\displaystyle O}{\underset{\displaystyle O}{\overset{\displaystyle \|}{\underset{\displaystyle \|}{S}}}} - ONa \qquad (II) $$

scavenged the acute urotoxic side effects of oxazaphosphorines in animal experiments as well as in humans (12). The inhibition of cyclophosphamide-induced urinary bladder tumors was dose-related. The experimental design and essential results are presented in Tables 1 and 2.

Administration of cyclophosphamide alone induced urinary bladder carcinomas or papillomas in 32% of the animals used, while concomitant administration of mesna or dimesna resulted in a reduction to 2% and 6%, respectively. To our knowledge, this was the first time that carcinogenesis induced by alkylating agents was inhibited in a specific organ by an antidote.

Table 1. Chronic oral administration of cyclophosphamide alone or in combination with mesna and dimesna to male Sprague-Dawley rats

Group	Initial number of animals	Individual doses Cyclophosphamide	(mg/kg b.w. 5x/week) Mesna	Dimesna
1	100	–	–	–
2	50	–	15	–
3	50	–	–	35
4	100	2.5	–	–
5	50	2.5	5	–
6	50	2.5	15	–
7	50	2.5	–	12
8	50	2.5	–	35

Table 2. Tumors observed in the urinary bladder after treatment according to Table 1

Group	Animals with tumors in the urinary bladder Total No	%	Papillomas No	%	Carcinomas No	%
1	0	0	0	0	0	0
2	0	0	0	0	0	0
3	0	0	0	0	0	0
4	32	32	15	15	17	17
5	12	24	6	12	6	12
6	1	2	0	0	1	2
7	7	14	2	4	5	10
8	3	6	2	4	1	2

Acute cystitis observed in the urinary bladder after administration oxazaphosphorines is predominantly caused by release of acrolein (III)

$$H_2C = CH - C \diagup^{O}_{\diagdown H}$$ (III)

in the urine (13). Reaction of the thiolic compound mesna with activated primary metabolites of oxazaphosphorines in the kidneys and the urinary tract largely inhibits the release of acrolein. This mechanism is also assumed to prevent tumorigenesis in the bladder (12).

It cannot be excluded that thiolen acts as a protector in the urine against bladder carcinogenesis. The inhibition of bladder carcinogenesis might, furthermore, be effected by reaction of activated cyclophosphamide with mesna, thus inhibiting the alkylation of genes in the bladder epithelium and, consequently, the malignant transformation. The mechanism of this process would be qualitively different from the inhibition of toxic cystistis by acrolein.

Thus, mesna proved to be a highly selective inhibitor of bladder carcinogenesis in the rat. In clinical trials it prevented acute cystitis after cyclophosphamide administration and it is to be expected that the experimental inhibition of bladder carcinogenesis by mesna will be confirmed in clinical studies.

C. SO_2 and NO_x

The influence of the common air pollutants SO_2 and NO_x on the tumorigenesis induced by chemical carcinogens has not yet been studied extensively. We may envisage the gases to act either inhibitorily, additively or synergistically during the induction of tumors by carcinogens. Preliminary studies did not show that lifelong inhalation of SO_2 in combination with benzo(a)-pyrene yielded an increased tumor rate in laboratory animals. The study on combination effects of the gases with another class of carcinogens, the N-nitrosamines, has not been studied so far.

In order to obtain basic information on the mechanisms of this type of combination, first in vitro experiments were performed with fetal hamster lung cells (FHLC) and rat hepatocytes (RH) (14,15). We studied the influence of SO_2 or NO_x exposure on toxicity (trypan blue exclusion, plating efficiency, activity of marker enzymes) and genotoxicity (DNA single strand breaks). For the assay of toxic effects, it was shown that the plating efficiency and measurement of lactate dehydrogenase (LDH) activity were more sensitive parameters than the assay of trypan blue exclusion. Thus, FHLC has a distinct reduced plating efficiency after two hours of of SO_2 exposure, whereas the staining with trypan blue revealed no differences between control and SO_2 exposed cells. SO_2 also reduced the LDH activity of both indicator cells after two-hour exposure. This effect may be due to a direct interference with the measurement, since the LDH activity of a serum standard is also reduced after SO_2 treatment. SO_2 application for one hour decreased the rate of DNA single strand breaks in FHLC and RH caused by - acetoxymethylmethylalmine, when compared to the effects this compound induced on its own or in combination with NO_x (16). In RH this may be due to conjugation of the carcinogen with SO_4 ions (formed by the sulfite oxidases from SO_3^-), as the addition of $MgSO_4$ also caused a reduction of genotoxicity. In FHLC, SO_2 application for 24 hours also decreased the induction of DNA single strand breaks induced by benzo(a)pyrene, as shown in Table 3 (17). Other explanations for the reduced genotoxicity of these two chemicals in the presence of SO_2 may be either that the energy of the cells is impaired by the pollutant or that SO_2 affects activating/deactivating enzymes. In comparison to the results in FHLC reported above, NO_x application for 24 hours was more effective than SO_2 in reducing the rate of induced single strand breaks by benzo(a)pyrene (Table 3) (18). At present we have no further explanation for

Table 3. Influence of SO_2 and NO_x on genotoxicity of benzo(a)pyrene (B(a)P) in vitro (fetal hamster lung cells)

Compound	Concentration μg/ml	ppm	SSB (rad equ.)
B(a)P	2		249
B(a)P	2		225
SO_2		35	-35
B(a)P + SO_2	2 +	35	160
B(a)P	2		248
NO_x		12.5	-14
B(a)P + NO_x	2 +	12.5	101

the mechanisms of inhibition of genotoxicity by air pollutants.

In addition to these in vitro studies, we investigated systemic effects of air pollutants after in vivo application. For this, Sprague-Dawley rats were treated with the pollutants (50 ppm, two weeks) and different susceptibility parameters (toxicity and genotoxicity) were analyzed in the blood and in explanted liver and lung cells (19). The parameters were determined after isolation and after incubation of the cells for one hour with and without carcinogens. SO_2 treatment resulted in an elevated LDH level in the blood serum. This LDH apparently did not arise from liver cells, however, as they failed to liberate LDH after explanation and cultivation. It was further found that the rate of DNA single strand breaks induced by carcinogens in vitro is lower in hepatocytes derived from SO_2-treated animals than in hepatocytes from untreated or NO_x-pretreated rats. An inhibition of the carcinogen-activating enzymes is probably not involved. The measurement of the foreign compound metabolizing enzymes aryl hydrocarbon hyroxylases (AHH), N-nitrosodimethylamine de-methylases (NDMA-D) and glutathione transferases (GST) demonstrated SO_2 increased the level of NDMA-D in the liver and decreased the level of GST in the lung. The pretreatment with NO_x instead revealed a decrease of AHH, NDMA-D and GST in liver and GST in lung.

These results are of general importance, because they show that both SO_2 and NO_x are systemically biologically active if taken up via inhalation. Whether or not SO_2 can inhibit the genotoxicity of NDMA in liver and lung is presently being determined by measuring the rate of DNA single strand breaks in cells of rats that inhaled both agents simultaneously. Furthermore, we are presently performing a long-term bioassay with NDMA and air pollutants to assess the role of these combinations effects in carcinogenesis. To which extent these effects may influence the carcinogenicity of N-nitrosodimethylamine is being investigated in this long-term bioassay.

CONCLUSION

The above examples demonstrated that in specific cases (cyclophosphamide-mensa) a defined carcinogenic risk can effectively be prevented. In other cases (disulfiram) the possibility of protecting a specific target organ against carcinogenic effects (liver carcinogenesis induced by N-nitroso-compounds) was theoretically and experimentally shown. This result is of no practical relevance up to date because the antidote disulfiram is not suited for human use and its carcinogenic effect was not inhibited altogether but only shifted to another organ. The mechanism responsible for this shift is still unknown. The third example discussed might acquire high practical relevance in the evaluation of possible enhancing or inhibiting effects of certain air pollutants on carcinogenesis. A definite conclusion can be drawn only after termination of the inhalation experiments in vivo.

It is notable that so far only a few research groups have dealt with the question of "anticarcinogenesis". Although the difficulties inherent in this field of activity are known to the experts, especially young scientists should be encouraged to take up this type of research.

REFERENCES

1 Schmahl, D., Preussmann, R., and Berger, M.R. (1988) Br. J. Cancer, submitted.
2 Schamhl, D. and Kruger, F.W. (1972) in Topics in Chemical Carcinogenesis (Nakahara, W., Takayama, S., Sugimura, T., Odashima, S. eds.) pp 199-211. University Press Tokyo
3 Schamhl, D., Kruger, F.W., Habs, M., and Diehl, B. (1976) Z. Krebsforsch.

4   Bertram, B., Schuhmacher, J., Frei. E., Frank, N., and Wiessler, M.
    (1982) Biochem. Pharmacol. 31, 3613-3619
5   Schuhmacher. J., Scherf, H.R., Bertram, B., Frei, E., Hauser, H., and
    Wiessler, M. (1985) J. Cancer Res. Clin. Oncol., 109, 16-22
6   Brown, D. A., Chatel, K.W., Chan, A.Y., and Knight, B. (1980) Chem. Biol.
    Interact., 32, 13-27
7   Sparnins, V.L., Venegas, P.L., and Wattenberg, L.W. (1982) J. Natl.
    Cancer Inst., 68, 493-496
8   Schmähl, D., Habs, M., Lorenz, M., and Wagner, I. (1982) Cancer Treat.
    Rev., 9, 167-194
9   Schmähl, D., and Kaldor, J.M (1986) Carcinogenity of Alkylating Cyto-
    static Drugs, IARC Sci. Publ. 78, Lyon
10 Schmähl, D., and Habs, M. (1979) Int. J. Cancer, 23, 706-712
11 Habs, M., Schmähl, D., and Lin, P.Z. (1981) Int. J. Cancer,
    28, 91-96
12 Brock, N., Habs, M., Pohl, J., Schmähl, D., and Stekar , J. (1982)
    Therapiewoche. 32, 4975-4996
13 Brock, N., Stekar, J., Pohl, J., Niemeyer, U., and Scheffler, G. (1979)
    Arzneim-Forsch. 29, 659-661
14 Pool, B.L., Klein, R.G., Schmezer, P., and Zeller, W.J. (1986) in 2.
    Statuskolloquim des PEF, Vol. 3, 959-975, Kernforschungszentrum,
    Karlsruhe
15 Klein, R.G., Pool, B.L., Scmezer, P., and Zeller, W.J. (1987) in 3
    Statuskolloquim des PEF, Vol. 4, 739-751, Kernforschungszentrum,
    Karlsruhe
16 Pool. B.L., Janowsky, I., Klein, R.G., Schmezer, P., Vogt-Leucht, G.,
    and Zeller, W.J. (1988) Carcinogenesis, in press
17 Zeller, W.J. Vogt-Leucht, G., Pool, B.L., and Schmezer, P. (1988) in
    4 Statuskolloquim des PEF, in press
18 Zeller, W.J., Vogt-Leucht, G., Pool, B.L., Janowsky, I., Schmezer, P.,
    and Klein, R.G. (1988), Proc. Am. Ass. Cancer Res., Abstract 440
19 Pool, B.L., Brendler, S., Klein, R.G., Monarca, S., Pasquini, R.,
    Schmezer, P., and Zeller, W.J. (1988) Carcinogenesis, submitted

CHEMOPREVENTION AND REDUCTION OF CANCER RISKS CAUSED BY NITROSATABLE DRUGS

T. Schramm and D. Ziebarth

Central Institute of Cancer Research
Academy of Sciences of the GDR
Berlin-Buch

ABSTRACT

As the mechanisms of action of many chemopreventive substances are up to now poorly understood, it is difficult to classify them comprehensively. One of the categories of such chemicals which can prevent the formation of N-nitroso compounds from reactions of precursor amines and amides with nitrate is a topic of interest. As more and more data suggest that N-nitroso compounds may be cancer risks for man, the possibilities of the formation of N-nitroso compounds from nitrosatable drugs currently used in the GDR and the possibilities to inhibit or to reduce nitrosation by adding ascorbic acid have been investigated. The results obtained should encourage this special type of chemoprevention of cancer using ascorbic acid in combination with selected drugs.

INTRODUCTION AND GENERAL CONSIDERATIONS

There is no doubt, that removal of carcinogenic agents and factors is the main goal in primary cancer prevention, but for the forseeable future this is an unlikely proposition. Therefore cancer prevention based on chemo-prevention in the broadest sense has gained importance. If we consider chemopreventive substances in connection with other factors which may inter-fere or modify chemical carcinogenesis, we can see, that chemoprevention is only one of several mechanisms involved in the modification of carcinogenic processess, either enhancement or inhibition (Table 1).

Although the mechanisms of action of many inhibitors of carcinogenesis are incompletely understood they can be divided into several categories according to their type of action or activities (Figure 1).

The first category consists of compounds that can prevent the formation of carcinogens from precursor substances. A special focus of this approach has been the prevention of the formation of N-nitroso compounds from the reactions of precursor amines or amides with nitrite as reported by Lijinsky et al. (6, 7). The endogeneous formation of N-nitroso compounds is theo-retically possible with all compounds that contain substituted amino groups, and many experiments in animals have shown, that nitrosation reactions can occur in vivo (9). The formation of N-nitrosoproline in humans has been demonstrated in a direct fashion by feeding nitrite and proline and measuring

Biochemistry of Chemical Carcinogenesis
Edited by R. Colin Garner and Jan Hradec
Plenum Press, New York, 1990

Table 1. Some factors and mechanisms which may interfere with the actions of chemical carcinogens.

MODIFYING FACTORS

a) Exogeneous

Exposure mode of application
Total dose, single dose, dose distribution
Duration of exposure or application
Syn-, co- and/or anticarcinogenic chemical, physical and biological agents and factors
Noncarcinogenic environmental agents, factors and health hazards
Therapeutic measures (hormones, radiation, immunosuppressives)
Chemopreventive substances

b) Endogeneous

Genetic constitution pattern
Hormonal and immunological patterns and activities
Enzymatic pattern and activities
Ability to detoxify and/or activate xenobiotics eg precarcinogens
Hormonal activities and disorders
Repair
Syn-, co- and/or anticarcinogenic factors/mechanisms

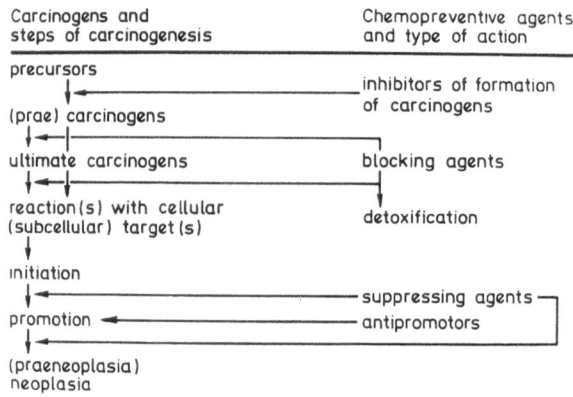

Figure 1. Classification scheme of chemopreventive agents according to their type of action in different steps of chemical carcinogenesis (Adapted and changed from 14)

the N-nitrosoproline excreted in the urine (12). A favoured organ for the formation of N-nitroso compounds is the stomach where the pH is close to the optimum for the reaction.

There may be a potential risk related to the endogeneous formation of N-nitroso compounds in patients taking nitrosatable drugs. The information available suggests that N-nitroso compounds are most effective by the oral route and when given as multiple small doses over a long period of time, which is the case for several drugs. A number of drugs, some of which are taken chronically and in substantial amounts contain nitrosatable groups and evidence has been presented that they can react with nitrite both "in vitro" and "in vivo" to yield N-nitroso compounds (2, 4). The carcinogenic risk linked to the use of nitrosatable drugs can be considered to depend either on the amount of N-nitroso compounds formed in the gastric environment by its reaction with nitrite or on the intrinsic genotoxic potency of the reaction products. In the special case of endogeneous formation of N-nitroso compounds chemopreventive agents can block formation presumably alter exposure (1).

Ascorbic acid is probably the most effective inhibitor of this reaction which was demonstrated first in 1972 (8). It is an effective scavenger of nitrite ions (13) and it can reduce the mutagenicity of human gastric juice as determined by the Ames test (11). It is very possible that the amounts of N-nitroso compounds synthesized intragastrically may differ over a wide range depending on the chemical reactivity of the active agents, the dose, their presence inside the stomach, the pH-conditions and on the nitrate concentrations.

Our investigations are correlated to experiments in which the nitrosation behaviour of 60 orally administered drugs used in the GDR (3) has been investigated. Only those pharmaceuticals were selected whose active ingredients are assessed as being nitrosatable by their chemical structures. Ten drugs have been found to react positively (15).

MATERIALS AND METHODS

Medium: Human gastric juice
Volume: 500 ml
Incubation temperature: $37^{\circ}C$
Time of incubation: 60 min, anaerobic conditions
Amount of drug used: Maximum therapeutic dose
Amount of nitrate; 320 u Mol sodium nitrate
pH-value: Gradual diminution from pH 7 to pH 2

RESULTS

Inhibitory action of ascorbic acid was verified under simulated conditions of the human stomach for drugs undergoing nitrosation at different rates which contain the following pharmaceutical substances: Aminophenazone, ampicillin, noramidopyrinemethane sulfonate, clomipramine, oxacillin, desipramine, piperazine, imipramine, ethambutol resp. phenoxymethyl-penicillin.

The inhibition rates of nitrosation by ascorbic acid for the ten drugs mentioned are shown in Table 2. An example of these investigations: Nitrosation of piperazine with and without ascorbic acid is demonstrated in Figure 2.

DISCUSSION ON CONCLUSIONS

The presence of chemicals causing inhibitions of carcinogenic processes points to the possibility to prevent cancer in the future not only by elimination of carcinogens or other carcinogenic risks. It must also be emphasized that most work in this area is still on a basic and exploratory level and not ready for regulatory action. Many of the anticarcinogenic substances only act in combination with specific carcinogens or in distinct species, specific organs and tissues or situations and different effects under other circumstances cannot be ruled out.

Table 2. Inhibition of nitrosation by ascorbic acid under simulated gastric condition.

Name of the drug in the GDR (3) / International nonproprietary name for pharmaceutical substance	Minimum ascorbic acid necessary for maximal inhibition of nitrosation per tabloid, dragee or capsule	Final yield of of N-nitroso compounds in % 1.)		Rate of inhibition in % 2.)
		without inhibitor	with ascorbic acid	
Aminophenazon tabloids (0.3 g) / aminophenazone	100 mg	87.5	7.0	92.0
Ampicillin capsules (250 mg /ampicillin	100 mg	5.5	4.5	18.2
Analgin tabloids /noramidopyrinemethanesulfonate sodium	500 mg	100.0	17.5	82.5
Hydiphen dragees /clomipramine	25 mg	< 1.0	n.d.	-
Oxacillin capsules	100 mg	49,0	3.0.	93.9
Petylyl dragees /desipramine	25 mg	< 1.0	n.d.	-
Piavermit tabloids /piperazine	100 mg	100	5.5	94.5
Pryleugan dragees /imipramine	25 mg	< 1	n.d.	-
Syntomen dragees /ethambutol	40 mg	42.0	1.0	97.6
V-Tablopen (400000 IE) /phenoxymethylpenicillin	25 mg	54.9	1.0	80.0

1.) = calculated on the basis of initial nitrate concentration

2.) = calculated on the basis of yield without inhibitor

n.d. = not detectable

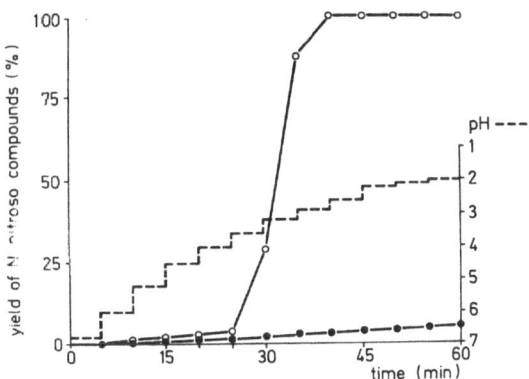

Figure 2.   Nitrosation of piperazine (Piavermit tabloids) under simulated
gastric conditions with ●-● and without o-o ascorbic acid.   Yields without
ascorbic acid 100% with ascorbic acid 5.5% (calculate on the basis of initial
nitrite concentration).
Conditions of the reaction: Gastric juice 500ml, piperazine adipate 4800 mg
(=16 Piavermit tabloids, advised maximal dose), ascorbic acid 1600 mg,
incubation temperature 37oC, anaerobic conditions, gradual diminution of the
pH from 6.8 to 2.0 amount of nitrate 320 u Mol (corresponding to a rather
high nitrite-containing meal).

    The results obtained verify the possibility to reduce drastically the
amount of nitrosation of drugs by means of inhibitory agents such as ascorbic
acid.   Due to the large number of theoretically nitrosatable drugs we should
adopt for prevention of this potential iatrogenic cancer risk a few simple
measures.

- To use alternative non-nitrosatable drugs whenever possible

- To avoid, as far as possible, foods rich in nitrites and nitrates although
   nitrate can be formed endogeneously.

- Not to take the drug close to meals whenever possible.

- To ingest the drug together with ascorbic acid preferably by combining
   nitrosatable drugs together with sufficient amounts of ascorbic acid (This
   should be regarded as the best way of inhibition of nitrosation because of
   its pronounced activity over a wide pH range and its lack of toxic
   effects).

    These approaches should be of value particulary because of observations
with regard to endogenous nitrosation by co-operation of macrophages (10) and
bacteria (5).

REFERENCES

1 Bertram J.S., Kolonel, L.N. and Meyskens, F.L (1987) Cancer Res., 47,
   3012-3031
2 Brambilla, G. (1985) Pharmocol. Res Commun., 17, 307-321
3 Gerecke, K. (Edit.) (1987) Arzneitmittelverzeichnis der DDR, VEB Verlag
   Volk und Gesundheit, Berlin
4 IARC Monographs on the evaluation of the carcinogenic risk of chemicals
   to humans - Some pharmacuetical drugs, Vol. 24 (1980) WHO/IARC, 297-314
   Lyon

5 Leach, S.A., Thompson, M. and Hill, M. (1987) Carcinogenesis 8, 1907-1912
6 Lijinsky, W., Conrad, E. and Van de Bogart, R. (1972) Nature, 239, 165-167
7 Lijinsky, W., Conrad, E. and Van de Bogart, R. (1972) formation of carcinogenic nitrosamines by interaction of drugs with nitrite. in: N-nitroso compounds: Analysis and Formation. (Bogovski, P., Preussmann, R., Walker, E.A. eds) IARC Sci. Publ. No. 3. 130-133, IARC, Lyon
8 Mirvish, S.S., Wallcave, L., Eagen, M. and Shubik, P. (1972) Science, 177, 65-68
9 Mirvish, S.S. (1975) Toxicol. appl. Pharmacol. 31, 325-351
10 Miwas, M., Stuehr, D.R., Marletta, M.A., Wishnok, J.S. and Tannenbaum, St. R. (1987) Carcinogenesis, 8, 955-958
11 O'Conner, H.J., Habibzedah, N., Schorah, C.J., Axon, A.T.R., Riley, S.E. and Garner, R.C. (1985) Carcinogenesis,6, 1675-1676
12 Oshima, H. and Bartsch, H. (1981) Cancer Research, 41, 3658-3662
13 Oshima, H., Bereziat, J.C. and Bartsch, H. (1982) Carcinogenesis, 3, 115-120
14 Wattenberg, L.W. (1985) Cancer Res., 45, 1-8
15 Ziebarth, D. (1982) Archiv Geschwulstforsch., 52, 429-442

INFLUENCE OF DIETARY CONSTITUENTS IN CHEMICAL CARCINOGENESIS IN RATS

M. Beth, M. R. Berger and D. Schmahl

German Cancer Research Center, Institute of Toxicology and
Chemotherapy, Im Neuenheimer Feld 280, 6900 Heidleberg FRG

The aim of this study was to separate the effects of calorie intake on
chemically methylnitrousourea (MNU) induced mammary carcinogenesis from those
of fat content and fat composition in female Sprague-Dawley rats. Furthermore we wanted to elucidate the influence of dietary vitamin A and E supplementation on carcinogenesis and correlate this with the fat content of the
respective diet.

The principal observations of this experiment are the following:

1) Decreasing the calorie level by 30% significantly inhibited tumor
development of MNU-treated rats as measured by all parameters of carcinogenesis. This influence was independent of the level and composition of fat.

2) The fat content of semisynthetic diets, although varying by 44.4%, did
not significantly influence mammary tumorigenesis; in fact, carcinogenic
expression was discontinously related to fat level. A plateau of tumor
incidence was observed at the level of 35 energy percentage of fat.

3) Fat composition did not influence tumorigenesis, even when the ratio of
poly-unsaturated fatty acids (linoleic acids) was doubled and that of
saturated fatty acids was lowered by one third.

4) Vitamin A and E supplementation showed no significant chemopreventive
effect against chemically induced mammary tumor development. This result was
independent of ingested fat level of the respective diet.

The role of caloric restriction is thus stressed in relation to possible
dietary prevention of cancer.

Breast neoplasia is still the most frequent lethal cancer in women. Thus
in the last few years there has been a growing interest in the concept of
prevention of mammary carcinogenesis by modifying the diet or by adding
micronutrients like hormones.

In a number of studies attempts have been made to identify dietary intake
of fat as an environmental factor related to the development of mammary
cancer. However, the title of a publication by Willet in 1986 summarizes the
absence of conclusive evidence from epidemiological and experimental studies
with the following provocation:

N – NITROSO – N – METHYLUREA

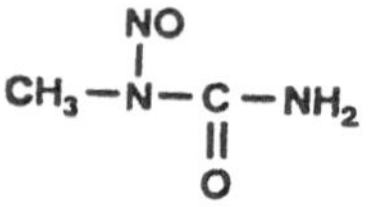

$$CH_3-\underset{\underset{\displaystyle}{|}}{\overset{\overset{\displaystyle NO}{|}}{N}}-\underset{\underset{\displaystyle O}{||}}{C}-NH_2$$

Figure 1

"Diet and Cancer: A whirlwind odyssey through a sea of inconsistency" (1).

Therefore the aim of our study (2) was to provide conclusive answers to the following questions:

1) Does the fat content of diet exert a specific effect on chemically induced mammary carcinogenesis, independently of its calorie content?

2) Has the dietary fatty acid composition an influence on mammary tumor development?

3) What is the dietary calorie content in MNU induced mammary tumorigenesis?

N-Nitroso-N-Methylurea (Fig 1) is a chemical carcinogen able to induce carcinomas that are exclusively localized in the mammary gland in different strains of rats (3), when administered at 25mg/kg bodyweight intravenously at day 50 of life.

The autochthonous MNU-induced mammary tumors are hormone-responsive, and their growth kinetics, histology, and chemotherapeutic sensitivity mimic the human counterpart (4). Because of the close resemblance with human mammary cancer the MNU-induced mammary tumor model was suitable to investigate the putative modulation or inhibition of chemical carcinogenesis by physiological factors such as diet.

To clarify these questions experiments based on the following separate dietary approaches were undertaken:

i) Animals received isocaloric diets with increasing levels of fat as in Table 1. The fat content of the diets was 25% of total calories in the ND25, which means normal diet with 25 energy % fat content. The fat content was 35% in the ND35 and 45en% in the ND45. The highest fat level was chosen to correspond to the fat content of the normal Western German diet, which is 45%. For this reason the abbreviation ND for normal diet is used.

The fatty acid composition reflected the average intake pattern (Table 2). The proportion of saturated fatty acids and mono- and polyunsaturated fatty acids was 46%, 38% and 16%, respectively, in the normal diets, the ND's.

ii) In addition the composition of fat was varied at the level of 35 en% of fat, according to recommendations aimed at the prevention of cardiovascular diseases, so called recommended diet, or RD (Table 1). The content of poly-unsaturated fatty acids was doubled whereas the amount of saturated fatty acids was decreased by one third. So the proportion of saturated fatty acids and mono- and polyunsaturated fatty acids was 33%, 34% and 32%, respectively, in the RD, the recommended diet (Table 2).

iii) These diets were fed at either at 50 kcals or at 35 kcals per rat per

190

Table 1. Composition of Diets

Ingredients	Designation of diets			
	ND 25	RD 35	ND 35	ND 45
Composition: g/100 g				
Casein	23.9	25.7	25.7	27.8
Corn starch	57.4	49.8	49.8	41.0
Palm Oil	7.8	6.24	11.7	16.28
Lard	1.46	3.12	2.18	3.04
Sunflower seed oil	1.14	6.24	1.72	2.39
Cellulose	5.8	6.2	6.2	6.7
Vitamins	0.39	0.41	0.41	0.45
Minerals	2.08	2.24	2.24	2.42
Protein (en %)[1]	24.0	24.0	24.0	24.0
Carbohydrate (en %)	51.0	41.0	41.0	31.0
Fat (en %)	25.0	35.0	35.0	45.0
Fiber (mg/kcal)	15.0	15.0	15.0	15.0
Vitamins (mg/kcal)	1.0	1.0	1.0	1.0
Minerals (mg/kcal)	5.4	5.4	5.4	5.4

[1] Percentage of total energy

Table 2. Profile of Fatty Acids in Semi-Synthetic Diets

Fatty Acid	Diets			
	ND25	ND35	ND45	RD35
C 14:0	1.1	1.2	1.1	0.7
C 16:0	39.1	37.6	39.8	25.8
C 16:1	-	-	-	0.9
C 17:0	0.1	0.2	0.1	0.1
C 17:1	-	-	-	0.1
C 18:0	5.4	5.2	5.4	6.1
C 18:1	37.9	36.6	37.5	33.0
C 18:2	15.5	15.6	15.3	31.5
C 18:3	0.2	0.3	0.2	0.3
C 20.0	0.3	0.3	0.3	0.3
C 20:1	0.3	0.3	0.2	0.3
C 20:2	-	-	-	0.1
C 20:3	-	0.1	-	-
C 20:4	-	0.8	-	-
C 22:0	0.1	0.3	0.1	0.4
C 22:1	-	0.2	-	-
C 24:1	-	1.2	-	-
Saturated fatty acids	46.1	44.8	46.8	33.4
Mono-unsaturated fatty acids	38.2	38.3	37.7	34.3
Poly-unsaturated fatty acids	15.7	16.8	15.5	31.9
Total	100.0	99.9	100.0	99.6
Fat content (weight %)	10.5	15.8	21.8	15.8

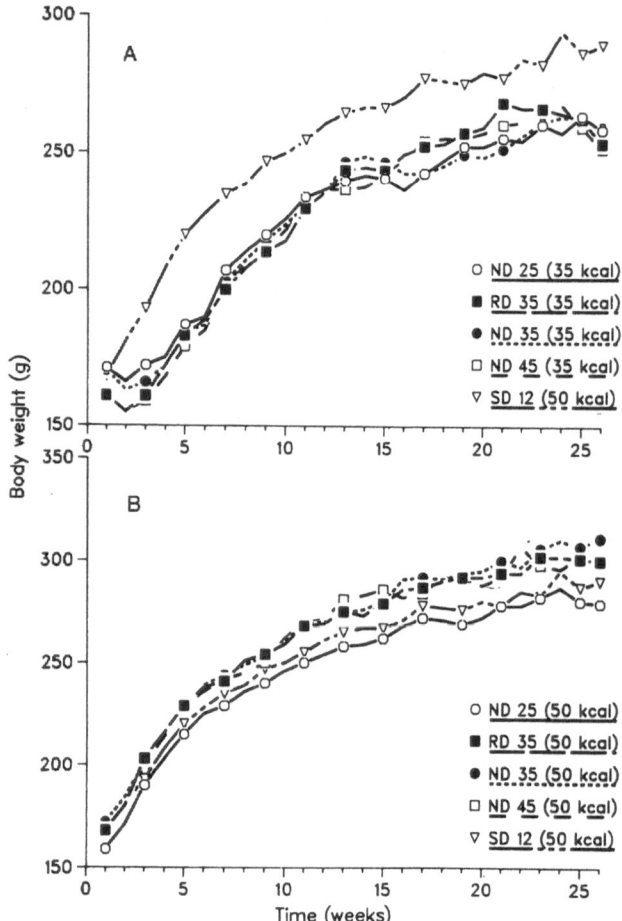

Figure 2. Mean body-weight curves of groups receiving different fatty diets
at either 35 kcal/day (a) or 50 kcal/day (b) in comparison to standard
laboratory chow (50 kcal/day).

day to offer either an <u>ad libitum</u> level of calories or at a caloric intake
reduced by 30%.

Thus, this experiment compared the effects of diets containing various
levels of fat versus those of 2 seperate levels of calories and different
compositions of fat.

In a further part of the experiment, which will be discussed later on, we
intended to elucidate the influence of dietary vitamin A and E supplementation on chemically induced mammary carcinogenesis in correlation to the
fat content of the respective diet.

To induce mammary carcinomas 270 female Sprague Dawley rats were injected
with 25 mg of MNU/kg intraveneously, on day 50 of life. One day after tumor
initiation, the rats were randomly divided into nine experimental groups to
receive on the described diets.

Figure 2 shows the influence of calories and fat level of diets on mean

Table 3: Influence of Different Fatty Diets on Tumor Manifestation Time in Female SD Rats Induced with Methylnitrosourea

Diet	Mean tumor manifestation time in affected rats[1]		Overall tumor manifestation time [1,2]	
	35 kcal	50 kcal	35 kcal	50 kcal
SD 12		14.2 + 5.1		20.9 + 3.02
ND 25	16.3 + 5.2	13.5 + 5.1	24.3 + 3.41	17.6 + 2.20
RD 35	17.8 + 4.8	15.4 + 5.3	22.6 + 2.01	17.8 + 1.68
ND 35	16.2 + 5.5	15.4 + 4.3	20.6 + 2.18	16.8 + 1.24
ND 45 [3,4]	17.2 + 6.5	12.5 + 3.5	20.5 + 2.15	14.4 + 1.27

[1]Weeks + standard deviation. [2]Maximum likelihood estimates, assuming a (right-censored) log normal distribution of tumor manifestation times.
[3]Significant difference between tumor manifestation times in rats fed with 35 or 50 kcal of diet ND 45 according to Kaplan Meier estimate (p=0.01).
[4]Significant difference between tumor manifestation times in rats fed with 50 kcal diet SD 12 or ND 45 according to the Kaplan Meier estimate (p=0.04).

body weight gain of MNU-treated female SD rats.

In the upper part of the picture it can be seen that all groups receiving their diets at a calorie level of 35 kcals/day exhibited very similar mean body weights during the whole experimental period, the maximum mean values not exceeding 270 grams. The slight increase in variation following week 13 is related to the occurrence of tumors.

As can be seen also, the dietary fat level did not influence body weight. Thus the question of a higher net energy intake resulting from fat is not positively supported by our results. In comparison with the lower part of the picture we see, that all groups receiving 35 kcals per day showed a distinctly lower weight gain than groups receiving their diets with 50 kcals, which proved to be highly significant.

The mean time of tumor manifestation of affected rats, presented in the left part of Table 3, was not related to the fat content of diets. Only marginal differences without any trend were seen at both caloric levels between the fatty diets, which was the same with the overall tumor manifestation time. On the other hand, when comparing the average time of tumor onset between high- and low-calorie fed groups, the latter were found to present a significantly later manifestation of tumors than the former.

In Figure 3 the incidence of tumor bearing rats after feeding the different fatty acids with the two different calorie-contents is expressed through the higher, striped columns, whereas mortality is presented by the smaller ones below.

The incidence of mammary tumors increased with rising dietary fat content up to the level of 35% of fat. The increase of 13% in the low-calorie group, and of 10% in the high-calorie group, however, was not significant with no further increase, or even a slight decrease in the low-calorie group, after feeding of 45 en% of fat.

Changes in the composition of fatty acids within the RD, the recommended diet, at the level of 35 en% of fat, did not influence tumor incidence.

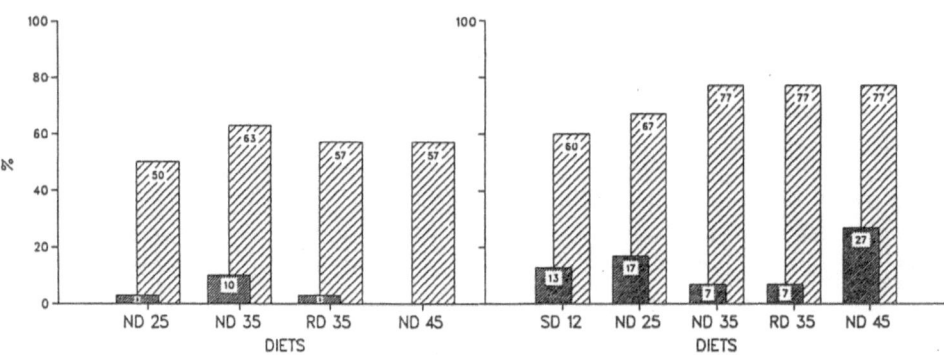

Figure 3. Influence of fat contents of different diets (abscissa),
administered either at a caloric level of 35 kcal/day and rat (left side)
on 50 kcal/day and rat (right side) on % tumor incidence (▨) and %
mortality (■) of MNU induced female SD-rats.

With respect to the amount of calories, however, the difference in tumor
incidence between groups receiving low- or high-calorie intake was
significantly greater, with 17% in the low-fat group, 14% in the ND35, and
with 20% difference between low and high caloric intake in the high-fat group
ND45.

Comparing the mortality following both calorie levels, as significantly
lower mortality was observed in rats fed the caloriclly restricted diets
(Figure 3). Mortality was not related, however, to the fat concentration of
diets.

Figure 4 shows the total number of tumors in the different dietary groups
with increasing fat level. The striped columns show the high-calorie groups,
the dotted areas the respective low-calorie groups.

As can be seen, the total number of tumors per group was less affected by
the fat level of diets than by the amount of calories, which is particularly
visible in period III, after termination of the six-months study.

Whereas the difference in fat content between the ND25 and ND45 in the
low- and in the high-calorie group, resulted in a number of tumors only
increased by 8, the number of tumors was enhanced by 16 in the low- and 16 in
the high-fat group, after feeding the respective diets with high-calories,
which was a significant influence of the calorie content of diets.

Comparisons at periods I, II, and III ie after 2,4 and 6 months,
respectively of tumor numbers per rat (Table 4) from all low-calorie groups
versus the respective high-calorie groups revealed significant differences,
whereas no significant differences in tumor number of rats were effected by
different fat levels of semisynthetic diets.

In Figure 5 the increase in mean tumor volumes of 4 groups fed the
various fatty diets in comparison to the standard diet is shown.

In all semisynthetic diets administration of 50 kcals/day produced a
significantly greater tumor volume than the feeding of 35 kcals. The

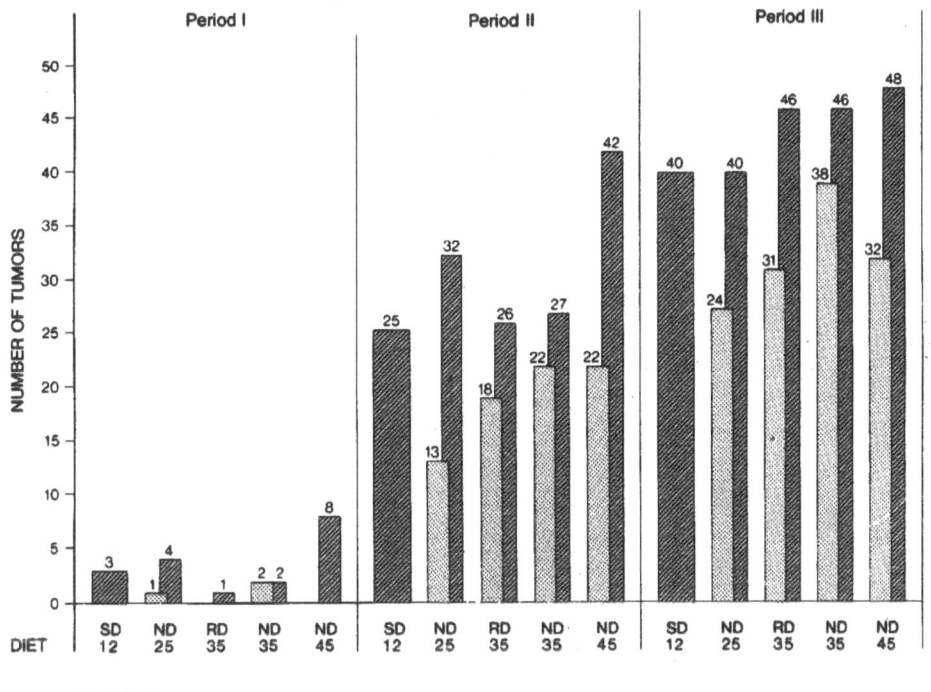

Figure 4. Cumulative tumor number of female SD rats in relation to time and calories following administration of 5 different fatty acids. The tumor number per group (ordinate) is given in relation to the respective diet at periods I, II and III (abscissa). ▢ , lower calorie level (35 kcal/day); ▨ , higher calorie level (50 kcal/day).

Table 4. Number of tumors per tumor-bearing rat in relation to time and calories following administration of five different fatty diets.

Diet	Period I		Period II		Period III	
	35 kcal^2	50 kcal^2	35 kcal^3	50 kcal^3	35 kcal^4	50 kcal^4
SD 12		1.0 ± 0		1.9 ± 1.8		2.2 ± 1.7
ND 25	1.0 ± 0^1	1.1 ± 0.4	1.3 ± 0.5	2.1 ± 1.1	1.6 ± 0.96	2.0 ± 0.9
RD 35	0	1.0 ± 0	2.0 ± 1.3	1.7 ± 0.5	1.8 ± 1.2	2.0 ± 1.5
ND 35	1.0 ± 0	1.5 ± 0.7	1.6 ± 0.6	1.7 ± 1.2	2.0 ± 0.9	2.0 ± 1.5
ND45	0	1.4 ± 0.7	1.6 ± 0.7	1.9 ± 0.9	1.9 ± 1.1	2.1 ± 1.2

[1] ± Standard deviation. [2] p = 0.012; all low calorie groups versus the respective high-calorie groups of semi-synthetic diets, according to Dunn. [3] p = 0.002; all low calorie groups versus the respective high-calorie groups of semi-synthetic diets, according to Dunn. [4] p = 0.008; all low calorie groups versus the repsective high-calorie groups according to Dunn.

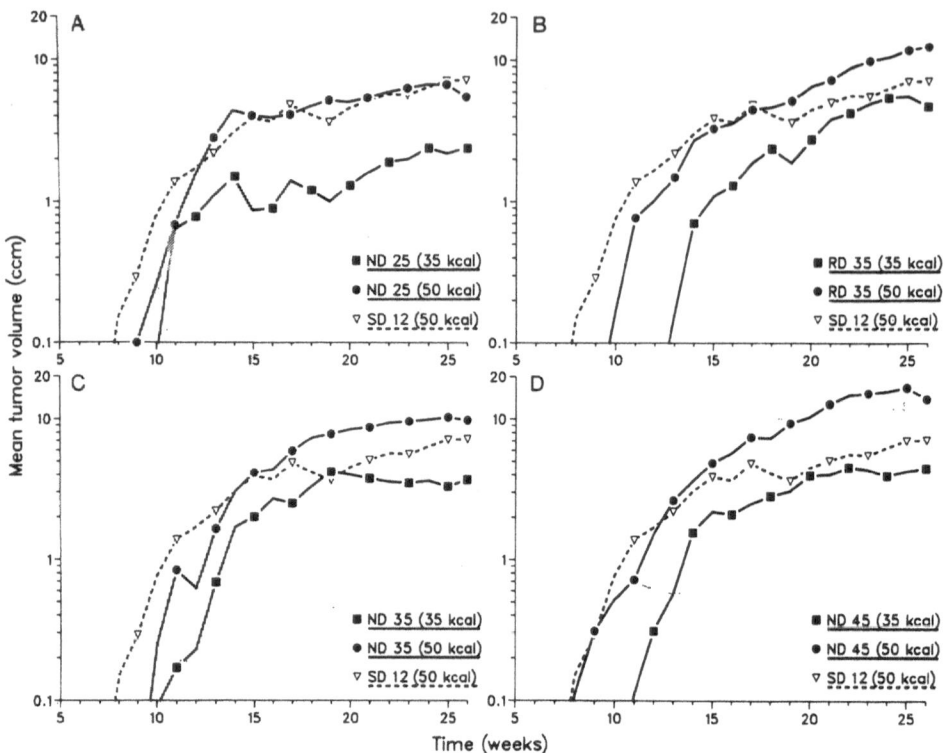

Figure 5. Tumor growth development of methylnitroso-urea-induced mammary carcinomas in SD rats: comparison of the influence of fatty diets ND 25 (a), RD 35 (b), ND 35 (c), and ND 45 (d) at 2 caloric levels, respectively, to standard laboratory chow (SD 12).

difference was still significant, when the tumor sizes were expressed as percentages of body weight.

There was no significant influence of dietary fat level on tumor growth development.

The same approach as already described was used for the assessment of the influence of dietary vitamins.

As found in a former experiment in our institute (5) both toxicity, as indicated by bone fractures, loss of weight, and peliosis cutis, and anti-tumor efficacy were found, when the high dose of 100 000 IU per 1000 kcals of vitamin A palmitate ester was added to a diet containing the very low fat level of 12 en% as fat. However, in a high-fat group with 45 en% as fat neither toxicity nor tumor-inhinitory activity following the same vitamin amounts were observed.

To elucidate this discrepancy we examined the influence of two levels of dietary fat in iso-caloric diets supplemented with normal or tenfold higher amounts of vitamin A and E on MNU-induced rat mammary carcinogenesis (6).

Vitamin E acts as biological antioxidant, inhibiting lipid peroxidation, and thus protects cell membranes and possibly nucleic acids against oxidative damage.

As shown in Table 5 the lower fat level was increased to 25 en% of fat, since the very low fat level of only 12% is not encountered in human diets. Accordingly the level of vitamin A was halved to a non toxic level of 50.000 IU per 1000 kcals.

Specifically we want to clarify the following questions:

1 Can dietary vitamin A and E supplementation at a non-toxic level suppress or block chemically induced mammary tumor development?

2 How does the influence of these fat-soluble vitamins on tumorigenesis correlate with the dietary fat level?

120 Sprague Dawley rats were initiated with 25mg of MNU/kg intravenously to induce mammary carcinomas. One day after tumor intiation the animals received one of the diets described before, offered at a daily caloric supply of 50 kcal per rat, five times a week.

The effect of the supplemented vitamins A and E on mean body weight gain of the chemical induced rats is shown in Figure 6.

Neither the dietary fat level nor the vitamin supplementation was found to have a significant effect on body weight gain, which confirms that vitamin A was given below the toxic level.

Regarding the mean time of tumor manifestation the vitamen A and E supplemented dietary groups showed a slight tendency of having a longer tumor free time, which, however, was not significant (Table V1).

Both, tumor incidence, which is expressed through the high columns of Figure 7, and mortality, represented by the lower ones, were not significantly influenced by the fat content of diets.

The effect of dietary vitamin A and E supplementation on the chemical induced mammary tumor incidence was non-homogeneous and not significant:

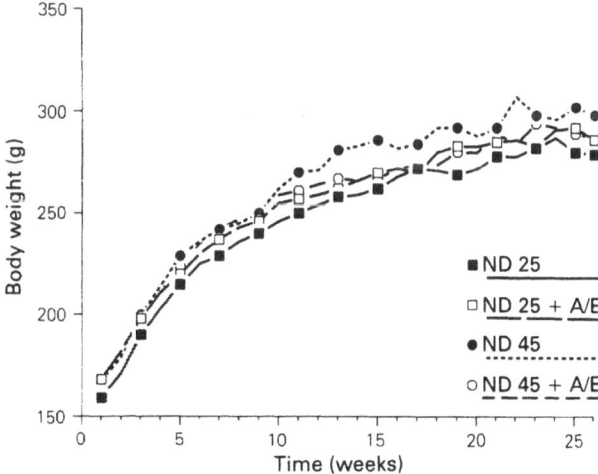

Figure 6. Effects of supplementation of vitamins A and E and different amounts of fat on body weight development in rats induced with methylnitrosourea.

Table 5. Composition of Diets

Experimental group	g/100g Diet						en %[a]					mg kcal^{-1}
	Proteins	Carbo-hydrates	Fat	Cellulose	Minerals	Vitamins	Protein	Carbo-hydrates	Fat	Cellulose	Minerals	Vitamins[b]
ND25	23.9	57.2	10.4	5.8	2.3	0.4	24.0	51.0	25	15.0	5.4	1.0
ND25+A/E	23.8	57.2	10.3	5.8	2.3	0.6	24.0	51.0	25	15.0	5.4	1.7
ND45	27.7	40.9	21.6	6.7	2.6	0.4	24.0	31.0	45	15.0	5.4	1.0
ND45+A/E	27.6	40.8	21.6	6.7	2.6	0.7	24.0	31.0	45	15.0	5.4	1.7

[a]Percent of total energy; [b]Contains 50.00 IU vitamin E per 1,000 kcal for the diets ND25+A/E and ND45+A/E; 2,500 IU vitamin A and 20 IU vitamin E for the diets ND25 and ND45.

Table 6: Influence of two different fat levels of diets supplemented with vitamins A and E on manifestation time and tumor number of female SD-rats induced with methylnitrosourea.

Diet	Mean tumor manifestation time of affected rats [a]	Overall manifestation time [a,b]	Number of tumors per tumor bearing rat	Total number of tumors per group.
ND25	13.5 ± 5.1	17.6 ± 2.2	2.1 ± 0.9[c]	43
ND25 + A/E	14.3 ± 4.6	17.7 ± 1.8	2.0 ± 0.9	41
ND45	12.5 ± 3.5	14.4 ± 1.3	2.1 ± 1.2	52
ND45 + A/E	13.4 ± 3.9	17.5 ± 2.0	3.0 ± 1.6	63

[a]Weeks ± s.d.; [b]Maximun likelihood estimates, assuming a (right-censored) log normal distribution of tumor manifestation times; [c] ± s.d.

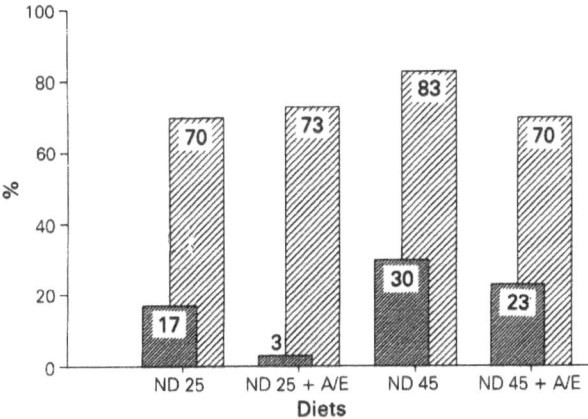

Figure 7. Influence of vitamins A and E supplemented to diets containing two different fat levels on % tumor incidence (▨) and % mortality (▧) of methylnitrosourea induced female SD-rats.

In comparison to ND25 the ND25 + A/E-group showed a slight increase in the number of tumor bearing rats whereas in the groups receiving the diet with 45 en% of at the vitamen supplementation decreased tumor incidence.

The diminishing effect of vitamin supplementation on mortality data was uniform but not significant in both groups fed with the different fat levels (Fig. 7). The mortality was significantly lower in the group.fed with the ND25 + A/E-diet compared to ND45 + A/E-group.as related to 5, 9 and 7 dead animals in the ND25, ND45, and ND45 + A/E-groups, respectively, this significance might be interpreted as a runaway-value rather than being a treatment effect.

In Table 6 it is visible that vitamin A and E supplementation effected a somewhat reduced tumor number in the low fat group and enhanced the number of tumors per rat as well as the total number of tumors in the groups fed the high fat diet which, however, was not significant.

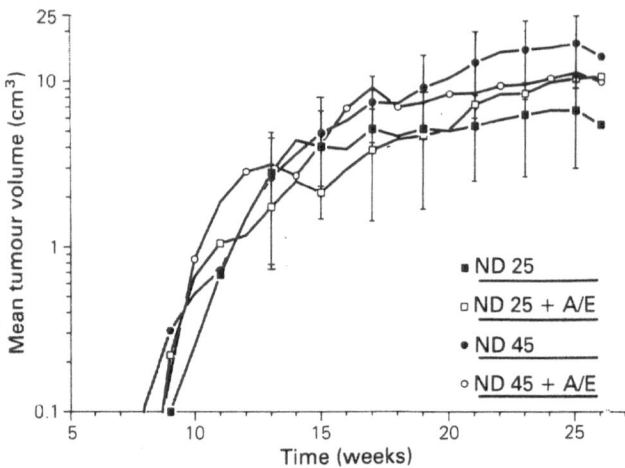

Figure 8. Tumor growth development of methylnitrosourea induced mammary
carcinomas in SD-rats; comparison of the influence of two different fat
levels with or without vitamen A and E supplementation. Indicated are the
mean tumor volumes and selected 95% confidence limits of diets ND25 and ND45.

    Concerning the tumor growth development (Fig. 8) the rats fed with 45 en%
of fat in their diets produced a little greater volume than the group fed
with 25 en% of fat, which was not significant.

    Vitamin administration resulted in a higher mean tumor volume when
supplemented to the low fat diet and conversely caused a smaller tumor volume
when given with the high fat diet, which was insignificant, as well.

    Summarizing the result of our experiments it can be stated, that the 4
principal observations are the following:

1 Decreasing the calorie level by 30% significantly inhibited chemically
with MNU-induced mammary tumor development in female SD-rats as evalurated by
all parameters of tumorigenesis. This influence was independent of the level
of fat.

2 The fat content of semisynthetic diets, although varying vy 44.4%, did not
significantly influence mammary tumorigenesis; in fact, carcinogenic
expression was discontinuously related to the fat contents of diets. Above a
dietary fat level of 35% of energy there was no further increase of tumor
incidence.

3 Fat composition did not influence tumorigenesis, even when the ratio of
poly-unsaturated fatty acids (linoleic acids) was doubled and that of
saturated fatty acids was lowered by one third.

4 Vitamin A and E supplementation showed no significant chemopreventive
effect against chemically induced mammary tumor development. This result was
independent of the supplied fat level of the respective diet.

    The role of caloric restriction is this stressed in relation to possible
dietary prevention of mammary cancer (7).

REFERENCES

1 Willett, W.C. (1986) Diet and Cancer: A whirlwind odyssey through a sea of inconsistency. Biblthca.Nutr.Dieta.,37, 121-129, Karger, Basel
2 Beth, M., Berger, M.R., Aksoy, M., und Schmahl, D., (1987) Int.J.Cancer 39, 737-744
3 Gullino, P M., Pettigrew, H. M., und Grantham, F.H., (1975) J. Natl. Cancer Inst., 54(2), 401-414
4 Berger M.R., und Zeller W.J., (1984) Wege zur rationalen praklinschen Testung antieoplastischer Chemotherapeutika. Beitr.Onkol.,18,274-286 Karger,Basel.
5 Aksoy, M., Berger, M.R., und Schmahl, D.,(1985) Arch. Geschwulstforsch. 55, 443-447
6 Beth, M.,Berger, M.R. Aksoy, M., und Schmahl, D.,(1987) Br.J.Cancer, 56, 445-449
7 Beth, M., Berger, M.R., und Schmahl, D., (1987) Influence of some dietary constituents in chemical carconogens in rats. In: Combination Effects. Schmahl, D., VCH Verlagsgesellschaft. In press.

LIVER GLUTAMATE AND GLUTAMINE CONCENTRATION IN HEPATOCARCINOGENESIS

EXPERIMENTAL: EFFECT OF GLUTATHIONE

C. Fernandez-Aguado, M. Minguez, Mª J. Miro,
C.J. Martinez-Honduvilla, and B. Feijoo

Dpto. Biogquimica y Biologia Molecular, Facultad de Farmacia
versidad Complutense, Madrid, Espana

SUMMARY

    Among the metabolic alterations involved in thioacetamide (TAA) induced
hepatocarcinogenesis, effects on ammonia metabolism have been demonstrated.
This study was undertaken to investigate the response of liver glutamate and
glutamine concentration after glutathione administration.

    After 20 days of TAA administration, liver glutamate and glutamine
concentration were 140% and 111% respectively of the control value. 2-oxo-
glutarate decreased to 25%. Glutamate dehydrogenase (GDH) as measured by
glutamate formation and γ-glutamyl transferase (GGT) activities increased to
278% and 128%. Phosphate dependent glutaminase and glutamine synthetase
decreased to 85% and 43%. Glutathione administration to the TAA treated
animals showed effects on glutamate and glutamine concentration whose values
decreased close to the control; 2-oxoglutarate increased. GDH activity
decreased to values close to the control. No effect on glutaminase and
glutamine synthetase activities was observed. The present results suggest
that metabolic adaptation of liver in TAA animals treated for 20 days, lead
to the synthesis of glutamate through GDH activity. Glutathione administration partially stops the enhanced content of glutamate and glutamine.

INTRODUCTION

Thioacetamide (TAA) is a well known hepatotoxin which has been much studied
since the first report of its toxic properties (1). Most experiments have
been carried out on the induction of hepatic necrosis (2, 3), cirrhosis (4),
hyperplastic nodulos and tumors (5, 6), cholangiomas and cholangiocarcinomas
(7). Schweitzer and Schaetz in 1957, reported that TAA-induced cirrhosis
resembles human cirrhosis more closely than other experimental forms.
Following the concept based on the focal nature of the cellular response to a
carcinogen, and on the inhibitory nature of chemical carcinogens on cell proliferation, a hepatocarcinogen induces as a first effect, an alteration of
metabolic response, in which a few hepatocytes became resistant to the
cytotoxic action of carcinogens (8).

    We have observed in liver from rats treated with TAA a considerable
decrease in carbamoyl-P-synthetase, ornithine transcarbamylase and arginase
activities as a consequence of which urea concentration decreased. More-

over, TAA treatment produces a high ammonia concentration. From this, it was concluded that the urea cycle is partially inhibited in TAA-treated rats (9, 10).

Intercellular reduced glutathione (GSH) plays an important role in detoxication of different exogenous compounds (11, 12). Moreover, several factors indicate that GSH probably plays a role in chemical carcinogenesis in rat liver (13).

There is substantial experimental evidence indicating that glutamine and glutamate are the major sources of respiratory energy for tumor cells (14). Intact Ehrlich cells oxidize glutamine to CO_2 at higher rates than any other amino acid present (15).

The aim of this paper is to study the main enzymes activities involved in the release and uptake of those metabolites, as well as the glutathione effect.

Material and Methods

The TAA hepatotoxic condition was induced by intraperitoneal administration of TAA to male Wistar rats weighing 180-220 g. TAA treated animals were injected daily for 20 days with a freshly prepared solution of TAA in saline (NaC1 0.15 mol/1) at a dose level of 100 mg/Kg body weight. Controls received the same volume of saline. Another group of animals received TAA (100 mg/Kg body weight) and glutathione at a dose level of 100 mg/Kg body weight/day during the same period. Controls of this group of animals were injected daily with a solution of glutathione in saline. All experiments were performed in metabolism cages; the animals had free access to food and water. The animals were killed 24 hours after their corresponding last dose. From 10 to 16 rats were used for each experiment. A liver tissue sample from each animal was immediately frozen clamped and the others homogenized.

γ-glutamyl transferase (EC 2.3.2.2.) was assayed with γ-glutamyl-p-nitroanilide as substrate and glycylglycine as acceptor using the procedure of Szasz (16). Phosphate dependant glutaminase (EC 3.5.1.2.) was measured by the method described in reference 17. Mitochondrial glutamate dehydrogenase was measured as described by Strecker (18) and glutamine synthetase was determined by the procedure of Iqbal and Ottaway (19). Ammonia, glutamate, glutamine, 2-oxo-glutarate and alanine were determined in neutralized perchloric acid extracts of freeze clamped livers obtained as described by Lagunas et al. (20). Ammonia was determined spectrophotometrically with 2-oxoglutarate and glutamate dehydrogenase (glycerol suspension). Glutamate was determined as described by Pfleiderer (21) and glutamine was assayed by measurement of glutamate after enzymatic hydrolysis. 2-oxoglutarate was determined by Wallenfels and Christian (22). Alanine was determined with L-alanine dehydrogenase by the method of Williamson (23). Values are given as μmoles per Kg. of fresh liver.

Measurements were made on a Unicam SP-1800 spectrophotometer with a scale expansion unit. Results are expressed as means + S.E.M. Fisher's P values where P is greater than 0.05 are considered as not significant (N.S.).

RESULTS

As can be seen in Table 1, glutathione administration to control animals caused a decrease of 10% in liver glutamine and 2-oxoglutarate concentration, while alanine and ammonia concentration increased 10% and 20% respectively.

Table 1. Free Ammonia, Glutamate, Glutamine, 2-oxoglutarate and Alanine in rat liver after 20 days in Thioacetamide treated rats and Thioacetamide + glutathione treatment.

Values are expressed as μmol/kg of fresh liver. Results are given as means $\pm$ S.E.M.

* = p 0.05; ** = p 0.01; *** = p 0.001;

( ) = % of the control value

	CONTROL	GLUTATHIONE	TAA	TAA+GLUTATHIONE
Ammonia	220 ± 32 (100%)	264 ± 20** (120%)	315 ± 35*** (143.6%)	320.5 ± 20*** (145.7%)
Glutamate	2.118 ± 120 (100%)	2.0756 ± 110 (98.9%)	2.965 ± 140*** (140%)	2.583 ± 135*** (122%)
Glutamine	3.980 ± 170 (100%)	3.570 ± 150*** (89.7%)	4.417,8 ± 185*** (111.5%)	3.753,2 ± 110** (94%)
2-Oxoglutarate	120 ± 2.2 (100%)	108.6 ± 1.2** (90.5%)	30 ± 3*** (25%)	48 ± 3.2*** (40%)
Alanine	1.230 ± 60 (100%)	1.353 ± 20* (110%)	1.759 ± 45** (143%)	1.857,5 ± 34*** (151%)

TAA administration showed an increase in concentration of glutamate (40%), glutamine (11%), and both ammonia (43%) and alanine (43%), while 2-oxoglutarate concentration decreased to 25% of the control value.

Glutathione administration to thioacetamide treated animals caused a slight decrease with respect to the latter, glutamate (122% against 140%) and glutamine (94% against 111%) values also fell compared with controls. 2-Oxoglutarate increased to 40% against 25% of the TAA value, as well as alanine. Ammonia remained almost at the same level of the TAA animals (145%).

In Table 2, the enzyme activities of our study are showed. Glutathione administration had effects on glutamine synthetase and glutamate dehydrogenase activities, which increased by 28% and 16% respectively.

Thioacetamide administration caused a decrease in glutamine synthetase and glutaminase activities, whose values were 42% and 85% of the control. Glutamate dehydrogenase and γ-glutamyl transferase increased 178% and 28%, respectively.

Glutamine synthetase and glutaminase activities decreased to 36% and 68% of the control value, when glutathione was administered to TAA-treated animals. γ-glutamyl transferase increased to 37% of the control value. GDH, however, showed a decrease (61% of TAA).

DISCUSSION

It is well known that glutamine and glutamate are the major respiratory fuels of tumor cells. The results presented in this paper show that these metabolites are increased in liver of TAA-treated animals. In this condition, the marked decrease of 2-oxoglutarate concentration seem likely to be a consequence of the elevated glutamate dehydrogenase activity observed, since glutamate concentration is also increased. This glutamate formed from 2-oxoglutarate through glutamate dehydrogenase catalytic activity, after transamination with pyruvate, releasing alanine (which is increased) and 2-oxoglutarate, which has two fates. One of them is to supply this metabolite to continue the tricarboxylic acid cycle, which is inhibited by the elevated ammonia content (24) at the level of isocitrate dehydrogenase enzyme, and the other recycling glutamate by the glutamate dehydrogenase activity. One source of glutamate, could be glutathione released by the hydrolytic reaction of γ-glutamyltransferase activity, which is enhanced in TAA treated animals. An increase in plasma glutamate and alanine content has been observed when TAA was admimistered (25, 26). In thioacetamide animals the increase in glutamine content could be interpreted as a consequence of glutaminase activity which is decreased. The increase in glutamine concentration is accompanied by a decrease in glutamine synthetase activity.

However, glutathione administration to control animals induces an increase in liver glutamine synthetase activity and a decrease in glutamine content.

These effects together seem to indicate that glutamine synthetase activity is enhanced by decreased glutamine concentration, and the increase in glutamine concentration induces an inhibition of glutamine sythetase activity.

The increase in activity of this synthesizing enzyme, has been seen in Chinese hamster cells (27) and in neuroblastoma cells (28) when the cells are deprived of glutamine. The last one had suggested that synthetase activity induction was due to increased synthesis of new enzyme.

Table 2. Catalytic concentrations of Glutamine synthetase, Glutaminase, Glutamate dehydrogenase and -glutamyl transferase in rat liver in 20 days of Thioacetamide administration and Thioacetamide + Glutathione treatment.

	CONTROL	GLUTATHIONE	TAA	TAA+GLUTATHIONE
Glutamine synthetase	30.3 ± 4.2 (100%)	38.96 ± 3.3** (128.6%)	12.98 ± 5.2*** (42.86%)	10.81 ± 4.7*** (35.7%)
Glutaminase	4.100 ± 0.44 (100%)	4.095 ± 0.20 (95%)	4.087 ± 0.35** (85%)	4.072 ± 0.33*** (68%)
Glutamate dehydrogenase	26.8 ± 4.5 (100%)	29.6 ± 3.6** (116.3%)	57.4 ± 4.2*** (278%)	46.9 ± 3.5*** (217%)
α -GGT	130 ± 0.05 (100%)	136.5 ± 0.03 (100.8%)	339.5 ± 0.01*** (128.6%)	400.5 ± 0.02*** (136.7%)

Values are expressed as umol. min^{-1}. Kg^{-1}. A unit of enzyme is defined as the amount catalyzing the formation of 1 umol of product per min. at 37°C.

* = p<0.05; ** = p<0.01; *** = p<0.001

( ) = % of the control value

Following glutaminase treatment, glutamine synthetase and γ-glutamyl-transferase activities, while remaining low, increased in resistant tumors but not in sensitive tumors, suggesting that the increase may be related to the insensitivity of the resistant tumor toward glutaminase treatment (29).

When glutathione was administered to TAA animals a decrease in glutamate and glutamine concentration and an increase of 2-oxoglutarate can be observed. These variations are accompanied by a marked decrease in glutamate dehydrogenase activity.

The ammonia concentration could be interpreted as a result of the decreased activities of glutaminase and glutamine synthetase.

In this treatment, the increase in γ-glutamyl transferase can be explained by the availability of glutathione, and its function in detoxication of different exogenous compounds (30, 31). Interfering with the metabolic adaption of liver in TAA-treated animals, glutathione treatment reduces glutamate and glutamine concentration and consequently the cells no longer have available an excess of these metabolites as a primary source of energy.

REFERENCES

1 Fitzhugh, O. and Nelson, A. (1984) Science 108, 626-631.
2 Trennery, P. and Waring, R. (1983) Toxicol. Lett. 19, 299-307.
3 Chieli, E. and Maldaci, G. (1984) Toxicology 31, 41-52.
4 Pap, A. and Varro, V. (1981) Acta Med Sci Hung 38/4, 381-384.
5 Anghileri, L., Heidreder, M., Weiler, G., Dermietzel, R. (1977) Exp. Cell. Biol. 45, 34-47.
6 Frederick, F. and Becker, M. (1983) INCI 71 (3), 553-558.
7 Praet, M. and Roels, H. (1984) Exp Pathol 20, 3-14.
8 Solt, D., Medline, A. and Farber, E. (1977) Ann. J. Pathol. 88, 595-609.
9 Cascales, M., Feifoo, B., Cerdan, S., Cascales, C. and Santos-Ruiz A. (1979) J. Clin. Chem. Clin. Biochem 17, 129-132.
10 Feijoo, B., Toledo, C., Aylagas, H. and Cascales, M. (1984) Rev. Esp. Oncologia 31, 15-20.
11 Reed D. J. and Beatty, P. W. (1980) Rev. Biochem. Toxic. 2, 213-220.
12 Ketterer, B., Coles, B. and Meyer, D. J. (1983) Envir, Hlth Perspect 49-59.
13 Mitchell, J. R., Jollow, D. J., Potter, W. Z. (1973) J. Pharmacol. Exp. Ther, 187, 211-217.
14 Reitzer, L., Wice, B., Keunell, D. (1979). J. Biol. Chem. 254, 2669-2676.
15 Lazo, P. (1981). Eur. J. Biochem. 117, 19-25.
16 Szasz, G. (1969). Clin. Chem. 15, 124-136.
17 Watford, M., Smith, E. M. and Erbelding, E. J. (1984) Biochem. J. 224, 207-214.
18 Strecker, H. J. (1953) Arch. Biochem. Biophys. 46, 128-140.
19 Iqbal, K. and Ottaway, J. H. (1970) Biochem. J. 119, 145-156.
20 Lagunas, R., McLean, P. and Greenbaum, A. L. (1970) Eur. J. Biochem. 15, 179-190.
21 Pfleiderer, G. (1965), in Methods of Enzymatic Analysis (Bergmeyer, H. U. ed) pp. 394-397. Academic Press.
22 Wallenfels and Christian (1986), in Methods of Enzymatic Analysis (Bergermeyer, H. U. ed) Vol IV, 20-24. Academic Press.
23 Williamson, D. H. (1986) in Methods of Enzymatic Analysis (Bergermeyer H. U. ed) Vol IV, 341-344. Academic Press.
24 Katunuma, N., Okada, M., Nishii (1966) in Advance Enzyme Reg. 4, 317-335.
25 Albrecht, J. and Hilgier, W. (1986) Acta Neurol. Scand. 73, 498-501.
26 Trennery, P. N. and Waring, R. H. (1983) Toxicol. Letters, 19, 299-307.
27 Tiemeier, D. C. and Milman, G. (1972) J. Biol. Chem. 247, 5722-5727.

28 Lacoste, L., Chandhary, K. D. and Lapointe, J. (1982) J. Neurochem. 39, 78-75.
29 Karen, C. Rosenspire, Alan S. Gelbard, Arthur J. L. Cooper, Franz A. Schmid and Roberts J. (1985) Biochem. Biophys. Acta. 843, 37-48.
30 Bellomo, G., Mirabelli, F., DiMonte, D., Richelmi, P., Thor, H., Orrenius, C. and Orrenius, S. (1987). Biochemical Pharmacology 36, 1313-1320.
31 Lindwell, G. and Boyer, T. D. (1987) The Journal of Biological Chemistry 262, 5151-5158.

MODIFICATION OF NITROSAMINE CARCINOGENESIS BY ASCORBIC ACID

P. Bogovski, R. Birk, L. Kildema, and L. Teras

Institute of Experimental and Clinical Medicine
Tallinn
U.S.S.R.

ABSTRACT

The modifying action of ascorbic acid (AA) on endogenous formation of N-Nitroso compounds is well known. We have studied the action of AA on the mortality of rats due to hepatoma induced by N-nitrosodiethylamine (NDEA) and on the activity of some marker enzymes of hepatocarcinogenesis. AA (50 mg per animal intraperitoneally three times a week during three months) accelerated the NDEA induced hepatocarcinogenesis. In the NDEA group one rat died after 400 days, in the NDEA+AA group 11 rats ($p < 0.01$). At later stages the mortality difference of both groups was less distinct. The macroscopic grading of hepatomas in necropsied rats of the NDEA+AA group was signifi-cantly higher (grade 4.41) than in the NDEA only group (grade 3.09, $P < 0.02$). Administration of AA augmented the increase of activity of glucose-6-phosph-ate dehydrogenase and the decrease of the activity of glucose-6-phosphatase compared with activities in the NDEA group. The activity of hexokinase, pyruvate kinase and glucose-6-phosphate dehydrogenase in the liver increased in the NDEA+AA group much earlier than in the NDEA group. The accelerating action of high doses of AA on nitrosamine hepatocarcinogenesis has to be taken into account in cancer chemoprevention and treatment.

INTRODUCTION

It has been repeatedly stated that basic research becomes increasingly important in cancer control, especially in cancer prevention. Though an essential part of primary cancer prevention is the control of exposure to such risk factors as tobacco, nutritional aspects of cancer control are very important. Epidemiological investigations have established associations of a wide range of cancer sites with various nutritional factors. Among the latter ascorbic acid as an anticarcinogenic modifying factor has deserved special attention. It has been found that the intake of fresh fruits and vegetables which contain ascorbic acid is negatively correlated with the incidence of cancer of the stomach, oesphagus, larynx, mouth and cervix (1, 2). Ascorbic acid efficiently inhibits the endogenous formation of N-nitroso compounds (NNC), from nitrite, amines and amides. Most likely the decreased risk of stomach cancer in people eating fresh fruits and vegetables is due to an inhibition of in vivo formation of NNC, as most in vivo nitrosation occurs in the stomach, where hydrochloric acid catalyzes the reaction. About 80% of nitrite entering the stomach is probably produced in the saliva by bacterial

Biochemistry of Chemical Carcinogenesis
Edited by R. Colin Garner and Jan Hradec
Plenum Press, New York, 1990

reduction of salivary nitrate, which in turn arises mostly from dietary nitrate (2).

The problem of the inhibiting action of ascorbic acid on the formation of NNC requires however, further investigation as it has been shown that in some situations ascorbic acid can increase nitrosation. Nair et al.(3) demonstrated, that the presence of ascorbic acid inhibited the increased nitrosation of proline in the saliva of persons chewing betel quid with tobacco in 4 chewers out of 10 and in 5 out of 10 betel quid chewers. In the rest of the chewers ascorbic acid increased nitrosation of proline. It was shown also that higher doses (5% in the diet) of ascorbic acid promoted in rats urinary bladder carcinogenesis induced by butyl-4-hydroxybutylnitro-samine or methylnitroso urea (4).

In cancer patients vitamin C deficiency is not uncommon. Administration of ascorbate was postulated to increase host resistance against cancer by a variety of mechanism, including enhancement of lymphocyte functions and increase of the resistance of the intercellular ground substance (5). The use of high doses (up to 10 g of ascorbic acid by mouth daily) advocated by L. Pauling was however found to be not justified, especially by the investigators at the Mayo Clinic. Vitamin C was not better than a placebo with respect to either amelioration of symptoms or survival time (6).

Controversial conclusions have been drawn from experiments carried out to clarify the role of ascorbic acid on the action of preformed carcinogens (2). The data are insufficient and mostly not comparable as different protocols, carcinogens and animal species have been used.

Apparently the action of ascorbic acid on the successive stages of carcinogenesis should be investigated in more detail. The important steps of nitrosamine carcinogenesis are formation of the carcinogen from precursors, its metabolic activation, the reaction of the ultimate carcinogen with the cellular target (DNA or other macromolecules i.e. RNA or proteins) and detoxication of the carcinogen (formation of noncarcinogenic metabolites, binding with glutathione, glucuronic acid and other low-molecular compounds). These events occur during the stage of initiation, which either with or without the stage of promotion results in neoplastic transformation of one or several cells. The following progression of transformed cells into a tumor is based on heterogeneity of tumor cells, their natural selection and gradual increase of malignant characteristics, including anaplasia, infiltrating growth formation of metastases, acquisition of drug resistance etc. A special field of investigations should apparently be the study of interaction of ascorbic acid with various classes of cytostatic drugs.

It has recently been shown that vitamin C deficiency decreases the metabolic activation of N-nitrosodimethylamine (NDMA) and N-nitrosodiethylamine (NDEA) in guinea pig liver and delays the blood plasma clearance of both NNC and the covalent binding of ^{14}C from (^{14}C) NDMA and (^{14}C)NDEA to DNA in the liver. Supplementation of ascorbate had the opposite effect – the cytochrome P-450 mediated microsomal metabolic activation of carcinogens in the liver was increased (7). In mice ascorbic acid decreased the incidence of lung adenomas induced by intraperitoneal administration of NDMA but increased the number of adenomas when cyclic NNC mononitrosopiperazine and nitrosomorpholine were given in drinking water (8). These data indicate that the modifying action of ascorbic acid on carcinogenesis depends on the liver metabolism of carcinogens and the resulting distribution in various organs and changes of the blood plasma clearance. The decrease of the frequency of lung adenomas induced by NDMA in ascorbate supplemented animals can be explained by more intensive metabolism and binding of NDMA in the liver whereas hydroxylation of cyclic NNC in the liver produces alkylating radicals to a smaller extent.

212

No data were found in the literature on the modifying action of ascorbic acid on the hepatocarcinogenicity of nitrosodialkylamines in long terms studies.

We investigated in chronic experiments the action of ascorbate on the mortality of rats due to hepatomas induced by NDEA and on the activity of some marker enzymes of hepatocarcinogenesis.

MATERIALS AND METHODS

Experiments were carried out in male Wistar rats with an initial body weight of 160-200 g. In both experiments 360 rats were used. In the mortality experiment in the first group (55 rats) NDEA was administered in drinking water at a dose of 2.5 mg/kg body weight daily six days a week for four months. The second NDEA + ascorbic acid group (55 rats) received in addition 50 mg sodium ascorbate per animal as a 5% solution administered intraperitoneally three times a week during the first three months of the experiment. Dead animals were necropsied and the number of rats with fatal liver neoplasms established in both groups. The life span and intercurrent mortality were taken into account using the method of Peto et al. (9), the Chi-square method was used for statistical analysis.

The extent of the neoplastic process in the liver was evaluated macroscopically using the grading scale proposed by Gurkalo and Zabezhinski (10) modified by us.

0   - liver with no alterations
I   - uneven surface (colour), no nodules,
II   - single tumor nodules (up to 5) 0.1-0.2 cm in diameter,
III - multiple tumor nodules (more than 5) 0.1-0.2 cm or single nodules
       0.3-0.5 cm,
IV   - multiple nodules 0.3-0.5 cm or single nodules 0.6-1.0 cm,
V   - multiple nodules 0.6-1.0 cm or single nodules more than 1 cm in
       diameter,
VI   - multiple nodules more than 1 cm in diameter,
VII - tumors more than 3 cm in diameter, tumor tissue predominates,

Tumors of grades V-VII were considered as fatal. The validity of the method of macroscopic evaluation of liver neoplasma is supported by the correlation of the grades with activity of the marker enzyme of hepatocarcinogenesis gamma-glutamyl transpeptidase (GGT) as shown in Fig. 1.

In the biochemical experiment NDEA was administered in the same way for four or six months and ascorbic acid during the first, two or four months respectively. Biochemical examinations were carried out between the fourth and ninth month. Animals were sacrificed by decapitation. Livers were removed and washed in icecold 0.15 M KCl, then 1 g of liver tissue was homogenized in a 10-fold dilution of 0.05 M tris-HCl buffer, containing 0.15 M KCl and 0.001 M EDTA (pH 7.5). The homogenate was centrifuged at 17 000 g and the resulting supernatant was used for enzyme assays. We studied the activity of the following enzymes:

1. Enzymes of glycolysis
   Hexokinase (HK)
   Glucokinase (GK)
   Pyruvate kinase (PK)
2. Enzymes of glyconeogenesis
   Glucose-6-phosphatase (G-6-Pase)
   Fructose-1, 6-diphosphatase (F-1, 6-DPase)
3. Enzymes of the pentose phosphate pathway

Glucose-6-phosphate dehydrogenase (G-6-PDH)
6-phosphogluconate dehydrogenase (6-PGDH)
4. Gamma-glutamyl transpeptidase (GGT).

RESULTS

Mortality of rats due to hepatomas

Neoplasms were found at necropsy mainly in the liver. Histologically the
tumors were hepatomas and hepatocarinomas of different stages of progression.
The observed (0) and expected mortality (E) due to hepatomas is presented in
Fig. 2. The observed mortality during the first 400 days in the NDEA group
was significantly lower than in the NDEA+AA group, respectively one rat and

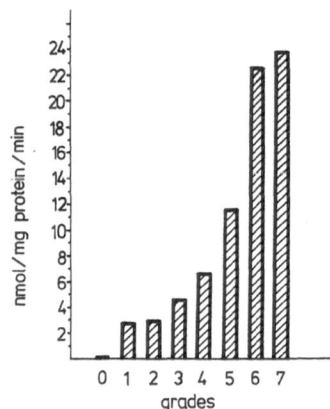

Figure 1. The activity of gamma-glutamyl transpeptidase depending on
neoplastic lesions (grades) in the liver.

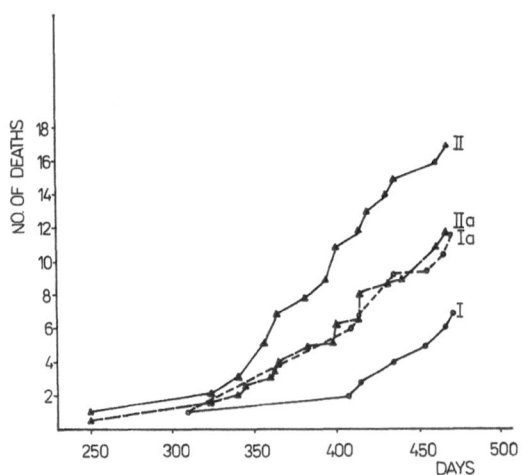

Figure 2. The mortality of rats due to hepatomas induced by NDEA. Observed
(I) and expected (Ia) mortality in NDEA group, observed (II) and expected
(IIa) mortality in NDEA+AA group.

Table 1. The macroscopic evaluation of hepatocarcinogenesis grades

Time of death (months)	NDEA		NDEA + AA		p
	N	M + m	N	M + m	
0 - 9	2	0	2	0	-
9 - 12	11	2.09 + 0.44	12	4.17 + 0.84	<0.05
12 - 14	11	3.45 + 0.28	11	4.81 + 0.54	<0.05
14 - 16	9	4.55 + 0.58	7	5.00 + 0.62	>0.06
Total	33	3.09 + 0.32	32	4.41 + 0.41	<0.02

Table 2. The activities of marker enzymes of hepatocarcinogenesis in rat liver

Groups	Activities (umoles/min . g protein)	
	G6PDH	G6Pase
Control	13.8 + 1.7	19.8 + 1.0
AA	16.4 + 1.5	21.3 + 3.0
NDEA	23.3 + 2.3	14.0 + 1.5
NDEA + AA	37.2 + 5.0	12.6 + 1.5

11 rats. (The expected mortality (E) was respectively 5.801 and 6.199, x^2 = 7.692, p<0.01). Later the mortality equalized: between the 400th and 470th day in both groups six rats died due to hepatomas (E in NDEA group - 5.675, in NDEA+AA group - 6.344 rats, x^2 = 0.034, p>0.7). The results indicate, that in our experimental conditions ascorbic acid accelerated considerably the progress of liver tumors.

Macroscopic evaluation of the neoplastic changes in all dead rats (including the intercurrent mortality) gave similar results (Table 1). The tumors progressed in the NDEA+AA group considerably faster than in the NDEA group. The difference was most marked in the earlier period of tumor development (9-12 months). In the NDEA group the average grading was 2.09, in the NDEA+AA group - 4.17 (p<0.05). Later the difference between the groups diminished steadily. The average macroscopic grading of all dead rats reached in the NDEA group 3.09, in the NDEA+AA group - 4.41 (p<0.02). On the 470th day the surviving rats were sacrificed. The grading of hepatocarcinogenesis reached 3.84 in the NDEA group and in the NDEA+AA group - 4.50 (p<0.3).

Biochemical experiments

The accelerating action of ascorbic acid on hepatocarcinogenesis induced by NDEA was revealed also in the biochemical study. In the liver tissue of rats receiving NDEA for six months characteristic enzyme activity changes were found after nine months. The activity of the enzymes of glycolysis (HK, PK) and the pentose phosphate cycle enzymes (G-6-PDH and 6-PGDH) was increased and the activity of GK and the enzymes of glyconeogenesis (G6Pase and F-1, 6-DPase) decreased. In the NDEA+AA group the changes were of the same character but in the majority of cases more expressed. In the NDEA+AA group the activity of G-6-PDH was 60% higher than in the NDEA group (p<0.02). At

Table 3. Morphological changes in the rat liver in hepatocarcinogenesis

	NDEA		NDEA+ AA		p
	M ± m	(N)	M ± m	(N)	
Grades	3.83 ± 0.49	(12)	4.83 ± 0.44	(12)	<0.05
Number of tumors	5.5 ± 1.6	(11)	11.9 ± 2.8	(11)	<0.1

Table 4. The content of ascorbic acid (mg%) in the rat tissues in hepato-carcinogenesis

Group of rats	M ± m	(N)	%	p
	Liver			
Control	40.0 ± 0.9	(11)	100	-
NDEA	28.4 ± 2.6	(10)	70	<0.01
	Blood Serum			
Control	0.90 ± 0.01	(11)	100	-
NDEA	0.76 ± 0.04	(11)	80	<0.01

the same time the activity of G6Pase decreased substantially as compared with the control group (p<0.01) (Table 2). The activity of PK was also markedly (42%) increased under the influence of ascorbic acid, whereas in rats receiving only NDEA the increase was lower (16%).

Morphological examination confirmed the more progressive neoplastic changes in rats receiving ascorbate. On the liver surface in rats receiving the carcinogen alone the average number of tumors was 5.5, in the group receiving NDEA and ascorbate twice as many tumors were counted on the liver surface. The macroscopic grading was 3.83 and 4.83 respectively (p<0.05) (Table 3).

The ascorbic acid content of the liver tissue during carcinogenesis was of interest. In rats, which received NDEA for six month, the liver ascorbic acid concentration was on the ninth month of the experiment 70% of the value in normal rats. The blood serum ascorbic acid content showed also a clear decrease (Table 4).

In rats receiving NDEA + ascorbate statistically significant changes of enzyme activity appeared earlier than in animals receiving only NDEA. Increase or respectively decrease of enzyme activity was found already after four months. The activity of HK was 140% (p<0.01), that of PK 137% (p<0.01), G-6-PDH - 155% (p<0.01) and the activity of GK 62% compared with the values in control rats (p<0.01) (Fig. 3). In rats receiving only NDEA no significant changes of enzyme activity could be found after four months.

DISCUSSION

These results demonstrate, that in our experimental conditions sodium as-

corbate enhanced the hepatocarcinogenic action of NDEA. The animals died
earlier due to fatal neoplasms in the liver, the tumors were more expressed.
In the literature numerous data can be found on the stimulating action of
ascorbic acid on the metabolism of xenobiotics and drugs in the liver (review
of Zannoni et al., 11). A possible explanation may be that ascorbic acid
increases the synthesis of cytochrome P-450 apoprotein and/or decreases its
catabolism protecting it from the destructive action of xenobiotics. Alternatively ascorbic acid may favour the binding of heme and apo-cytochrome P-
450 to assemble into active enzymes (12) or affect the transfer of electrons
from NADPH to cytochrome P-450 (13).

Recently it was shown that in mutant rats unable to synthesize ascorbic
acid the dietary requirement of ascorbic acid for the maximum induction of
hepatic drug metabolism was increased several fold (from 300 up to 1000 mg
per kg of diet) by the administration of xenobiotics (14). On the other
hand, administration to mice of NDMA in drinking water (50 ug/ml during 6
days, total dose per mouse 1.2-1.5 mg) produced a fourfold decrease of the
activity of N-demethylase of NDMA diminishing the metabolic activation of
this carcinogen by the cytochrome P-450 mediated microsomal monooxygenase
system (15).

The enhancing action of ascorbate on NDEA hepatocarcinogenity in our
experiments can be explained by higher NDEA metabolizing activity in the
livers of AA treated rats compared with rats receiving the carcinogen only.
More NDEA is metabolized in the liver and less passes the liver barrier. At
the same time the blood plasma clearance of carcinogen is accelerated.

It is supposed that microamounts of NDMA absorbed into the portal blood
from the gut are almost completely metabolized by the liver and do not enter
the general circulation (16). In that case it is not excluded, that the
improvement of microsomal metabolic activation increases the binding of alkylating radicals to numerous nucleophilic groups in the cytoplasm decreasing
the alkylation of DNA in the nucleus of the liver cell. For example, the
half-life of the alfa-hydroxymetabolite of NDEA reacting with cysteamine is
only 39 seconds (17). For this short time the diffusion in the cell is
apparently limited.

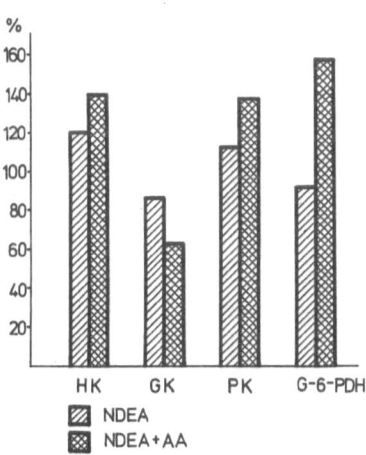

Figure 3. The activity of enzymes in the liver - administration of NDEA and
NDEA + ascorbic acid during 4 months of experiments, percentage of normal
values.

Obviously the modifying action of ascorbic acid on NNC carcinogenesis requires further investigation. From the point of view of cancer prevention in humans it is important to clarify the action of ascorbic acid on small amounts of NNC absorbed from the gut.

It is possible that the role of ascorbic acid in preventing stomach cancer is connected solely with the inhibition of endogenous formation of NNC, whereas in areas exhibiting a high liver cancer frequency the high concentration of ascorbic acid in food may appear as a risk factor. Until more detailed data is obtained the accelerationg action of ascorbic action on nitrosamine carcinogenesis has to be taken into account in cancer chemoprevention and treatment.

REFERENCES

1 Doll, R. and Peto, R. (1981) J. Natl. Cancer Inst. 66, 1191-1308
2 Mirvish, S.S. (1986) Cancer 58, 1842-1850
3 Nair, J., Nair, U.J., Ohshima, H., Bhide, S.V., Bartsch, H. (1987) in Relevance of N-nitroso compounds to human cancer: Exposures and mechanism. IARC Sci. Publ., 84, p. 465-469
4 Fukushima, S., Imaida, R., Sakata, T., Okamura, T., Shibata, M. and Ito, N. (1983) Cancer Res., 43, 4454-4457
5 Cameron, E., Pauling, L., Leibovitz, B. (1979) Cancer Res. 39, 663-681
6 Moertel, C.C., Fleming, T.R., Creagan, E.T., Rubin, J., O'Connel, M.J. and Ames M.M. (1985) N. Engl. J. Med. 312, 137-141
7 Ton, C.C.T., Fong, L.Y.Y. (1984) Carcinogenesis 5, 533-536
8 Mirvish, S.S. (1981) in Inhibition of tumor induction and development. (eds M.S.Zedeck and M.Lipkin) Plenum Press, NY a. London, pp. 101-126
9 Peto, R., Pike M.C., Day, N.E., Gray, R.G., Lee, P.N., Parish, S., Peto, J., Richards, S., Wahrendorf, J. in Long-term and short-term screening assays for carcinogens: a critical appraisal. IARC Monographs, suppl. 2. - Lyon, 1980 pp. 311-426
10 Gurkalo. V.K., Zabezhinski, M.A. (1982) Neoplasma 29, 301-307
11 Zannoni, V.G., Holsztynska, E.G., Lau, S.S. (1982) in Ascorbic Acid: Chemistry, Metabolism and Uses. 2nd Chem. Congr. N. Amer. Contin. Washington, D.C., pp. 349-368
12 Omaye, S.T., Turnbull, J.D. (1980) Life Sci. 27, 441-449
13 Rikans, L.E. (1982) J. Pharmacol. Exp. Ther. 204, 702-713
14 Horio, F., Ozaki, K., Kohmura, M., Yoshida, A., Makino S. and Hayashi, Y. (1986) J. Nutr., 116,2278-2289
15 Rubenchik, B.L., Juchimenko, M.D., Boim, T.M., Mihhailenko, V.M. (1988) Eksperimentalnaya Onkologiya (Russ.) 10, 29-32
16 Diaz-Gomez, M.I., Swann, P.F., Magee, P.N. (1977) Biochem. J., 164, 497-500
17 Keeman T.H., Weinkam, R.T. (1985) Toxicol. and Appl. Pharmacol. 78, 316-320

GENETIC DAMAGE IN PERIPHERAL LYMPHOCYTES OF CHRONIC ALCOHOLICS

R.J. Sram, J. Topinka, B. Binkova, J. Kocisova, V. Kubicek, and
J.A. Gebhart

Psychiatric Research Institute, 18103 Prague 8
Czechoslovakia

SUMMARY

A possible impact of long-term overexposure to ethanol was studied in two
groups of chronic alcoholics in a psychiatric hospital. The hypothesis that
acetaldehyde may act as a mutagen for somatic cells and induce free radicals
in alcoholics tissues was tested. To monitor the amount of injury resulting
from alcohol abuse the following markers were used: chromosome aberrations
and unscheduled DNA synthesis (UDS) in peripheral lymphocytes and lipid
peroxidation (LPO) in plasma and peripheral lymphocytes. The following
results were observed: 2.05% aberrant cells (AB.C) in alcoholics vs. 1.29%
in the control group; 12 weeks abstinence decreased the frequency of chromosome aberrations from 2.13 X AB.C to 1.52% AB.C. If the patients were supplemented by 1000 mg of ascorbic acid and 400 mg of α-tocopherol per day in
the course of 12 weeks, original UDS T/C increased 3.74 to 4.12 (patients
without antioxidants = 3.48, controls = 4.72), LPO in plasma decreased from
1.57 to 1.37 nmol MDA/ml (untreated patients = 1.89, controls = 1.12), LPO in
lymphocytes decreased from 1.72 to 1.34 nmol/mg protein (untreated patients =
1.96, controls = 1.14).

Key words: ethanol, chromosome aberrations, unscheduled DNA synthesis, lipid
peroxidation, ascorbic acid, α-tocopherol

INTRODUCTION

New data indicates that long-term overconsumption of ethyl alcohol may
create genetic risk for alcoholics. The most significant injury is probably
induced in somatic cells, affecting in this way various target tissues. The
impact of this injury may be cancer, atherosclerosis, high-blood pressure and
speeding up the process of ageing.

Ethanol is eliminated from the body by oxidation, it is metabolized
enzymatically first to acetaldehyde (AA) and than to acetate (1). Alcohol
consumption leads to accelerated ethanol metabolism by different mechanisms:
the alcohol dehydrogenase pathway, the microsomal ethanol oxidizing system
and the catalase pathway. During chronic alcohol consumption ethanol is
oxidized mostly by adaptive increase of the cytochrome P-450 dependant microsomal ethanol oxidizing system (MEOS) (2). Also a mechanism involving
hydroxyl radicals has been proposed (3). The MEOS speeds up blood alcohol

Biochemistry of Chemical Carcinogenesis
Edited by R. Colin Garner and Jan Hradec
Plenum Press, New York, 1990

clearance and thus contributes to the accelerated rate of ethanol metabolism.
However, simultaneously with increased MEOS activity due to chronic alcohol
intake, there is an increased activity of drug metabolizing enzymes. Ethanol
is acting as a cocarcinogen through its effects on enzyme systems in
carcinogen activation and detoxification. Since ethanol is a potent inducer
of the cytochrome P-450 dependent microsomal biotransformation system (4), it
may be speculated that microsomal cytochrome P-450 dependent carcinogen activation is faster in alcoholics. It could mean a higher probability of increased mutagen and carcinogen activation and therefore a higher possibility
of active metabolites reaching the DNA and inducing mutations. How this
process proves effective is simultaneously the result of effective DNA repair
mechanism.

    Adverse health effects of alcohol overconsumption on somatic cells DNA
are generally determined by:

1 the rate of conversion of ethanol to acetaldehyde,

2 acetaldehyde activity which induces mutations, and

3 organism capability to repair induced DNA damage

    AA is the ethanol-active metabolite. The genetic resk of chronic
alcoholism and DNA damage may be related to the mutagenic and carcinogenic
activity of AA (5, 6). AA is highly reactive: its electrophilic properties
cause it to react with nucleophilic groups of various macromolecules, like
DNA. Its condensation with amino groups of nucleic acid bases may be related
to effects on DNA synthesis (adducts, cross-links) (7, 8) or chromosomal
aberrations (9). AA has been shown to be a clastogen, inducing chromosome
aberrations in human peripheral lymphocytes and other cells in vitro and in
vivo, as well as sister chromatid exchanges (20). Its carcinogenic properties have been proven in Syrian hamsters (11) and rats (12); it may also
carry a promoting activity as it increases the frequency of benzo(a)pyrene -
induced tumors (11).

    An induced MEOS pathway in chronic alcoholics could enhance free oxygen
radical production. AA binding with glutathione may contribute to a
depression of liver glutathione (13), and as such increase lipid peroxidation. Increased activity of microsomal NADPH oxidase could result in
enhanced H_2O_2 production (14). If free radicals, produced as the result of
chronic ethanol ingestion, are not eliminated by protective enzymes and/or
natural antioxidants, they may induce oxidative damage of DNA and act as
mutagens (15), or react with polyunsaturated fatty acids and induce lipid
peroxidation (16). Enhanced lipid peroxidation has been proposed as a
mechanism for ethanol-induced fatty liver (17). LPO leads to formation of
malondialdehyde (MDA), which has been reported to be an inhibitor of
mitochondrial aldehyde dehydrogenase, thus contributing to the elevation of
AA (17).

    In our study, we analyze the following questions: a) the frequency of
chromosome aberrations in peripheral lymphocytes of chronic alcoholics; b)
the effect of abstinence on the frequency of chromosome aberrations; c) the
level of UDS and LPO in chronic alcoholics; d) the effect of antioxidant
supplementation on UDS and LPO level.

MATERIALS AND METHODS

    Subjects for study were chosen from in-patient male alcoholics undergoing
therapy at the psychiatric hospital. The first part of the study, (A), analyzed the chromosome aberrations in peripheral lymphocytes. In the second

220

part of the study, (B), unscheduled DNA synthesis and lipid peroxidation processes were analyzed in other groups of patients. Patients in group A came to the hospital voluntarily for the first time. They were examined in two consecutive groups. A pilot study was performed on 24 patients aged 48.2 $\pm$ 2.1 years. The matching control group consisted of males, aged 46.4 years $\pm$ 2.0 years, with the same place of residence and the same smoking habits. The other group comprised 49 patients aged 46.8 $\pm$ 2.3 years. Blood samples were taken upon patients entering the hospital. 23 patients from the above group were also examined after 12 weeks of therapy, to check the effect of abstinence upon analyzed changes. There was no proven exposure to any known mutagen in any of the examined subjects.

Patients in group B came to the hospital voluntarily for the first time, or after recidivism. 52 patients, aged 38.4 $\pm$ 7.6 years, were chosen for the study. Blood samples were taken at the beginning of therapy. Their therapy with disulphiram was then supplemented for 12 weeks with antioxidants in the form of Celaskon efferevescens Spofa (L-ascorbic acid) at 1000 mg per day, and with Vitamin E Spofa (-toco-pherol) at 400 mg per day. The supplementation therapy terminated with 36 patients. The control group included 23 healthy volunteers, aged 37.9 + 8.0. The effect of antioxidant therapy was checked after 12 weeks of hospitalization in 20 patients not treated with antioxidants.

Cytogenetic analysis

Cytogenetic analysis of peripheral lymphocytes was carried out on short-term lymphocyte cultures essentially as described by Hungerford (18). Lymphocytes were stimulated for 52 hours by phytoheamaglutinin (PHA 15 Wellcome), when according to BUdR checking most cells are in the first division, and cultivated in medium RMPI 1640. Slides were coded and examined by two observers. 100 metaphase were analyzed from each subject. Four basic categories of chromosome aberrations were evaluated: chromatid and chromosome breaks, plus chromatid and chromosome exchanges. Cells bearing breaks and/or exchanges were scored as aberrant cells (AB.C.).

Isolation of Lymphocytes

Lymphocytes were isolated from 20 ml of heparinized blood on Ficoll 400 (Pharmacia, Sweden)- Verografin (Spofa, Czechoslovakia) gradients by the modified method described by Harris (19). Cells were washed three times in buffer (NaCl 14 mmol/1; Tris 14.5 mmol/1; KCl 5.4 mmol/1; $MgCl_2.6.H_2O$ mmol/1; $CaCl_2.2H_2O$ 0.05 mmol/1 and anhydrous D-glucose 5 mmol/1; pH 7.2) and resuspended in RPMI 1640 medium for determination of UDS and LPO.

Unscheduled DNA synthesis (UDS)

Excision DNA repair was estimated by measuring in vitro unscheduled DNA synthesis (UDS) in peripheral lymphocytes (20). UDS was induced by 1-methyl-3-nitro-1-nitrosoguanidine (MNNG). After isolation from blood the cells were incubated with 10 mmol/1 hydroxyurea and 2 umol/1 5-fluor-deoxyuridine to inhibit semiconservative DNA synthesis. All cell cultures were divided into two aliquots. The first aliquot of cells from each donor (treated cells -T) was incubated in the presence of MNNG (0.1 mmol/1) dissolved in DMSO and 400 kBq/ml of (methyl-^{3}H) thymidine (2 TBq/mmol) for 3 hours. The second aliquot (control cells - C) was incubated in the same conditions but without MNNG. These incubations were followed by the chemical isolation of DNA from lymphocytes and measuring the incorporated radioactivity in DNA. The concentration of DNA in all samples was also estimated by the diphenylamine method. The specific activity of samples was expressed as CPM/ug of DNA and calculated for treated (T) and control (C) samples. Ratio T/C gives information about the increasing incorporation of radiolabelled nucleoside 3H-thymidine as a consequence of standard damage of DNA by MNNG. This ratio is therefore called UDS.

Lipid peroxidation

The LPO level in lymphocytes and plasma was measured by modified thiobarituric acid (TBA) assays (21, 22). These involve an acid hydrolysis of lipoperoxides to malondialdehyde (MDA). which subsequently reacts with TBA producing a MDA-TBA adduct suitable for a sensitive spectrophotometric or fluorometric measurement. The maximum formation of the reaction products is attained in 20% acetic acid solution at pH 3.5 after raising the temperature to 95^0C for 60 minutes. TBA- active products of lipid peroxidation were measured spectrophotometrically, after extraction in n-butanol, as the differences in absorbance at 532 nm and 580 nm (incase of the lymphocytes) or fluorometrically at 554 nm with excitation at 515 nm (plasma). The obtained values were expressed as nmoles of MDA/mg proteins or nmol MDA/ml plasma using malondialdehyde-bis diacetal as a standard. The content of proteins in the suspension of lymphocytes was determined by the Bradford (23) method using Coomasie Blue G reagent.

Statistical analysis

The inter-group differences were statistically evaluated by unpaired Student's t-test (24).

Chemicals. Ficoll 400 (Pharmacia, Sweden), 60% Verografin (Spofa, Czechoslovakia), MNNG, hydroxyurea, 5-fluordeoxyuridine (all Serva, FRG), thiobarbituric acid (Merck, FRG), medium RPMI 1640 (USOL, Czechoslovakia), methyl-^{3}H) thymidine (UVVVR, Czechoslovakia), all other chemicals had a minimal degree of the purity necessary for analysis.

RESULTS AND DISCUSSION

The cytogenic analysis of peripheral lymphocytes (Table 1) proved the greater increase of aberrant cells (p$\frac{1}{4}$0.05) in alcoholics as compared to matching control groups. Comparing the number of patients carrying 4 and more % AB.C. with the control, which is usually formed in a mutagen untreated population in 3.22% of lymphocyte cultures (25), we observed 18.1% of such lymphocyte cultures among alcoholics and 2.7% in the controls.

Analyzing the spectrum of chromosome aberrations and comparing alcoholics with matching controls and "historical controls" (26) in the group Patients I, we found an increase of dicentric chromosomes. This result corresponded to the data of Obe et al. (27), which reported that the frequency of exchange-type aberrations are positively correlated with the duration of the dependency on alcohol. However, we were not able to repeat this result in the group Patients II. There was also significant effect of smoking habits in alcoholics on the frequency of chromosome aberrations.

In the group of 23 patients (from Patients II) we followed the effect of 12 weeks' therapy (usually by disulphiram) and abstinence (Table 2). After three months in hospital, the frequency of chromosome aberrations in alcoholics decreased to the control level (p$\frac{1}{8}$0.05). This decrease of chromosome aberrations in chronic alcoholics after 3 months' therapy corresponds to the idea that chromosome aberrations induced by mutagens with clastogenic activity may be taken as indicators of lymphocyte injury by mutagens in the three or four months before the blood for cytogenetic analysis is collected (25).

AA is believed to be responsible for the clastogenic activity of ethanol

Table 1. Cytogenetic analysis of peripheral lymphocytes in chronic alcoholics.

Group	N	number of meta- phases	AB.C. %	B'	B''	E'	E''	B/C
Patients I	24	2400	1.83*	21 0.87	20 0.83	1 0.04	3 0.12	0.020*
Patients II	49	4800	2.21*	75 1.56	32 0.66	2 0.04	–	0.023*
Summe I+ II	73	7200	2.05*	96 1.33	52 0.72	3 0.04	3 0.04	0.022*
Controls I	24	2400	1.21	16 0.67	13 0.54	–	–	0.012
Controls II	44	4400	1.34	48 1.09	21 0.48	–	–	0.016
Summe I+II	68	6800	1.29	64 0.94	34 0.50	–	–	0.014
Historical	234	23400	1.44	219 0.94	114 0.49	3 0.01	1 0.04	0.015

$p < 0.05$, % AB.C = percentages of aberrant cells; B'= chromatid breaks B''= chromosome breaks; E' = chromatid exchanges; B/C = number of breaks per cells; number in the second line indicate the number of aberrationss per 100 cells.

Table 2. Effect of 12 weeks' therapy on the frequency of chromosome aberrations in chronic alcoholics

Group	N	Number of metaphases	AB.C.	B'	B''	E'	E''	B/C
Before	23	2300	2.13	32 1.39	16 0.70	1 0.04	–	0.022
After	23	2300	1.52*	26	10	1	–	0.016

$p < 0.05$, legend see Table 1

(28). As AA may induce oxidative damage of DNA (15) and lipid peroxidation (16) its effect could be supressed using antioxidants. Our experience indicated the anticlastogenic activity of ascorbic acid, if supplied to a

group of workers occupationally exposed especially to polycyclic aromatic hydrocarbons (29). As AA also affect cell membranes and induce lipid peroxidation the other antioxidant of choice became α-tocopherol. This can act as an oxygen free radical scavenger, as well as terminate the chain reaction of lipid peroxidation and structurally stabilize the membrane bilayer.

Therefore, in the second part of our study, (B), alcoholics were treated for 12 weeks simultaneously with ascorbic acid and with α-tocopherol. UDS

223

Table 3. Effect of 12 weeks' antioxidants supplementation on UDS and LPO level in chronic alcoholics.

Group	Alcoholics			Controls (K)
	before	after	without	
		supplementation of antioxidants		
Length of therapy (weeks)	0	12	12	0
U D S				
n	50	36	20	23
(T/C)	3.73 **	4.12	3.48	4.72
s.d.	0.73	0.81	0.64	0.59
$(x - 2s)_K$	37%	20%	50%	0%
PLASMA LPO				
n	51	36	20	23
nmol MDA/ml	1.57 **	1.37	1.89	1.12
s.d.	0.29	0.26	0.23	0.14
$(x - 2s)_K$	77%	37%	85%	4%
LYMPHOCYTES LPO				
n	40	28	19	23
nmol MDA/ml	1.72 **	1.34	1.96	1.14
s.d.	0.40	0.18	0.33	0.19
$(x - 2s)_K$	67%	11%	100%	4%

* $p < 0.01$ ** $p < 0.005$

and LPO levels were measured before and after the antioxidate supplementation (Table 3). UDS was decreased in chronic alchoholics as compared to the controls by more than 20% ($p < 0.05$). As for the patients having a UDS level lower than the average which was determined as the double of the control group, i.e. $< (x - 2s)_K$, 37 patients had a decreased UDS level. In patients supplemented for 12 weeks with antioxidants, no change in the UDS level during abstinence was observed. The LPO level in plasma and lymphocytes. Similarly, LPO in plasma $> (x+2s)_K$ increased in 77% of the patients, LPO in lymphocytes in 67% of the patients. The antioxidant supplementation decreased LPO level in plasma and lymphocytes ($p < 0.005$), but the observed value were still higher than in the controls. If patients were not supplemented with antioxidants, 12 weeks' abstinence did not decrease LPO level in plasma nor in lymphocytes.

The data for alcoholics can be related to UDS and LPO level across an ageing male population (Table 4), (30). Comparing the observed results from both groups, it may be speculated that chronic alcohol consumption speeds up the process of ageing in males aged 38.4 $\pm$ 7.6 years by at lease 15 years, if we accept the UDS and LPO levels as indicators of ageing. Particularly noticeable is the increase of lipid peroxidation in plasma and lymphocytes. Our data seems to indicate the advantage to using ascorbic acid and α-tocopherol during alcoholic therapy in psychiatric hospitals. Data on UDS and LPO levels in alcoholics without antioxidant supplementation shows that abstinence is probably not sufficient for the improvement of all enzyme activities in an alcoholic liver. The health effect of damage induced by long-term overconsumption of alcohol and its mutagenic metabolite AA may be

Table 4.   Effect of age and alcohol abuse on unscheduled DNA synthesis and lipid peroxidation in men.

Group	Alcoholics	Geriatric Persones			Control
		60 - 69	70 - 79	>80	
Age	38.4	60 - 69	70 - 79	>80	37.9
N	52	9	20	20	23
U D S					
T/C	3.77	3.48	3.41	2.94	4.72
s.d.	0.60	0.62	0.77	0.77	0.59
L P O					
PLASMA					
nmol MDA/ml	1.55	1.38	1.33	1.41	1.12
s.d.	0.14	0.30	0.26	0.35	0.14
LYMPHOCYTES					
nmol MDA/mg	1.79	1.60	1.47	1.43	1.14
s.d.	0.29	0.51	0.28	0.34	0.19

observed as an increased susceptibility to several diseases, such as cancer of the respiratory and gastrointestinal tracts, atherosclerosis, essential hypertension and senile dementia of the Alzheimer type. If antioxidant supplementation is used as part of the complex therapy and care for alcoholics, it seems to improve the long-lasting impact of chronic alcoholism as expressed by the DNA damage in somatic cells.

REFERENCES

1   Von Wartburg, J. P. (1971) in Biology of Alcoholism (Kissin, B., and Begleiter, H., eds.) Vol.1, pp. 63-102, Plenum Press, New York
2   Lieber, C. S. (1977) in Metabolic Aspects of Alcoholism (Lieber, C.S., ed.), pp. 1-29. University Park Press, Baltimore
3   Winston, G. W., Cedersbaum, A. I. (1983) J. Biol. Chem., 258, 1514
4   Lieber, C. S., Baraona, E., Leo, M. A. Garro, A. (1987) Mutation Res., 186, 201-233
5   Dellarco, V. L. (1988) Mutation Res., 195, 1-20
6   IARC Monographs on the Evaluation of the Carcinogenic Risk of Chemicals to Humans (1985) Vol. 36, pp. 101-132. IARC Lyon
7   Tuma, D. J., Sorrell, M. F. (1985) Progr. Clin. Biol. Res., 183, 3-17
8   Hemminki, K., Suni, R. (1984) Arch. Toxicol., 55, 186-190
9   Korte, A., Obe, G. (1981) Mutation Res., 88, 389-395
10 Obe, G., Anderson, D. (1987) Mutation Res., 186, 177-200
11 Feron, V. J., Kruysse, A., Wontersen, R. A. (1982), Evr. J. Cancer Clin. Oncol., 18, 13-31
12 Wontersen, R. A., Appelman, L. M., Feron, V. J., Lieber, C. S. (1981) J. Lab. Clin. Med., 98, 417-425
13 Show, S., Jayatilleke, R., Ross, W. A., Gordon, E. R., Lieber, C. S. (1981) J. Lab. Clin. Med., 98, 417-425
14 Reitz, R. C. (1975) Biochem. Biophys. Acta, 380, 145-154
15 Ames, B. N. (1986) in Antimutagenesis and Anticarcinogenesis Mechanism (Shankel, D. M., et al., eds) pp. 7-36, Plenum Press, New York
16 Diansani, M. V. (1987) Proc. Nutr. Soc, 46, 43-52
17 Hjelle, J. J. Grubs, J. H., Peterson, D. R. (1982) Alcoholism: Clin. Exp. Res., 6, 145-149
18 Hungerford, D. A. (1965) Stain. Technol., 40, 333-338

19 Harris, R., Ukaijiofo, E. O. (1970) Brit. J. Haematol., 18, 229-235
20 Martin, C. N., McDermaid, A. C., Garner, R. C. (1978). Cancer Res., 38, 2621-2627
21 Okhawa, H, Ohiski, N., Yagi, K. (1979) Anal. Biochem., 95, 351-358
22 Yagi, K. (1976) Biochem Med., 15, 212-216
23 Bradford, M. M. 91976) Anal. Biochem., 72, 248-254
24 Dixon, W. J., (1985) BMDP Statistical Software 1985
25 Sram, R. J. (1981) in Industrial and Environmental Xenobiotics (Gut, I., Cikrt, M. Plaa, G. L. eds.) pp. 187-194, Springer Verlag, Berlin, Heidelberg
26 Sram, R. J., Kocisova, J., Marecek, P., Nerad, J. (1986) Cs. Psychiat., 82, 113-117
27 Obe, G., Goebel, D., Engeln, H., Herka, J., Nataragan, A. T. (1980) Mutation Res., 73, 377-386
28 Obe, G., Jonas, R., Schmidt, S. (1986) Mutation Res., 174, 47-51
29 Sram, R. J., Cerna, M., Hola, N. (1986) in Genetics Toxicology of Environmental Chemicals, Part B: Genetics Effects and Applied Mutagenesis, pp. 327-335, A. R. Liss, New York
30 Sram, R. J., Binkova, B., Topinka, J., Kotesovec, F. (1988) Cas.lek.ces., 127, in press

BIOCHEMICAL MARKERS OF CELL HETEROGENEITY IN RAT MAMMARY TUMORS INDUCED BY

7,12-DIMETHYLBENZANTHRACENE (DMBA) AND N-METHYL-N-NITROSUOREA (MNU)

V. Dolezalova[1], A. Rejthar, R. Neunutil, M. Cernoch,
J. Zaloudik, Z. Kolar[2], and V. Chylek

[1] Research Institute of Oncology, Institute of Medical Research
656 01 Brno PB63
[2] Dept of Pathology, Medical Faculty University of Palacky
775 15 Olomouc, Czechoslovakia

ABSTRACT

Two carcinogens, DMBA and MNU were compared for their ability to affect both epithelial and fibrous cell components of rat mammary tissue. Under given dosage regimens epithelium was more susceptible to both carcinogens than connective tissue in the early phase of tumor induction, whereas adenocarcinomas prevailed with a variable content of normal stroma. Repeated local doses of MNU evoked sarcomas of fibrous origin long after the epithelial tumors. The use of biochemical markers of tumor cell heterogeneity contributed to the knowledge about the role of fibrous elements in morphological alterations of primary and transplantable rat mammary tumors. Enzymatic activity of -glutamyltransferase (GGT) discriminated between epitherial and sarcomatous tumor components. Lectin binding capacity signaled a transition from epithelial to anaplastic tumor forms but did not reflect tissue origin in stained cells.

INTRODUCTION

A great deal of fundamental research in the field of mammary carcinogenesis, tumor growth and hormone responsiveness has been obtained with rat mammary tumors induced by DBMA and MNU. There are difficulties in working with the primary tumors because of variations in the main tumor categories viz. fibroadenomas (FA), adrenocarcinomas (CA) and sarcomas (SA) induced during the process of chemical carcinogenesis. Heterogeneity exists not only among tumors but also within one tumor containing transitory cells of various degrees of cell differentiation. These cell phenotypes respond to further morphological alterations during tumor progression and in the course of repeated transplantations. Therefore biochemical and biological studies are of value only when tumors are historically examined.

In the present work we tested selected biochemical markers for their discriminating ability between epithelial and connective tissue cells as well as other spindle cell elements which are indistinguishable by simple histology. For this reason -glutamyltranspeptidase (GGT), lectin-binding sites, DNA and reticulin content were analysed. Our results contribute to the knowledge of the role of fibrous tissue elements in the carcinogenic process, tumor growth and progression of primary and transplantable mammary tumors.

METHODS

Groups of female 50 day old Wistar rats were administered either DMBA using a standard dosage regimen (1) or MNU using the method of Guillano et al. (2) modified according to Thompson (3). The tumors appeared 11-35 weeks after the first carcinogen dose in 97 and 100 percent of treated rats respectively.

Tumors used for transplantation were finely minced and inoculated in 200 mg aliquots of tumor tissue per rat as described in a previous paper (4).

Enzymic activity of GGT was determined in a 105 000 x g supernatant of tissue homogenate treated overnight with Triton X-100 (1% v/v), using a kinetic procedure (measuring the increase in absorbance at 410 nm) with γ-glutamyl-4-nitranilide (4 mmol/1) and glycyl-glycine (40 mmol/1) as substrates.

Historical examination was performed with tumor sections elaborated by current method, stained with hematoxylin and eosin. Histoloical types of tumors were determined using two classifications established for rat mammary tumors (5,6).

The binding capacity for lectins was estimated histochemically using avidin-biotin complex for biotinylated UEA (Ulex europaeus), and WGA (Tritium vulgaris - Wheat germ); and an indirect immunoperoxidase method for DBA (Dolichos biflorus - Horse gram), Con-A (Concanavalin A) and PNA (Arachis hypogaea - Peanut). Semi-quantitative evaluation was made by two persons independently and was graded stepwise from - to ++.

The presence of reticulin was detected by Gomori special staining in histological sections.

DNA content was measured by flow cytometry from paraffin embedded 50 μ thick tumor slices (7).

RESULTS

Two carcinogens, DMBA and MNU were compared for their carcinogenicity in both epithelial and connective tissue cells in rat mammary gland. Under given dosage regimens epithelium was more susceptible to carcinogens than connective tissue since adenocarcinogens predominated in the early phase of the process (day 60-150) after the first dose. Sarcomas originated from fibrous component long after the epithelial tumors (day 160-300) (Table 1.)

Table 1 Distribution of the main tumor categories

Carcinogen	Rat Strain	Total Tumors N	CA N	%	FA N	%	SA N	%	Tumor Detection day
DMBA	Wistar	7	5	(72)	2	(28)	-	-	190
MNU	Wistar	13	10	(59)	-	-	7	(41)	162

Table 2 Phenotypic alterations of DMBA and MNU induced tumors

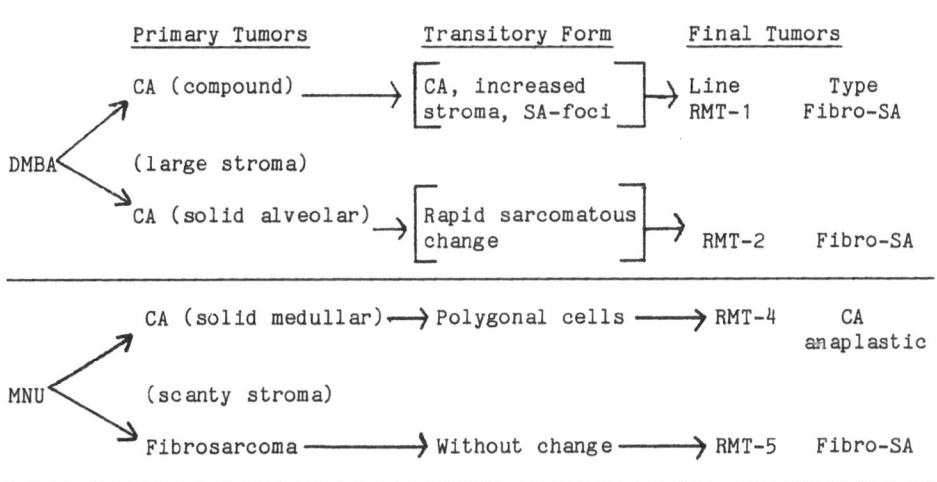

Transplantable tumor lines were established from both DMBA- and MNU-induced tumors as described in a previous paper (4) (Table 2). Comparing the morphology of all four transplantable rat mammary tumor lines (RMT-1, RMT-2, RMT-4, RMT-5) we found the following: relatively large stroma in DMBA-induced carcinoma increased with further increasing number of transplanted generations with signs of malignancy and sarcomatous foci. The fibrosarcomatous phenotype replaced original epithelial tumor gradually, in the case of RMT-1, or rapidly in RMT-2 (Fig 1, 2, 3).

MNU-induced carcinomas differed from those of DMBA in several features: they contained only scanty stroma without signs of malignancy or atypia and maintained a carcinomatous character in transplanted form (RMT-4). Transition from solid medullary form proceeded fluently to the anaplastic carcinoma. Its proportion in the tumors was variable. This tumor type could be hardly distinguished from sarcoma using light microscopy (Fig 4). In addition MNU induced primary fibrosarcoma providing evidence for the susceptibility of mammary connective tissue to the carcinogen. Absence of rudimentary epithelial ducts in tumor RMT-5 was the only difference compared to DMBA-derived sarcomas.

Staining for reticulin discriminated between the epithelial and fibrous cell origin of tumors showing no stain in epithelial parts of initial carcinomas, while all cells in fibrosarcomas were lined by reticulin (Fig 5, 6).

Reticulin formation distinctly outside of the groups of tumor cells provided evidence for the epithelial origin of this anaplastic tumor phenotype (Fig 7).

The enzyme activity of GGT was proposed recently as a marker of mammary carcinoma in DMBA system (8). In this work it was used for the first time to discriminate between transitory forms of adenocarcinomas and sarcomas. Statistical analysis has shown significantly lower values in all sarcomas when compared to carcinomas or normal adult virgin mammary tissue (Table 3). The dynamics of GGT activities in the course of transplanted passages of tumor (RMT- 2, RMT-4) in syngeneic rats has shown an initial increase above the controls when epithelial cells predominated. Consequent decreases correlated with the transitory states of tumors; then followed consistently low values in sarcoma RMT-2 and alternatively higher values in anaplastic carcinoma RMT-4 (Fig 8).

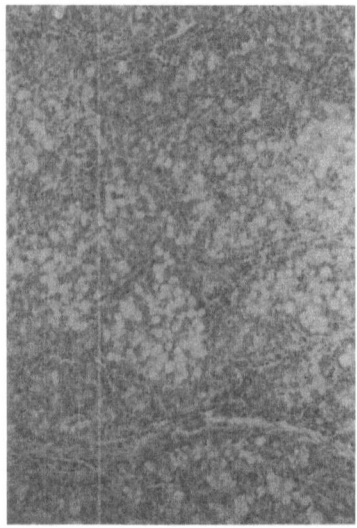

Fig 1. DMBA-induced primary
adenocarcinoma (compound type)

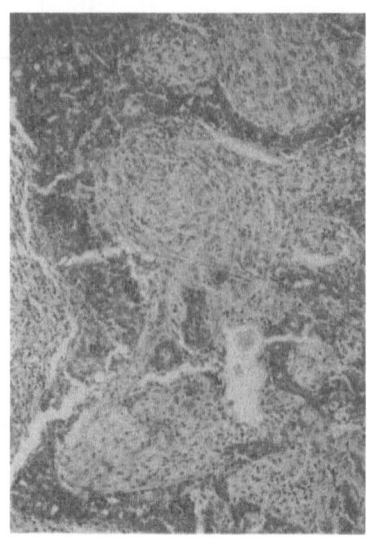

Fig 2. Carcinoma (phylloid formations
with sarcoma foci). RMT-2, passage 4

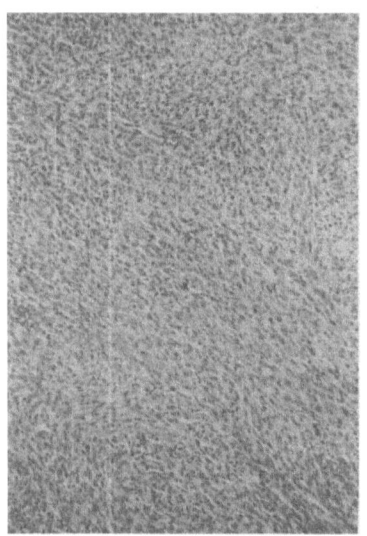

Fig 3. Fibrosarcoma replacing
original tumor (RMT-2, passage 9).

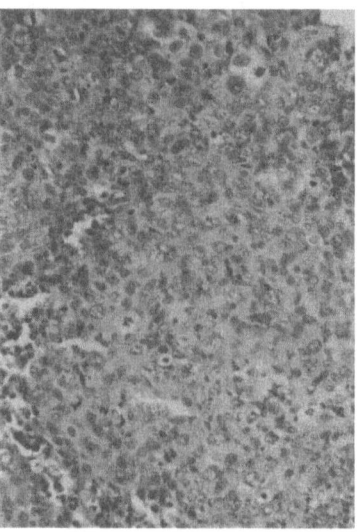

Fig 4. MNU-induced anaplastic
carcinoma (RMT-4, passage 41).

All four lectins tested have shown binding affinity to initial carcinomatous forms in tumor lines RMT-2 and RMT-4, but PNA was negative in MNU-
induced tumor. Binding capacity decreased during the course of passages up
to passage 13 being negative in passage 30 and 40. When the four tumor lines
were compared at the point when they reached stable characteristics, additional binding capacity appeared in sarcoma RMT-1 and RMT-5. That means that
other phenomena (9) than morphological phenotype responded by a decrease of
glycosylation during development of transplantable tumor.

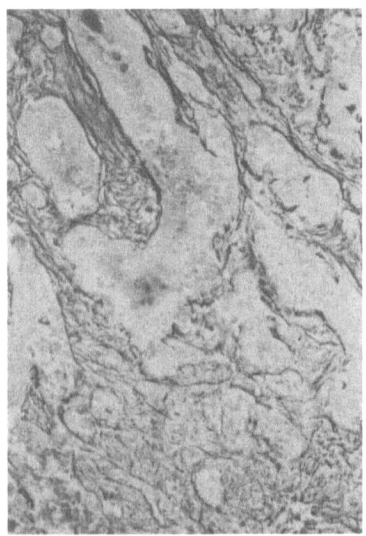

Fig 5. DMBA-induced carcinoma.
Note absence of reticulin stain in
epithelial component (RMT-2, passage
3).

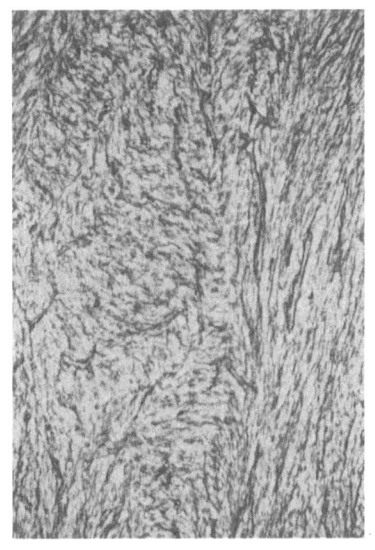

Fig 6. MNU-induced primary fibrosarcoma. Note reticulin fibres
surrounding every cell.

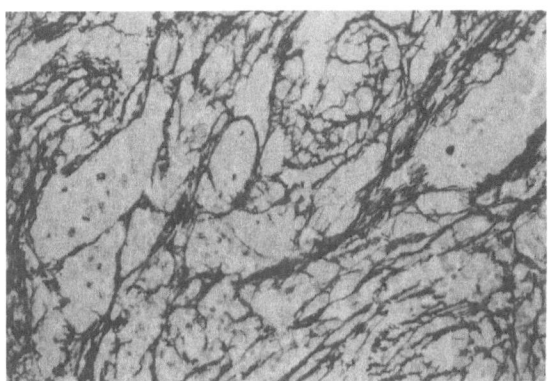

Fig 7. MNU-induced anaplastic carcinoma without reticulin. (RMT-4, passage
61)

Table 3 GGT activities in mammary tumor and control tissue

Tumor Type	Number of rats	GGT n kat / g protein $\bar{x}$		S.D.
RMT-1 (SA)	11	21.98*	+	9.54
RMT-5 (SA)	5	55.26*	+	19.57
RMT-4 (CA)	8	175.50**	+	41.23
Control	7	125.70	+	62.11

* = 1% level of significance (tumors - controls)
** = Nonsignificant

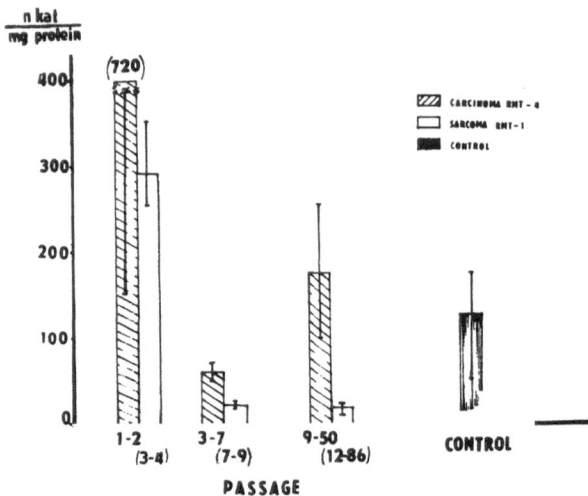

Fig 8. GGT activities following a series of transplantations (mean of 5-11 values).

Table 4   Detection of lectin-binding sites in transplanted tumors.

	Tumor Line							
	Transitory form			Stabilized form				
	RMT-2 Passage				RMT-4 Passage				RMT-1	RMT-2	RMT-4 Passage	RMT-5
Lectin	2	4	13	30	1	7	10	14	40	13	13	13	13
DBA	+	+	+	-	+	+	-	-	-	-	-	-	-
CoN-A	-	+	-	-	+	+	++	+	-	++	-	+	±
PNA	++	+	-	-	-	-	-	-	-	++	-	-	-
UEA$_1$	+	-	-	-	+	+	-	-	-	-	-	-	+

Flow cytometry measurement of the DNA content in the course of passages of tumor line RMT-2 has shown diploid character in all stages having no ability to discriminate among the transitory tumor stages.

DISCUSSION

It is well documented in the literature that the initial dose regimen of carcinogen can influence the carcinogenic process in rat mammary gland (10). The present paper reports the relationships between the early carcinogenic effect and post-transplantation tumor morphogenesis. DMBA- and MNU-induced carcinomas differed in the proportion and quality of fibrous stroma reflecting the severity of carcinogenic effect in mammary fibrous cell component. Both carcinogens produced transplantable tumor lines which underwent dramatic morphological changes after repeated transplantation in agreement with some recent papers (11,12). The cytogenetic basis of resulting tumor lines remains still questionable because stromal-epithelial interactions cannot be distinguished histologically in transitory and anaplastic tumor types. For that reason we examined biochemical markers to extend the histological findings. In our hands GGT discriminated between sarcoma and anaplastic carcinoma. Staining for reticulin provided a complementary marker supporting the identification of sarcomatous and stromal components from the tumor nodules of epithelial origin. Lectin binding capacity revealed a correlation with developmental stages of transplantable tumors but without referring to the tumor phenotype. Our results can be summarised as follows: 1) Stromal components of DMBA-induced carcinoma acquired a malignant pattern after several transplanted generations. Clonal selection of fibroblast like cells which appeared for the sarcomatous tumor conversion. 2) MNU-induced carcinoma was converted to anaplastic tumor of epithelial origin with normal stroma without signs of malignancy. Sarcomatoid conversion has also been observed in rat and mouse mammary carcinomas induced by intermediate cultivation of tumor cells in vitro (13,14). Disaggregation of tumor cells for the innoculation seemed to be an important step for this process. Epithelial-stromal interactions have to be taken into consideration in complex environment in vivo. GGT in conjunction with other biochemical markers provides a suitable tool for a better understanding of regulation of mammary tumor growth and morphogenesis.

REFERENCES

1   Hilf, R., Goldenberg, H., Michel, I., Carrington, M.J., Bell,C., Gruenstein, M., Meranze, D.R. and Shimkin, M.B. (1969) Cancer Res., 29, 977-983
2   Guillamo, P.M., Pettigrew, H.M. and Grantham, F.H. (1975) JNCI, 54, 401-414
3   Thompson, H.J. and Meeker, L.D. (1983) Cancer Res., 43, 1628-1629
4   Dolezalova, V., Nemecak, R., Toth, J. and Sugar, J. (1987) In Vivo, 1, 257-264
5   Komitowski, D., Sass, B. and Laub, W. (1982) JNCI, 68, 147-156
6   Williams, J.C., Gutherson, B., Humphreys, J., Monaghan,P., Coombes,R. C., Rudland, P. and Neville, A.M. (1981) JNCI, 66, 147- 152
7   Hedley, D.W., Friedlander, M.L., Taylor, I.W., Rugg, C.A. and Musgrove, E.A. (1983) J. Histochem. Cytochem., 31, 1333-1335
8   Sulakhe, S.J. (1987) Int. J. Biochem., 19, 509-515
9   Leathem, A. and Atkins, N. (1983) J. Clin. Pathol., 36, 747-750
10 Issacs, J.T. (1986) Cancer Res., 46, 3958-3963
11 Horn, H., Erlichman, I., Geier, A., and Levij, I.S. (1976) Eur. J. Cancer, 12, 189-194
12 Yoshida, H., Yoshida, A., Yamada, M., Yamanaka, K., Kiamura, Y and Matsumoto, K. (1985) Jpn. J. Cancer Res. (Gann) 76, 771-778
13 Hoon, D.B., Wang, H.Ch. and Ramshaw, I.A. (1984) Eur. J. Cancer Clin. Oncol., 20, 1517-1526
14 Kliss, R., Devleeschouwer, N., Paridaens, R,J., Danguy, A., Heuson, J.C. and Atassi, G. (1986) Anticancer Res., 6, 752-760
15 Haslam, S. (1986) Cancer Res., 46, 310-316

FATE OF GLUTATHIONE, RELATED ENZYME ACTIVITIES AND IN VITRO LYMPHOCYTE

TRANSFORMATION IN PATIENTS WITH BRONCHIAL CARCINOMA

C. Bluhm[1] and D. Greshuchna[2]

[1]Department of Dermatology, University of Dusseldorf, FRG
[2]Ruhrlandklinik, Essen, FRG

ABSTRACT

Glutathione and glutathione S-transferase (GST) are important for the detoxification of foreign compounds and thus for the prevention of chemical induced tumors.

No significant decrease in the specific activity of glutathione S-transferases was observed in lung tissue of patients with advanced bronchial carcinoma, but there was a reduction of glutathione reductase activity and total glutathione content, as well as a decrease in lymphocyte transformation rates after stimulation with concanavalin A.

Age-dependent reduction of GST activity and of in vitro lymphocyte transformation, but not of glutathione reductase activity and glutathione content in human lung tissue seems to be of significance in the development of malignant tumors.

The data presented here indicates a loss of detoxification efficiency due to a shift in the intercellular ratio of GSH and GSSG, and an impairment of lymphocyte function. Age-specific phenomena, like reduction of GST activity and reduced lymphocyte transformation rates could be indicators of an impaired defense system which may facilitate tumor growth.

Abbreviations

BP = benzo(a)pyrene, Con A = concanavalin A, GSH = reduced glutathione, GSSG = oxidized glutathione, GST = glutathione S-transferases, IL-2 = interleukin 2, PAH = polycyclic aromatic hydrocarbon, PHA = phytohemagglutinin, PWN = pokeweed mitogen, SD = standard deviation, SEM = standard error of the mean, ST = stimulation index.

INTRODUCTION

The lung is a target organ for a number of environmental pollutants, e.g., polcyclic aromatic hydrocarbons, which need enzymatic and non enzymatic activation to form the ultimate carcinogen (1-5). A well-known example is (BP), a carcinogen present in the environment and a constituent of cigarette smoke. The inhalation of this smoke is associated with an

Biochemistry of Chemical Carcinogenesis
Edited by R. Colin Garner and Jan Hradec
Plenum Press, New York, 1990

increased risk of bronchial carcinoma (6-10). Carcinogenic metabolites of PAHs are inactivated preferably by conjugation with glutathione, catalyzed by glutathione S-transferases (11-16). Detoxification is necessary to prevent a long retention time of reactive metabolites, especially epoxides.

Another important factor in tumor-defense is the immune system. From recent data in mice it has been concluded that BP in vivo causes immunomdulation by effects on T cells, B cells, and macrophages (17). BP suppresses IL-2 responsiveness and reduces IL-2 efficacy on T-cells (18, 19) probably by a reduced capacity of BP-treated macrophages to stimulate T-cells. All studies indicate that BP has various effects on several components of the humoral immune response. Similar regulation mechanisms may be present in humans. Data from several groups have shown an impairment of the immune system with advancing age (20-22). The age-dependent loss of immune function may facilitate tumor development.

MATERIALS AND METHODS

Patients (n = 190)

All patients included in this study underwent surgical resection of malignant lung tumors. Tumor stage was confirmed according to TNM classification (23). Immediately after removal of the tissue, some pieces without macroscopic signs of malignancy from the periphery of the lung were at once frozen in liquid nitrogen and stored at -80° until use (24). There was no significant loss of enzyme activities after storage for more than one year.

Preparation of subcellular fractions

Lung tissue was minced and homogenized in ice-cold 1.15% KC1 containing 3 mM EDTA and 10mM potassium phosphate buffer, pH 7.4. The homogenates were centrifuged at 100000 x g to remove cell debris, and the resulting super-natants were again centrifuged at 100000 x g for 60 minutes (24). The 100000 g supernatant fraction was used for measurement of gluthathione-S-tranferases, glutathione reductase, and total glutathione.

Assays

Glutathione-S-transferases were measured according to Habig et al. (25), using 2, 4-dinitrochlorobenzene (Merck) as substrate. Activity of glutathione reductase was measured according to Mize and Langdod (26) with a change of the final NADPH concentration to 0.11 mM. Total glutathione was determined according to Tietze (27). Protein content was measured as described by Lowry et al. (28).

In vitro lymphocyte transformation assays were carried out according to Grosse-Wilde et al (29). The mitogens used were concanavalin A (Difco, USAO, phytohemagglutinin P (Difco, USA) and pokeweed mitogen (Gibco, GB). Each sample was tested in triplicate at three different concentrations. Lymphocytes were cultured for 68 hours at $37^{\circ}C$, 90% humidity and 5% CO_2. After this time, [^{3}H]-thymidine was added and the cultivation was continued for another four hours. The lymphocytes were then harvested and counted in a beta-counter (Kontron). Lymphocytes stimulation rates were calculated as a stimulation index (SI): cpm with mitogen/cpm without mitogen.

Statistical Analysis

From the individual data a mean, standard deviation, and standard error of the mean was calculated for each parameter investigated. The variances were tested by an F-test for homogenous distribution (30). In case of homogeneity, data were compared by an unpaired Student's t-test, in case of homogenous distribution a Wilcoxon rank-test was performed. Correlations were tested using the correlation test according to Spearman. Differences were considered statistically significant at a value of p <0.05.

RESULTS

190 patients with primary bronchial carcinoma were included in this study (Table 1). Most of the patients were smokers or former smokers. Activity of glutathione S-transferases and in vitro lymphocyte transformation showed a significant correlation with the patient's age. No relation was observed for these parameters with the exception of lymphocyte transformation with Con A to the tumor stage. Glutathione reductase, however, was significantly correlated with the actual tumor stage (Table 2).

Table 1. Patients

Patients	190
Male	165
Female	25
Smokers	137
Former smokers	35
Non-smokers	18
Age (years)	32 - 77
Mean age (x $\pm$ SD)	57 $\pm$ 9

Table 2. Correlation of age and tumor stage with selected parameters (Spearman correlation)

Parameter	age significance	tumor stage significance
Glutathione S-transferases	0.005	n.s
Glutathione reductase	n.s	0.05
Glutathione	n.s	n.s
Concanavalin A (SI)	0.02	0.05
Phytohemagglutinin (SI)	0.05	n.s
Pokeweed mitogen (SI)	0.02	n.s

Table 3. Glutathione reductase, GST, glutathione, and _in vitro_ lymphocyte transformation with Concanavalin A at different tumor stages $(x \pm SEM)$

Tumor stage	GSSG reductase (nmol/mg/min)	GST	Glutathione $(\mu g/ml)$	Con A (SI)
I	22.1 ± 1.6	99.3 ± 6.5	3.1 ± 0.3	34.0 ± 4.5
II	22.0 ± 1.0	102.8 ± 4.0	3.9 ± 0.4	28.3 ± 2.1
III	$16.5 \pm 1.5**$	90.7 ± 7.8	$2.3 \pm 0.4*$	$20.4 \pm 2.5**$

* $p < 0.05$,   ** $p < 0.02$

As shown in Table 3 there was a significant reduction of total gluta-thione content and of glutathione reductase activity for tumor stage III. No significant differences between stages I and II were seen for glutathione content of lung tissue. Stage III with the greatest tumor expansion and the most favourable prognosis had a significant lower activity of glutathione reductase, but showed only a tendency of a decreased GST activity for stage III. Concanavalin A was the only mitogen that was significantly influenced by tumor spread.

An influence of age on specific activity of glutathione S-transferases (Figure 1) and on _in vitro_ lymphocyte transformation rates of all mitogens tested (Figure 2) was observed.

Glutathione S-transferases showed a slight decrease of activity with age. This finding was significant for patients over 70 years (Figure 1).

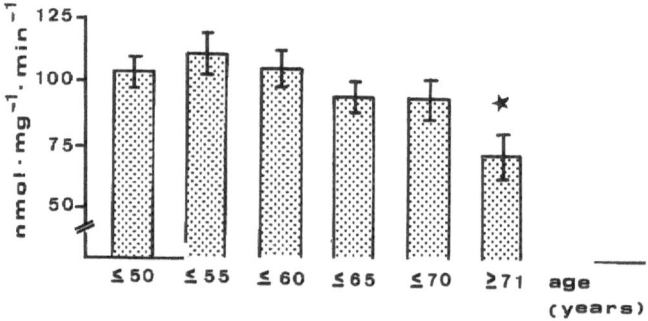

*p < 0.05

Figure 1. Specific activity of glutathione S-transferases as function of age $(x \pm SEM)$

A reduction of mitogenic responses was seen with advancing age, particularly for those mitogens preferentially causing blastogenic transformation of T-lymphocytes, although there was a considerable interindividual variation in [^{3}H]-thymidine incorporation.

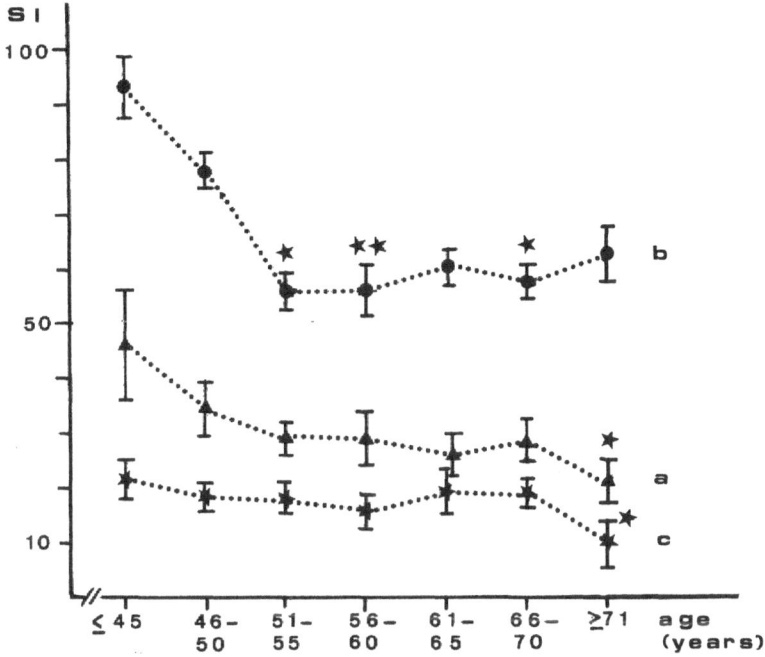

Figure 2. In vitro lymphocyte transformation with concanavalin A (a), phyto-hemagglutinin (b), and pokeweed mitogen (c) at different age groups (x ± SEM).

DISCUSSION

It is generally accepted that tumor development is a multistep process that can be divided into tumor initiation and tumor promotion (31,32.). While the first step is thought to be irreversible, tumor promotion is thought to be at least partially reversible and tumor manifestation can be reduced or prevented by an effective detoxification system. With advancing age the activity of GST and the mitogenic response was decreased (Figures 1 and 2). This indicates an influence on tumor defense (20, 33, 34).

One finding of our studies was a reduction of the specific activity of glutathione reductase and of total glutathione content in stage III, whereas glutathione S-transferase activity was not significantly influenced by the progression of the tumor (Table 3). These data can be interpreted as impairment of detoxification potency for reactive metabolites of PAH's due to a reduced GSH content. As GST activity did not change significantly with tumor growth, the decreased detoxification potency seems to be due to the depletion of reduced glutathione due to an impairment of glutathione reductase activity. Depletion of glutathione below a critical concentration allows an enhancement of lipid peroxidation evoked by endogenous promoters. This indicates that GSH is of great importance for the detoxification system of lipid peroxides. Therefore depletion of GSH may be one of the mechanisms causing peroxidative damage of cell membranes (35). Our data is reduced by a decreased activity of GST in patients with advanced bronchial carcinoma. Previous studies have shown that glutathione and GST were necessary to prevent binding of carcinogenic metabolites to DNA as has been found for different animal species and various organs (11, 14).

Another important factor of tumor prevention is the integrity of the immune system (36-38). The significant reduced mitogenic response to concanavalin A (Table 3) in stage III indicates an impairment of T-lymphocyte function.

Age-specific phenomena play an important role in the development and clinical manifestation of malignant tumors (20, 33, 34), as supported by response (Figure 2) in advanced age. This is thought to be due to an age-related decrease of lymphocyte function representing an impairment of the immune system and the lower activity of GST in the elderly are a condition for the development of bronchial carcinomas. Progression of malignant tumors seems to be made possible by depletion of GSH due to a decreased activity of glutathione reductase, rendering cell membranes susceptible to damage due to lipid peroxidation.

ACKNOWLEDGEMENTS

This work was supported by a grant of the Landesamt fur Forschung, NRW, and of Sandoz Therapeutische Stiftung.

REFERENCES

1 Daly, J.W., Jerina, D.M., Witkop, B. (1972) Experientia, 28, 1129-1149
2 Sims, P., Grover., P.L. (1974) Adv. Cancer Res., 20, 165-274
3 Gelboin, H.V., (1980) Physiol. Rev., 60, 1107-1166
4 Pelkonen, O., Nebert, D.W. (1982) Pharmacol Rev., 34, 189-222
6 Wynder, E.L., Graham, E.A. (1950) J.Am. Med. Assoc., 143, 329-336
7 Auerbach. O., Stout., A.P., Hammond., E.C., Garfinkel, L. (1961) N. Engl. J. Med., 265, 253-267
8 Doll, R. and Peto., R. (1976) Br. Med. J., 2, 1525-1536
9 Doll, R., (1981) Can. J. Public Health, 72, 372-381
10 Nou, E. and Hillerdal, O. (1981) Eur. J. Respir. Dis., 62, 152-159
11 Glatt, H. and Oesch, F. (1977) Arch. Toxicol., 39, 87-96
12 Jakoby, W.B. (1978) in Advances in Enzymology and Related Areas of Molecular Biology (Meister, A. ed), pp 383-414. Wiley & Sons, New York
13 Reed, D.J. and Beatty, P.W. (1980) Rev. Biochem Toxicol., 2, 213-242
14 Hesse, S., Jernstrom, B., Martinez, M., Moldeus, P., Christodoulides, L., Ketterer, B. (1982) Carcinogenesis, 3, 757-760
15 Meister, A. (1983) Science, 220, 472-477
16 Orrenius, S. and Moldeus, P. (1984) Trends Pharmacol. Sci., 5, 432-435
17 Blanton, R.H., Myers, M.J., Bick, P.H. (1988) Toxicol Appl. Pharmacol 93, 267-274
18 Lyte, M., Blanton, R.H., Myers, M.J., Bick, P.H. (1987) Int. J. Immunopharmacol., 9, 307-312
19 Myers, M.J., Blanton, R.H., Bick, P.H. (1988) Int J Immonopharmacol., 10, 177-186
20 Flood, P.M., Urban, J.L., Kripke, M.L., Schreiber, H. (1981) J Exp Med., 154, 275-290
21 Lighthart, G.J., Vlokhoven, P.C. van, Schuit, H.R.E., Hijmans, W. (1986) Immunology, 59, 353-357
22 Nagel, J.E., Chopra, R.K., Chrest, F.J., McCoy, M.T., Schneider, E.L., Holbrook, N.J., Adler, W.H. (1988) J. Clin. Invest., 81, 1096-1102
23 UICC (1982) TNM Classification of Malignant Tumors. /rd edn., Geneva
24 Oesch, F., Schmassmann, H., Ohnhaus, E., Athaus, U., Lorenz, J. (1980) Carcinogenesis, 1, 827-835
25 Habig, W.H., Pabst, M.J., Jakoby, W.B. (1974) J. Biol. Chem., 249, 7130 -7139
26 Mize, C.E. and Langdon, R.G. (1962) J. Biol. Chem, 237, 1589-1595
27 Tietze, F. (1969) Anal Biochem, 27, 502-522

28   Lowry, O.H., Rosebrough. N.J., Farr, A.L., Randall, R.J. (1951) J. Biol.
      Chem., 193, 265-275
29   Grosse-Wilde, H., Baumann, P., Netzel, B., Kolb, H.J., Mempel, W.,
      Wank, R., Albert. E.D. (1973) Transplant. Proc., 5, 1567-1571
30   Miller, E.C. and Miller, J.A. (1981) Cancer, 47, 1055-1064
32   Harris, C.C., Willey, J.C., Saladino, A.J., Grafstrom, R.C. (1985) in
      Carcinogenesis - A Comprehensive Survey, 8, (Mass, M.J. et al. eds),
      pp 159-171. Raven Press, New York.
33   Teller, M.N., Stohr, G., Curlett, W., Kubisek, M.L., Curtis, D. (1964)
      J Nat Cancer Inst., 33, 649-656
34   Geddes, D.M. (1979) Br.J. Dis. Chest., 73, 1-7
35   Siegers, C.P., Hubschner, W., Younges, M. (1982) Res. comm. Chem. Patol.
      Pharmacol., 37, 163-169
36   Kerman, R.H., and Stefani, S.S. (1977) Oncology, 34, 10-12
37   Reese, J.C., Rossio, J.L., Wislon, H.E., Minton, J.P., Dodd, M.C. (1975)
      Cancer., 36, 2010-2015
38   Gottesman, S.R.C. (1987) Rev. Biol. Res. Aging, 1, 103-162

TUMOR MARKERS IN BRONCHIOALVEOLAR LAVAGE

Munoz Calvo, R., Barbero E. M. A., Calloll L., Gomez de Terróros
F. J., and Blasco R.

Seccion Departamental de Fisiologia Animal
Falcultad de Farmacia. Hospital del Aire. Universidad
Complutense
28040 Madrid, Spain

ABSTRACT

Tumor markers may be defined as a series of chemical products that can be
produced by a tumor or that can accompany it during its development and
evolution reflecting the different changes occurring when the cells enter
upon a neoplastic transformation.

Bronchioalveolar lavage (BAL) consists of the introduction of some
quantity of sterile isotonic serum in the airways with posterior aspiration
of it.

By using radioimmunoassay techniques it has been possible to quantify
four tumor markers: tissue polypeptide antigen (TPA), carcinoembryonic
antigen (CEA) neuron specific enolase (NSE) and ferritin in the cell-free
supernatant of cells coming from 17 BAL of patients with etiopathogenetic
processes showing tumoral characteristics (oat cells, epidermoid carcinoma,
adenocarcinoma).

The result was the finding of higher concentrations of these parameters
in comparison with the ones found in 22 BAL of a reference group, being
possible to define different statistically significant relations.

INTRODUCTION

An antigen with a tumor can be considered as a malignant linked substance, whose clinical relevance is related to its properties as a diagnostic
and prognostic tool for inflammatory and neoplastic processes (1).

The bronchioalveolar lavage (BAL), also called alveolar lavage, consists
of the introduction of some quantity of sterile isotonic serum in the airways
with posterior aspiration of it (2). It should be regarded as a useful
technique for studying different biochemical parameters, giving the opportunity, not possible in other organs, of carrying out a simple and repetitive
analysis of the soluble pulmonary substances.

The BAL is considered as an important advance in the knowledge of pro-

cesses occurring in alveolar interstitial structures and permits one to study
phenomena during inflammatory-neoplastic activity (3) (4).

The study of different inmunoglobulins and certain enzymes in the BAL has
been the object of investigation in a great number of cases (5, 6, 7, 8, 9,
10), nevertheless, the quantification of substances directly related to neoplastic processes is a more restrictive field.

MATERIALS AND METHODS

A study on 39 BAL from subjects ages $50.52 + 17.8$ has been undertaken.

After radiological and analytical assays the following distribution was
made: 17 BAL from subjects with etiopathogenetic processes of tumoral origin
(6 epidermoid CA, 6 Oat cells, 5 Adenocarcinoma) and 22 BAL from normal subjects (reference or control group).

The BAL technique was based on the studies of different authors (11, 12,
13, 14) using a fibroscope with a cold light source.

The patients were asked not to smoke 48 hours before the BAL and analytical parameters as well as arterial blood gases were controlled. In all the
cases an electrocardiogram was performed.

A dose of 0.6 mg. of intramuscular atropine was injected half an hour
before the physical examination, using topical anaesthesia with tetracaine in
upper airways and cricothyroid anaesthesia with lidocaine (2%).

The BAL was made with a sterile solution of NaC1 (0.9%) at room temperature in bolus injections of 50 ml., making a total of 250 ml. and beginning the aspirations 30 seconds after the instillation of each bolus.

The material obtained was centrifuged at 200 g in order to separate the
cell fraction of the supernatant free of cells. Concentrations of TPA, CEA
NSE and ferritin were determined using radioimmunoassay techniques (RIA)
(inmunocompetition with I-125 monocloned antibodies) as well as total proteins (15).

Statistical methods used were students test and the linear correlation
coefficient.

RESULTS

The results obtained (TPA, CEA, NSE and ferritin), expressed in terms of
total proteins were higher in the BAL from patients with tumoral etiopathogenetic processes (TPA = 718.3 u/g P.T., CEA + 17.1 ng/mg P.T., NSE = 1.67
ng/mg P.T y Fer = 16.9 ng/mg P.T) in comparison with the reference group
(control), being statistically significant for the TPA ($p < 0.05$) (Table 1).

Linear correlations among the different parameters were set, obtaining
the following statistically significant differences: CEA/NSE ($p = <0.01$),
CEA/TPA ($p = <0.09$) y TPA/NSE ($p = <0.013$). (Table 2).

DISCUSSION

There are many studies demonstrating the relation between the plasma concentrations of TPA, CEA, NSE and ferritin and different tumoral processes
(16) (17).

Table 1

		x	min V.	max V.	
TPA	Tumoral	718.3	0	3.263	
(u/g P.T)	Control	29.6	0	125.1	(p<0.05)
CEA	Tumoral	17.1	0.25	149.6	
(ng/mg P.T)	Control	6.3	0.16	51.6	(p=0.22)
NSE	Tumoral	1.67	0.2	4.8	
(ng/mg P.T)	Control	1.5	0.3	6	(p=0.696)
FERT	Tumoral	16.9	0.5	71	
(ng/mg P.T)	Control	8.8	0.08	46	(p=0.182)

Table 2

	CEA	TPA	NSE	FER
CEA	1.000 -	0.5832 (p=0.009)	0.6887 (p=0.001)	0.3351 (p.0.094)
TPA	0.5832 (p=0.009)	1.000 -	0.5562 (p=0.013)	0.2730 (p=0.153)
NSE	0.6887 (p=0.001)	0.5562 (p=0.013)	1.000 -	0.0072 (p=0.489)
FER	0.3351 (p=0.091)	0.2730 (p=0.153)	0.0072 (p=0.489)	1.000 -

Plasma increases of TPA and CEA in patients with lung tumors have been described (18) (19), and these have been defined as tumor markers in oat cell (20), adenocarcinomas, etc. (21).

In lung pathology, a relation between NSE and lung cancer (22, 23) is observed, typifying this relation as a tumor marker for oat cells (24).

Many studies have been reported on the plasma increase of ferratin and the presence of lung tumors (25) as well as the relations between the increase of this protein and CEA in patients suffering from lung cancer and its posterior remission after certain therapy (26).

Therefore it is evident there is an increase of plasma levels of TPA,

CEA, NSE and ferratin in lung pathologies. If the BAL may contribute to understanding the possible pathologic processes developing in the lungs, it appears to be logical to think that these parameters would be increased in the BAL from patients with etiopathogenetic tumoral processes.

The study and quantification of parameters such as CEA, calcitonin, creatine-kinase-BB and DNA in BAL from patients with tumoral processes carried out by different authors (27) (28) confirm above, as these authors observed an increase in the concentration of these parameters in comparison with a reference group.

The development of radioimmunoassay techniques using monoclonal antibodies makes it possible to quantify the different parameters in certain biological fluids making this method very useful in the BAL.

The above findings are confirmed by our results in which an increase of TPA, CEA, NSE and ferratin obtained in the BAL from subjects with etiopathogenetic of tumoral origin is obtained, it was also observed that some correlation exist among the different parameters; the CEA concentrations are similar to those obtained by other authors using other analytical techniques (27).

REFERENCES

1   Galofre P, Ruibal A, Oncologia en Medicina Nuclear; aplicaciones diagnosticas de los isotopos radiactivos. Ed. Cientifico Medica, cap. 20.
2   Myrvick Q N, Leake ES, Farijs B, (1961) J. Inmunol 86; 133-139.
3   Hunninghake Gw, Crystal RG, (1981) N. Engl. J. Med. 305; 429-434.
4   Crystal R, Bitterman P, Renhard S, Hance A, Keogh B, (1984). N. Engl. J. Med. 310; 154-166.
5   Lawrence E, Blaeses R, Martin R, Stevens P, (1978) J. Clin. Invest. 62; 832-835.
6   Mandel MA, Dvorak KJ, Worman LW, De Cosse JJ, (1970) N. Engl. J. Med. 295; 694-698.
7   Danielle R, Elias J, Epstein P, Rossman P, (1985) Ann. Inter. Med. 102; 93-108.
8   Mordelet Ms, Stanislas GM, Huchow GJ, Baumen FC, Marsac JH, Chratien J. (1982) Am. Rev. Respir. Dis. 126; 472-475.
9   Brambilla C, Fourcy A, Stoebner P, Paranielle B. (1977) Chretien J. Ed. Paris Colloques Inserm. 373-380.
10 De Vust P, Jeward J, Dumortier P, Vandermorten B, Vandermeyer R, Yernault JC. (1982) Am. Rev. Dis. 126: 972-976.
11 Gee JBL., Fick RB, (1980) Thorax 35; 1-8.
12 Velluti G, Capelli O, Lusuardi M. (1983) Resp. 44; 403-410.
13 Basset F, Bernaudin J, Soler P, Chollet S. (1983) Lavage bronchoalveolaire d'exploration, Encycl. Med. Chir. Paris. Pouman. 6000-M-50, 11.
14 Callol L, Laguna R, Caro M, Gomez de Terreros FJ, (1984) Enf. Tor. 33, 2; 135-139.
15 Lowry OH, Rosebrough NJ, Farr AL. (1951) J. Biol. Chem. 193; 265-269.
16 Bjprklund B. (1980) Tumor Diagnostik 1; 9-20.
17 Nemoto T, Constantine R, Chu TM, (1979) J. N. Clin. 63; 1347-1350.
18 Vincent RG, Chu TM. (1972) J. Thorac. Cardiovas, Surg. 66; 320-328.
19 Cocanon J, Dalbow M, Liebler G. (1974) Cancer 34; 184-192.
20 Ruibal A, Encabo G, Getael R, Lafuerza A, Bodi R, Mtez Millares E. (1983) Rev. Esp. Med. Nucl. 3; 111-116.
21 Vincent RG, Chu TM, lane WW. (1979) Cancer. 44; 685-691.
22 Tapia FJ, Barbosa AJA, Marangos PJ, Polak JM, Bloom SR, Dermody C. (1981) Lancet 11; 808-811.
23 Carney D, Ihde D, Cohen M, Marrangos P, Bunn P, Minna (1982) Lancet 583-586.

24 Ariyoshi Y, Kato K, Isihiguro Y, Ota K, Sato T, Suchi T. (1983) Gann 74; 219-225.
25 Niitsu Y, Goto Y, Kohgo Y, Adach C, Unodera Y, Urushizakie I (1980) In radioimmunoassay of hormones, proteins and enzymes. Ed. Albertini A. Excert. Med. (1980) 256-266.
26 Groop C, Haveman K, Gunter L. (1978) Cancer 42; 2802-2808.
27 Goldstein NG, Lippman ML, Goldberg SK, Fein Am, Shapiro B, Leon SA (1985) Am. Rev. of Respir. Dis. 132; 60-64.

PHOSPHORYLATION AND DEPHOSPHORYLATION OF THE CYTOCHROME P-450-DEPENDENT

MONOOXYGENASE SYSTEM: REGULATORY DEVICES BY POSTTRANSLATIONAL MODIFICATION

Walter Pyerin and Hisaaki Taniguchi

German Cancer Research Center
Institute of Experimental Pathology
D-6900 Heidelberg, F. R. G.

ABSTRACT

Cyclic AMP-dependent protein kinase and unspecific phosphatase are
described as posttranslational modifiers of the hepatic microsomal cytochrome
P-450-dependent monooxygenase system. Kinase and phosphatase affect the
system not by a protein phosphorylation-dephosphorylation mechanism but
rather independently of each other at different sites of the monooxygenase
system. The protein kinase introduces a phosphoryl group into cytochrome P-
450, the terminal monooxygenase itself, at a critical serine residue causing
a conformational change which converts P-450, into its inactive form P-420.
Since only certain P-450 isoenzymes carry the kinase recognition sequence -
Arg-Arg-X-Ser-, the phosphorylation may act as an isoenzyme-specific form of
regulation of the monooxygenase system through the control of the degradation
of P-450. In contrast, the phosphatase digests FMN, one of the two prosthetic flavins of the NADPH-cytochrome P-450 reductase, thereby interrupting
the electron transport chain which leads to an unspecific decrease of the
overall monooxygenase activity. Neither kinase nor phosphatase affect significantly the rest of the protein components of the monooxygenase system.

INTRODUCTION

The hepatic microsomal cytochrome P-450-dependent monooxygenase system
has a fundamental role in the metabolism of a large variety of exogenous and
endogenous hydrophobic compounds including chemical carcinogens and co-
carcinogens. The system's activity may result in beneficial detoxification
or detrimental activation of such compounds (1). Understanding of the regulation of this system, therefore, is of major concern in cancer research. The
system is composed of the following protein components, all of which are
tightly bound to the endoplasmic reticulum (ER) membrane interacting with
each other in the plane of the membrane: The hemeprotein cytochrome P-450,
which occurs in various isoenzymic forms, is the component of catalyzing the
monooxygenase reaction in which a substrate S is converted to a product SO in
the presence of oxygen and electrons. The flavoproteins NADPH-cytochrome P-
450 reductase and NADH-cytochrome b_5 reductase and the heme-protein cytochrome b_5 are the components transferring electrons from NADPH and/or NADH to
cytochrome P-450 (1; Fig. 1).

A well-known and extensively studied form of regulation of the mono-

Biochemistry of Chemical Carcinogenesis
Edited by R. Colin Garner and Jan Hradec
Plenum Press, New York, 1990

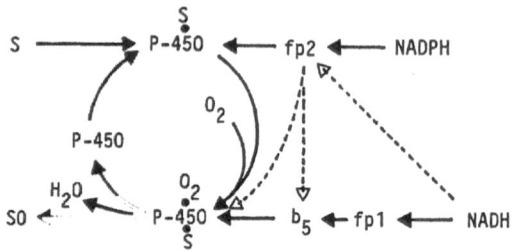

Figure 1. Cytochrome P-450 reaction cycle and electron transport pathways of the hepatic microsomal cytochrome P-450-dependent monooxygenase system. P-450, cytochrome P-450; fp2, NADPH-cytochrome P-450 reductase; b_5 cytochrome b_5; fp1, NADH-cytochrome b_5-reductase; S, substrate; SO, product.

oxygenase system is the induction of the P-450 isoenzymes, a process requiring de novo synthesis of RNA and protein (1). Other possible regulatory mechanisms such as posttranslational and modifications of monooxygenase components, however, have received much less attention. We describe here some of our recent studies on the phosphorylation of cytochrome P-450 isoenzymes which suggest the possibility of an isoenzyme specific regulation at the posttranslation level and those on the effect of dephosphorylating enzymes acting on NADPH-cytochrome P-450 reductase which show a further regulatory possibility.

Phosphorylation of cytochrome P-450

Various treatments which cause the increase of the cellular concentration of cyclic 3',5'-adenosine monophosphate (cAMP), such as the administration of catecholamines, have been known to depress hepatic drug metabolism (2). The direct involvement of cAMP in the process was shown when $N^6,0^{2'}$-dibutyryl-cAMP, was found to decrease microsomal monooxygenase activities, the effect being rapid in onset (10 min) and long-lasting (over 24 hours) (3,4). The target for the action of cAMP and the mechanism by which cAMP affects the monooxygenase system, however, remained unknown. Our observation that incubation of purified P-450 isoenzyme 2 (LM_2), the major phenobarbital-inducible isoenzyme in rabbit liver (phenobarbital is a hepatic tumor promoter; 5), with the catalytic subunit of cAMP-dependent protein kinase in the presence of ATP results in a phosphorylation of this P-450 isoenzyme (6), therefore, revealed for the first time one of the targets of the cAMP-action and, at the same time, opened the way to investigations of mechanisms.Phosphorylation of proteins is a frequently-observed powerful control device employed by cells to regulate numerous of their physiological functions (7).

The phosphorylation of P-450 LM_2 was found to take place at a serine residue located in position 128 of the amino acid sequence of the molecule (8,9). This serine residue belongs to a part of the molecule with the sequence -Arg-Arg-X-Ser-, the typical recognition sequence of the cAMP-dependent protein kinase (7). This sequence occurs only once in the molecule providing a single phosphorylation site per molecule. P-450 isoenzymes such as isoenzyme 4 (LM_4) of rabbit liver lacking this sequence was consequently not phosphorylated. From this it became clear that the phosphorylation of a certain P-450 isoenzyme can be predicted from its amino acid sequence. To check this, several P-450 isoenzymes of known sequence were purified from rat liver and phosphorylated with the catalytic subunit of cAMP-dependent protein kinase. As expected, the phenobarbital-inducible rat isoenzymes b (PB 3a) and e (PB 3b), the equivalents to rabbit P-450 LM_2, which both contain the kinase recognition site, were phosphorylated while isoenzymes c (MC 1b) and d (MC 1d), lacking the site, were not (10; Table 1).

250

Table 1. Amino acid sequence and phosphorylation of hepatic cytochrome P-450 isoenzymes by catalytic subunit of cyclic AMP-dependent protein kinase. Cytochrome P-450 isoenzymes were isolated from liver microsomes of rabbits and rats pretreated with phenobarbital (isoenzymes LM_2, PB 3a, and PB 3b) or with methylcholanthrene (isoenzymes LM_4, MC 1b, and MC 1d) and incubated with catalytic subunit of cAMP-dependent protein kinase in presence of [-^{32}P]ATP. The phosphorylation was estimated semiquantitavely by densitometer tracing following polyacrylamide gel electrophoresis and autoradiography. For experimental details see refs.6, 10, 11. Kinase recognition sequence underlined.

P-450 isoenzyme source	form	Primary structure[a]	Phosphorylation predicted	observed
rabbit	2 (LM 2)	121 -WRALRRFSLATMRDFGMGKR-	+	+
rabbit	4 (LM 4)	132 -WAARRRLAQDSLKSFSIASN-	-	-
rat	b (PB 3a)	121 -WKALRRFSLATMRDFGMGKR-	+	+
rat	c (MC 1b)	135 -WAARRRLAQNALKSFSIASD-	-	-
rat	d (MC 1d)	132 -WAARRRLAQDALKSFSIASD-	-	-
rat	e (PB 3b)	121 -WKALRRFSLATMRDFGMGKR-	+	+

[a] as given in ref. 20

The phosphorylation of P-450 LM_2 was found to have physiologically "meaningful" characteristics, i.e. a K_m value of 2-8 uM P-450 (11), the cellular concentration of P-450 being in the order of 10-50 uM depending on the treatment of the animals. Similar K_m values as that obtained with solubilized P-450 were measured with P-450 in presence of phosphatidylcholine or incorporated into model membranes of ER (phosphatidylcholine/phosphatidylethanolamine/phosphatidic acid, 2/1/0.06) (11), and also with P-450 bound to microsomes (9). This shows that phosphorylation of P-450 occurs also in its membrane-bound form, the state of occurrence in living cells. To test whether phosphorylation in living cells can really take place, primary cultures of hepatocytes were prepared from phenobarbital-induced rats and the cellular level of cAMP was increased appropriately to activate endogenous kinase. As a result, phosphorylated P-450 could be identified in the microsomal fraction by several different methods, the only amino acid residue phosphorylated being a serine residue located at the same site as in the P-450 molecules phosphorylated in vitro (12).

What is the effect of the phosphorylation of P-450 and what is the mechanism of the effect? Concomitant with the phosphorylation, the enzymatic activity of both rabbit liver P-450 (11) and rat liver P-450 (12) decreased. Phosphorylation of P-450, therefore, has an inactivating effect. The light absorption spectrum, a sensitive indicator of the conformational status of P-450 showed that the reason for this effect is the change in the conformation of P-450. In the CO difference spectrum of reduced cytochrome P-450,

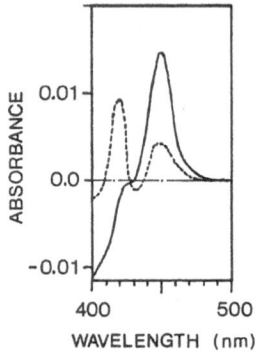

ABSORBANCE

0.01

0.0

-0.01

400   500
WAVELENGTH (nm)

Figure 2. Effect of phosphorylation on the CO-difference absorbance spectrum
of cytochrome P-450.
P-450 isoenzyme LM$_2$ was incubated for 60 minutes in presence of catalytic
subunit of cAMP-dependent protein kinase and ATP followed by dilution with
potassium phosphate buffer/glycerol/emulgen, bubbling with CO, and reducing
with dithionite as described in refs. 12 and 13. Given are the spectra
before (----) and after (- - - -) phosphorylation.

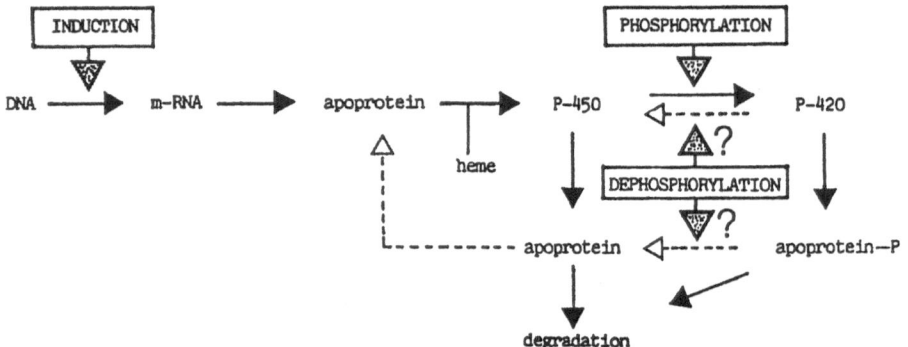

Figure 3. Schematic representation of the isoenzyme-specific control system
of the cellular cytochrome P-450 level.

the absorption peak at 450 nm decreased upon phosphorylation and concomit-
antly, the absorption shoulder at 420 nm developed into a peak (13; Fig. 2).

    The change in absorption strictly paralleled the phosphorylation degree
which reached approximately 0.6 mol phosphate/mol P-450 LM$_2$. This means that
phosphorylation converts cytochrome P-450 to cytochrome P-420 which is known
to possess no enzymatic activity. This mechanism can easily explain the
above mentioned rapid decrease in the hepatic monooxygenase activity observed
upon treatment of animals with dbcAMP.

    We therefore suggest that there exists, in addition to induction, an
isoenzyme-specific posttranslational modification of P-450 through protein
phosphorylation. This phosphorylation serves as a "down-regulating" mechanism in the isoenzyme-specific control systems of cytochrome P-450 (Fig. 3).

252

Effects of phosphatase on NADPH-cytochrome P-450 reductase

   In addition to the phosphorylation, the hepatic monooxygenase system was
found to be also affected by dephosphorylating enzymes. Treatment of microsomes obtained from livers of rabbits and rats with unspecific phosphatases
caused a marked decrease in monooxygenase activity (6). So the dephosphorylation has the same effect as the phosphorylation. How can this apparent
dichotomy be explained?

   Further analysis showed that the treatment with phosphatase affected not
only the overall monooxygenase activity but also NADPH-cytochrome c reductase
activity catalyzed by NADPH-cytochrome P-450 reductase while, in contrast,
the cytochrome P-450 was not affected (14-16). The latter was confirmed by
two different methods. Firstly, the CO-difference spectrum of reduced cytochrome P-450 was measured and no significant spectral change or increase in
cytochrome P-420 was observed. Secondly, cumene hydroperoxide-dependent N-
demethylase activity, which is catalyzed by P-450 alone and independent of
other protein components, was determined using N,N-dimethylaniline as substrate and, again, no significant change was noted (16). This indicates that
the sites of action of protein kinase and phosphatase differ, the main target
of the phosphatase being not cytochrome P-450 but rather NADPH-cytochrome P-
450 reductase.

   The time course of the decrease of the NADPH-dependent monooxygenase
activity of microsomes was similar to that of the decrease of NADPH- cytochrome P-450 reductase, and addition of phosphatase inhibitors abolished the
phosphatase effect (16) showing that it is in fact the result of the catal-
lytic activity of the phosphatase. In order to examine whether the phosphatase directly inactivates the reductase, NADPH-cytochrome P-450 reductase
was purified from liver microsomes of phenobarbital-induced rabbits and incubated with phosphatase under conditions similar to those employed in the
experiments with microsomes. A decrease of NADPH-cytochrome c reductase
activity with a similar time course was obtained (16). These results make
clear that phosphatase directly attacks NADPH-cytochrome P-450 reductase and
that the decrease of microsomal monooxygenase activity caused by the phosphatase is mainly due to the inactivation of the reductase by the phosphatase.

   What is the mechanism of the phosphatase effect? Dephosphorylation of
the enzyme protein can be excluded because NADPH-cytochrome P-450 reductase
becomes only marginally phosphorylated by protein kinases (10; see also
below). Measuring ferricyanide reduction in comparison to cytochrome c reduction gave the first hint to the mechanism, namely, that the phosphatase
probably inactivates one of the two prosthetic flavins of NADPH-cytochrome P-
450 reductase, the FMN, because cytochrome c reduction, which requires both
FMN and FAD, was affected more than ferricyanide reduction which requires
only FAD, and because FMN is known as a good substrate for the phosphatase
(16). To test this possibility, flavin analysis of phosphatase-treated
reductase was carried out. The ratio of FMN-to-FAD in untreated NADPH-
cytochrome P-450 reductase was approximately 0.8, i.e. both flavins were
present in comparable amounts (Table 2). Treatment with phosphatase resulted
in the appearance of a new peak at the expense of FMN, FAD remaining unchanged, This new peak was identified as riboflavine (16). Its appearance
can easily be explained by the splitting-off of the terminal phosphate group
of FMN.

   Phosphatase, therefore, interrupts the electron transport chain from
NADPH to cytochrome P-450 by digesting FMN and thereby inactivating NADPH-
cytochrome P-450 reductase activity. As a consequence, the activity of the
overall monooxygenase activity decreases. Since the presence of unspecific
FMN-phosphatase in liver cels has been demonstrated (17), the observed effect
of phosphatase on the reductase in another potential possibility for post-

Table 2. Flavin analysis of phosphatase-treated hepatic NADPH-cytochrome P-450 reductase.
Purified reductase was incubated without (control) or with (treated) alkaline phosphatase, flavins were extracted and chromatographed on a FPLC Mono Q column followed by fluorescence determination of flavinadeninedinucleotide (FAD), flavin-mononucleotide (FMN), and riboflavin (Rf) as described in ref. 16.

Incubates	Relative contents		
	FAD	FMN	Rf
control	1.0	0.8	0.0
phosphatase-treated	1.0	0.15	0.65

translational control of monooxygenase activity. In contrast to the iso-enzyme-specific phosphorylation of P-450, the phosphatase-effect concerns the total monooxygenase activity.

Phosphorylation and dephosphorylation of other monooxygenase components

Significant phosphorylation of components of the monooxygenase system other than cytochrome P-450 was not observed. NADPH-cytochrome P-450 reductase was a very poor substrate of cAMP-dependent protein kinase while NADPH-cytochrome b_5 reductase and cytochrome b_5 were not phosphorylated at all (10). Hence the kinase effect seems to be restricted to cytochrome P-450.

When the same phosphatase treatment of microsomal samples was carried out as in the case of examining the effect on monooxygenase activity, no change as observable in NADH-dependent ferricyanide reductase activity or in NADH-dependent cytochrome c reductase activity (16). Since the former is cata-lyzed by NADH-cytochrome b_5 reductase and the latter by NADH-cytochrome b_5 reductase togehter with cytochrome b_5, phosphatase has obviously no effect on these two monooxygenase components as it has on cytochrome P-450 (see above).

Conclusion

The destructive effects of the cAMP-dependent protein kinase and the un-specific phosphatase on the activity of the hepatic microsomal cytochrome P-450-dependent monooxygenase system described here are potential rapid control devices of the system at the posttranslational level. Their effect is based on covalent modifications of certain of the component proteins of the system, namely cytochrome P-450 by the kinase and NADPH-cytochrome P-450 reductase by the phosphatase, whereas the rest of the protein components are not affected. This is in contrast to those control devices acting by changes of the pro-tein-lipid interactions, in which the membrane of the endoplasmic reticulum plays an active role in determining diffusion, orientation, and interactions of the component proteins of the monooxygenase system as well as by exerting a substrate condensation effect (18,19). Detailed studies on the protein modifications of the system will shed more light on the expression and regu-lation of the cytochrome P-450 dependent monooxygenase system, which is one of the key host factors in chemical carcinogenesis.

REFERENCES

1   Schenkman, J. B. and Kupfer, D. (eds.) (1982) Hepatic Cytochrome P-450
    Monooxygenase System, Pergamon Press, Oxford/New York/Toronto/Sidney/
    Paris/Frankfurt.
2   Fouts, J. R. (1962) Fed. Proc. 21, 1107-1111
3   Weiner, M., Buterbaugh, G. G. and Blake, D. A. (1972) Res. Commun. Chem.
    Pathol. Pharmacol. 3, 249-263
4   Weiner, M., Buterbaugh, G. G. and Blake, D. A. (1972) Res. Commun. Chem.
    Pathol. Pharmacol. 4, 37-50
5   Kaufmann, W. K., Ririe, D. G. and Kaufman, D. G. (1988) Carcinogenesis 9,
    779-782
6   Pyerin, W., Wolf, C. R., Kinzel, V., Kubler, D. and Oesch, F. (1983)
    Carcinogenesis 5, 573-576
7   Krebs, E. G. and Beavo, J. A. (1979) Ann. Rev. Biochem. 48, 923-959
8   Muller, R., Schmidt, W. E. and Stier, A. (1985) FEBS Lett. 187, 21-24
9   Pyerin, W., Marx, M. and Taniguxhi, H. (1986) Biochem, Biophys. Res.
    Commun, 134, 461-468
10 Pyerin, W., Taniguchi, H., Horn, F., Oesch, F., Amelizad, Z., Friedberg,
    T. and Wolf, C. R. (1987) Biochem, Biophys. Res. Commun. 142, 885-892
11 Pyerin, W., Taniguxhi, H., Stier, A., Oesch, F. and Wold, C. R. (1984)
    Biochem. Biophys. Res. Commun. 122, 620-626
12 Pyerin, W. and Taniguchi, H., submitted
13 Taniguchi, H., Pyerin, W. and Stier, A. (1985) Biochemical Pharmacology
    34, 1835-1837
14 Pyerin, W., Jochum, C., Taniguchi, H. and Wolf, C. R. (1986)
    Res. commun, Chem, Pathol. Pharmacol, 53, 133-136
15 Pyerin, W., Horn, F. and Taniguchi, H. (1987) J. Cancer Res. Clin. Oncol.
    113, 155-159
16 Taniguchi, H. and Pyerin, W. (1987) Biochim. Biophys. Acta. 912,
    295-302
17 Lee, S. S. and McCormick, D. B. (1983) J. Nutr. 113, 2274-2279
18 Taniguchi, H., Imai, Y. and Sato, R. (1987) Biochemistry 26, 7084-7090
19 Taniguchi, H. and Pyerin, W. (1988) J, Cancer Res. Clin. Oncol., in
    press
20 Black, S. D. adn Coon, M. J. (1986) in Cytochrome P-450, Structure,
    Mechanism and Biochemistry, (Ortiz de Montellano, P. R., ed.) Plenum
    Press, New York/London

EFFECT OF S-ADENOSYL-L-METHIONINE ON SERUM LIPOPROTEIN IN EXPERIMENTAL

HEPATIC CARCINOGENESIS

M.J.Miro, C.Arce, H.Aylagas, C.Fdez-Aguado and E.Palacious-Alaiz

Instituto de Bioquimica. Facultad de Farmacia. Universidad
Complutense,
28040 - Madrid, Spain

SUMMARY

Plasma lipoprotein concentration and composition were investigated in
experimental hepatic carcinogenesis induced by thioacetamide (TAA). The
effect of S-adenosyl-L-methionine (SAM) on lipoprotein alterations during
hepatic injury was also tested.

After 30 days administration of the weak carcinogen i.p. (50 mg/Kg/day)
to male Wistar rates, a decrease of serum VLDL concentration and HDL
accumulation in plasma was observed. Consequently, the HDL/VLDL ratio was
elevated. Any increase of LDL/VLDL ratio was also observed. Total plasma
phospholipids were lower than reference values. Sphingomyelin/phosphatidyle-
thanolamine (SPH/PE) ratios/significantly decreased in plasma, as well as in
VLDL-LDL and HDL fractions, as a consequence of the reduced SPH level and
simultaneous increase of PE.

Treatment of experimental animals with SAM (2 mg/Kg/day) as well as with
TAA for 30 days resulted in a normal level of serum VLDL and the HDL/VLDL
ration. Phospholipid concentration in plasma was also normalized as a consequence of elevated level of these compounds in the VLDL-LDL fraction.
Administration of SAM also prevented the decrease of SPH/PE ratio in plasma
and different lipoproteins.

INTRODUCTION

The central role of the liver in lipoprotein metabolism is consistent
with an abnormal electrophoresis pattern and lipid composition of the plasma
lipoprotein in different liver diseases (1-3).

Human hepatomas, rat primary hepatocellular carcinomas and livers of rat
treated with hepatocarcinogenic agents lack feed back control of cholesterol
regulation (4, 5). Alterations in serum lipids and lipoproteins have been
described in cancer patients (6, 7) and in experimental carcinogenesis (8).

Between the different experimental models of parenchymal liver damage,
the hepatoxic compound thioacetamide (TAA) given to rats (9-12) offers a
suitable animal model to investigate disturbed lipid metabolic processess in
cirrhotic and preneoplastic tissue (9). This weak carcinogen reproduces in

rats necrosis of hepatocytes (10, 11) cirrhosis (13), hyperplastic nodules, tumors, cholangiomas and cholangiocarcinomas (14).

In previous investigations we observed that administration (i.p.) of TAA to rats results in alterations of subcellular organells composition (15) and hepatic phospholipid biosynthesis (16, 17). To obtain more information on these lipid changes we have investigated these parameters in experimental animals treated with S-adenosyl-L-methionine (SAM) as well as with TAA, because of the demonstrated contribution of this methyl group donor compound to the rapid adaptative response of liver to TAA-induced necrosis (11) and the various pharmacological effects reported for this compound (11, 18).

Abbreviations

ALAT, L-alanine: 2-oxoglutarate aminotransferase; ASAT, L-aspartate: 2-oxoglutarate aminotransferase; HDL, high density lipoproteins; LDL, low density lipoprotiens; LPC, lysophosphatidylcholine; PC, Phosphatidyl-choline; PE, phosphatidylethanolamine; PI, phosphatidylinositol; PS, phosphatidylserine; RER, rough endoplasmic reticulum; SAM, S-adenosyl-L-methionine; SPH, sphingomyelin; TAA, thioacetamide, VLDL, very low density lipoproteins.

MATERIALS AND METHODS

Chemicals

TAA, silicagel G type 60 and organic solvents were obtained from Merck (Darmstad). SAM was a donation from Europharma (Madrid). Phospholipid standards were supplied by Sigma Chemical Co. (St. Louis). Dextran sulfate was from Sochibo, S. A. (Bologne). Kits for cholesterol and triglycerides determinations were purchased from Boehringer Mannheim, S. A. (Barcelona) and reagents for aminotransferase enzymes from Technicon (Tarrytown).

Animals

Two month old male Wistar rats weighing 200-220 g were divided in the three following groups: 1) Rats which received a daily intraperitoneal injection of TAA (50 mg/Kg body weight) in 0.15 M CaC1 solution for 30 days; 2) Rats that received an (i.p.) injection of TAA as indicated above and a daily dose of S-adenosyl-L-methionine (2 mg/Kg body weight) for 30 days; 3) Control rats which were daily injected (i.p.) with 0.15 M NaC1 solution. Animals were fed with water and food ad libitum; 18 hours before sacrifice, control and treated rats were fasted (water ad libitum). For each experiment six animals were used. Blood samples were obtained from the abdominal aorta under light ether anesthesia and collected in tubes containing 0.1% Na_2 EDTA and centrifuged for plasma at 3,000 r.p.m. at 40^0C for 10 minutes. The liver was removed, weighed and scored for macroscopic alterations. Liver specimens were used for histological examinations.

Serum enzymes assays

Catalytic activities of serum aspartate aminotransferase and alanine aminotransferase were determined as previously described (11).

Morphological examinations

Each liver of the TAA, TAA + SAM-treated and control rats was examined at a macroscopic light level as indicated elsewhere (11).

Polyacrylamide gel electrophoresis

Electrophoresis of serum aliquots was carried out in a discontinuous concentration gradient flat polyacrylamide gel electrophoresis system according to standard laboratory techniques.

Isolation of serum lipoproteins

The dextran sulfate-Mg^{+2} precipitation procedure (19) was applied to achieve the HDL from VLDL-LDL separation.

Determination of serum and lipoprotein lipids

Isolation and identification of phospholipids were performed by two dimensional thin layer chromatography on silicagel G-60 coated glass plates (20). Lipid spots were visualised by iodine vapours and scraped into tubes for phosphorous analysis by the method of Rouser et al. (21). In serum and in each isolated lipoprotein fraction, triacylglycerol content (22, 23), total cholesterol (24, 22) and free cholesterol (22, 25) were determined. Esterified cholesterol was calculated as the difference between total and free cholesterol. Total and individual phosphalipids were quantified as indicated above and expressed as phosphatidylcholine (mg/100 ml plasma) and as ug of total P_i lipid/100 ml. plasma, respectively.

Statistical analysis

The results are given as means $\pm$ S.D. of six treated rats in duplicate experiments. For the statistical analysis Student's t-test was used. No significant differences were considered when p was equal to or greater than 0.05.

RESULTS AND DISCUSSION

Liver damage

The liver/body weight ratio increased about 68% in TAA-treated rats with respect to controls. Both the in vivo weight and liver weight changes contributed to this increase. The mean value of the former was 30% lesser in TAA-treated rats than in controls, while the increase of the absolute liver weight was 25% over reference value. Superficial examination of the liver showed a homogeneous micronodular appearance and histological analysis demonstrated a loss of liver architectural arrangement of the hepatocytes with dense inflamatory infiltrations and connective septa, limiting parenchymal nodules. Appearance of livers from rats treated with SAM as well as with TAA were normal.

Serum transaminases did not show changes in TAA-treated rats in any of the experimental conditions, which is consistent with our previous observations (11) which reported that serum ALAT and ASAT are not suitable parameters to monitor liver damage by TAA in long-term treatment. Serum ASAT and ALAT catalytic activities were 81 ± 9 IU/1 and 35 ± 10 IU/1 respectively, in control rats. Values for TAA-treated animals were 87 ± 10 IU/1 for ASAT and 37 ± 5 IU/1 for ALAT. In the same range were serum aminotransferases catalytic activities of TAA + SAM-treated rats.

Poliacrylamidegel electrophoresis

Figure 1 shows changes in the percentage distribution or serum lipoprotein classes from control, the TAA and TAA+SAM-treated rats, as measured densitometrically after its electrophoretic separation. The alfa-migrating band was found to increase in TAA-treated group, while a decrease was observed in the pre-beta-migrating lipoproteins. No changes were apparent in the beta-migrating band. Increased capacity for serum lipoprotein synthesis has been reported for the primary rat hepatoma induced by N-2-fluorenylacetamide (26) and in the Morris hepatoma 7777 and its host liver. Due to greatly increased $(U-{}^{14}C)-$ leucine incorporation in the predominant lipoprotein of rat serum HDL_2 the utilization of these lipoproteins by dividing cells has been suggested for the assembly of cellular membranes (26).

Present results show that in the experimental carcinogenesis by TAA increases of serum HDL occur even when signs of malignancy are not yet evident and the pathological substrate is compatible with a micronodular liver cirrhosis. The observed diminution of serum pre-beta lipoprotein may be a consequence of the decrease in hepatic VLDL formation and secretion. Reduction of the number of Golgi VLDL particles and impaired formation has been demonstrated on TAA-induced acute liver injury (27, 28). Changes in the serum lipoprotein pattern including decrease of the pre-beta-migrating band and increase of the alfa-lipoproteins as quantitated densitometrically after separation on agarose gel electrophoresis has been referred to incirrhosis provoked by long-term oral administration of TAA (9). As shown in figure 1, serum from rats that received SAM as well as TAA had a normal VLDL percentage, probably because of the participation of SAM in phospholipid methylation (29, 30) and possibly contributes to the preservation of integrity of

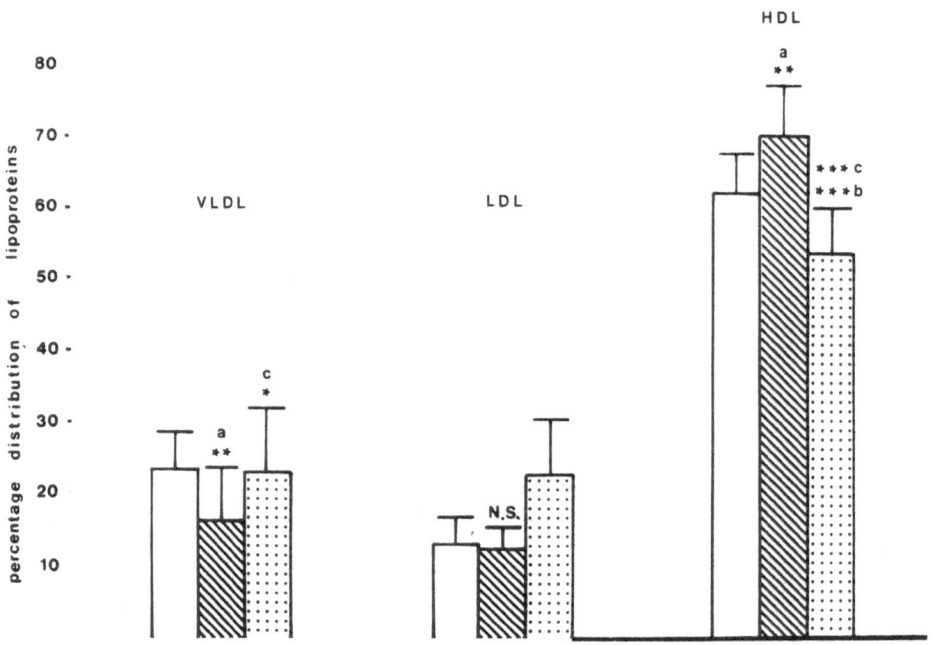

Figure 1: Polyacrylamide gel electrophoresis of serum lipoproteins from control ☐ ; TAA-treated ▧ ; and TAA+SAM-treated ⊡ rats. TAA was injected i.p. to rats at daily doses of 50 mg/Kg body weight in 0.15 M NaCl solution. SAM was similarly given in 0.15 M NaCl solution at daily doses of 2 mg/Kg body weight. Values, expressed as percentage distribution of lipoproteins bands after densitometric measure, are mean ± S.D. of 6 rats except controls (n = 20). The asterisks indicate a significant difference (a) TAA vs. control; (b) TAA + SAM vs. control; (c) TAA + SAM vs. TAA (p < 0.05 *. p< 0.01 ** and p > 0.001 ***)

membrane bound compartments that constitute the secretory pathway (31, 32). Depletion and disorganisation of the RER has been previously reported by effect of TAA (27).

The LDL/VLDL ratio was elevated by the effect of TAA, (0.5 for control and 0.8 for TAA-treated rats). SAM did not counteract this effect. Increase LDL concentration and abnormal composition, when isolated by ultracentrifugation, has been previously detected during TAA-treatment (33) and the suggestion of a possible covalent modification by TAA of amino acid residues in the basic region of functional domains of E and B apolipoproteins (34) has been proposed (33, 35) considering the implication of these residues in LDL and VLDL remnants binding to apo B.E (36) and hepatic apo E receptors (37, 38). Retarded LDL elimination from plasma may also be the consequence of defects in the lipoprotein receptor number. In contrast to other cancer diseases such as haematological malignancies (7), premalignant liver (39) and rat hepatoma grown in vivo or in vitro expresses a lower apo E receptor number (4, 5).

Lipids in plasma and in different lipoprotein classes

In figure 2 are presented the total plasma cholesterol and the content of the same in HDL and VLDL-LDL plasma fractions.

No changes by TAA administration were observed in plasma nor in the HDL fraction, but an increase of cholesterol content in VLDL-LDL fraction was evident, expressed as 100mg of plasma. This increase could be explained by

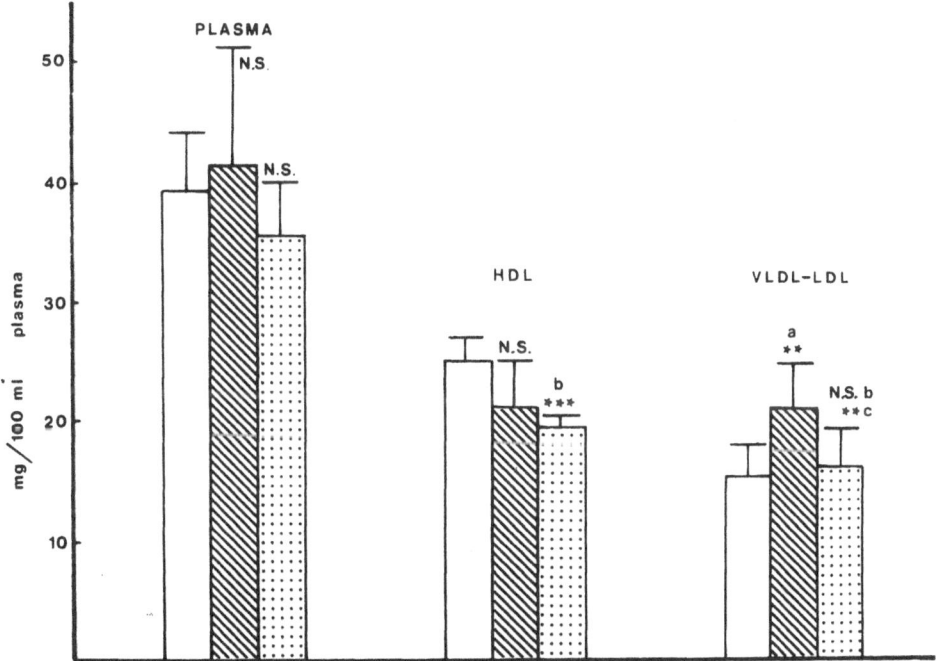

Figure 2. Serum and lipoproteins total cholesterol in control☐, after 30 days of i.p. TAA administration (50 mg/Kg body wt/day)◩, and after 30 days of TAA+SAM i.p. administration (2 mg/Kg body wt/day)▦. Values (expressed as mg/100 ml of plasma) are mean ± S.D. of six rats except controls (n=20) The asterisks indicate a significant difference (a) TAA vs. control; (b) TAA+SAM) vs. control; (c) TAA+SAM vs. TAA (p < 0.05 *, p < 0.01 ** and p < 0.001 ***).

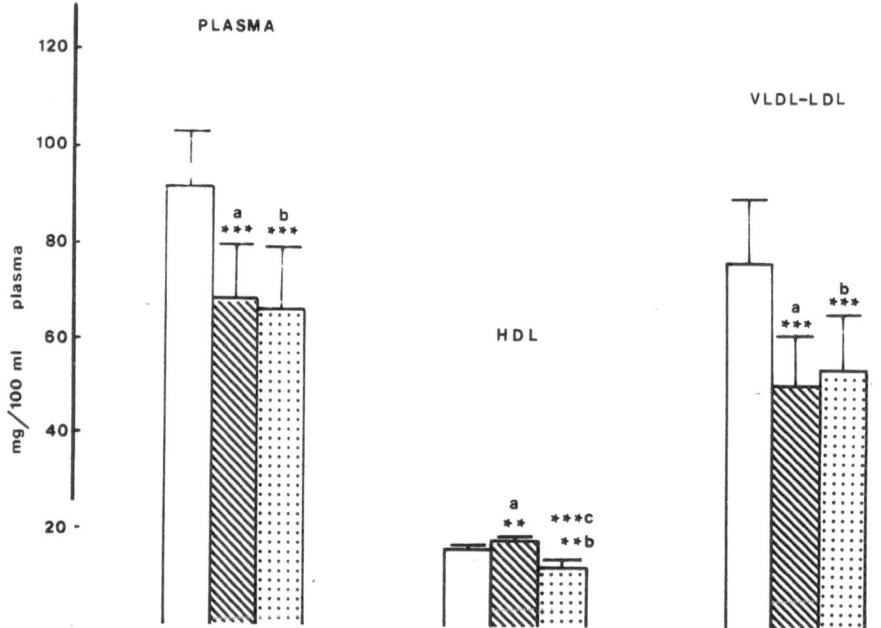

Figure 3.   Serum and lipoproteins total triglycerides in control ☐ , after
30 days of i.p. TAA administration (50 mg/Kg body weight/day) ▨ , and after
30 days of TAA + SAM i.p. administration (2 mg/Kg body weight/day) ⊡ .
Values (expressed as mg/100 ml of plasma) are mean + S.D. of six rats except
controls (n=20).   The asterisks indicate a significant difference (a) TAA vs.
control; (b) TAA + SAM vs. control; (c) TAA + SAM vs. TAA (p < 0.05 * p <0.01
** and p < 0.001 ***).

the possible limited (hepatic uptake of) VLDL-remnants due to fewer receptors
for apo E-rich lipoproteins in premalignant than in normal liver and because
of the limited access of lipoproteins to the pre- and neoplastic tissue (5,
39).

Absence of serum cholesterol changes was reported by Zimmermann after
chronic TAA application to rats (9).

Serum an VLDL-LDL fraction triglyceride content (figure 3) were lowered
about 25% and 35% respectively in TAA- and TAA+SAM-treated rats.

These results are consistent with findings reported in human liver
cirrhosis (40).   They are also compatible with serum triglyceride concentration and hepatic output behaviour in the chronic TAA-application (9).

Figure 4 shows plasma, HDL and VLDL-LDL total phospholipid content.   All
three presented lower values in TAA-treated animals than in corresponding
controls.   The increased phospholipid level in the apo B-containing lipoproteins by the effect of SAM may account for the normal levels found in plasma
from TAA+SAM-treated animals.

Evaluation of individual serum phospholipids showed that phosphatidyl-
coline (PC) represents about 60% of total lipid phosphorous; other identified
phospholipids were lysophosphatidylcholine (LPC), spingomyelin (SPH),

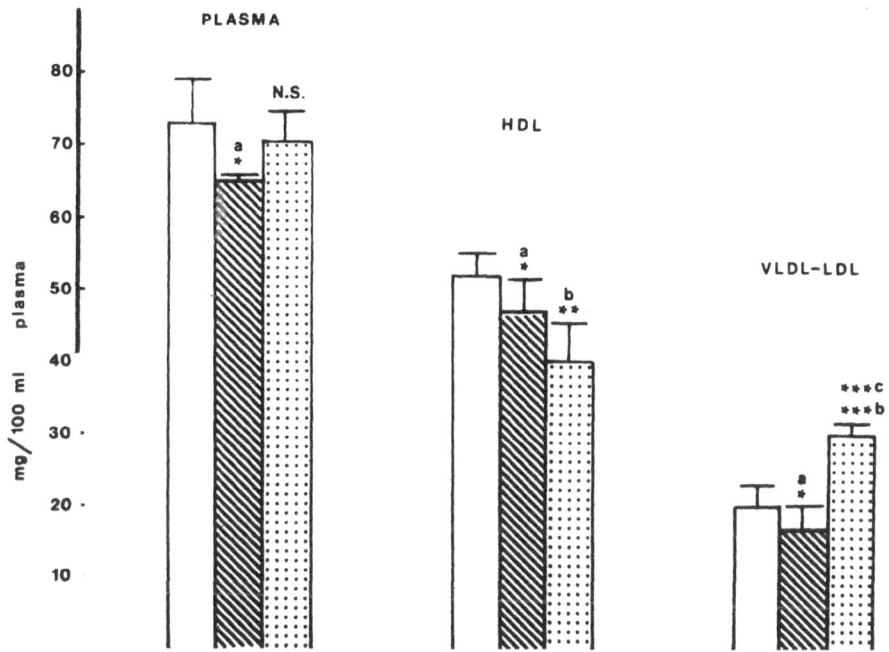

Figure 4: Serum and lipoproteins total phospholipids in control ☐ , after 30 days of i.p. TAA administration (50 mg/Kg body weight/day) ▨ , and after 30 days of TAA + SAM i.p. administration (2 mg/Kg body weight/day) ⦂⦂ . Values (expressed as phosphatidylcholine mg/100 ml of plasma) are mean ± S.D. of six rats except controls (n=20). The asterisks indicate a significant difference (a) TAA vs. control; (b) TAA + SAM vs. control; TAA + SAM vs. TAA ($p < 0.05$ *, $p < 0.01$ ** and $p < 0.001$ ***).

phosphatidylinositol (PI) + phosphatidylserine (PS) and phosphatidylethanolamine (PE), their levels decreasing.

Figure 5 shows the content of phosphatidylethanolamine and sphingoyelin in plasma and in HDL and VLDL-LDL fractions. As can be seen the concentration of PE in plasma from TAA-treated rats increases while the sphingolipid concentration decrease. As a consequence, the SPH/PE ratio (Table 1) presented values two times lower in plasma from TAA treated rats than in controls. Administration of SAM, as well as TAA, prevents this elevation. Values are also in the normal range for the apo B lipoproteins taken together (Table 1).

The mechanism(s) for the observed effect of exogeneous SAM in preventing the SPH/PE apo B lipoproteins alterations and SPH level in plasma and VLDL/LDL is unknown. Ceramide-phospho-ethanolamine has been recently demonstrated in liver (41). SAM is involved in the methylation process of this sphingomyelin analogous of which the direct precursor is PE. Even when this SPH way of formation may be minor in liver, it could be implicated in supplying SPH for hepatic lipoprotein formation, considering the specific metabolic origin of phospholipids for hepatic intracellular membranes and for lipoprotein secretion (32, 42, 30). The effect of SAM in this experimental hepatocarcinogenesis model could be in relation to the hepatic restoration of S-adenosyl-L-methionine level. Reduced SAM-synthesis activity has been

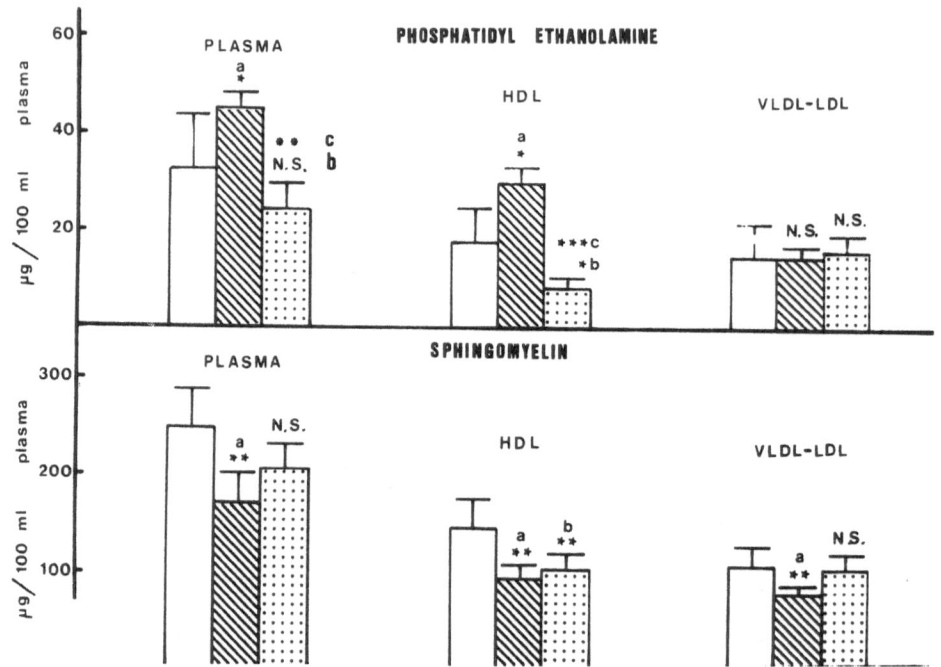

Figure 5. Serum and lipoproteins phosphatidylethanolamine and spingomyelin
in control ☐ , after 30 days of i.p. TAA administration (50 mg/Kg body
weight/day) ◪ , and after 30 days of TAA + SAM i.p. administration (2 mg/Kg
body weight/day) ▨ . Values (expressed as ug p_i/100 of plasma) are mean ±
S.D. of six rats except controls (n=20). The asterisks indicate a
significant difference (a) TAA vs. control; (b) TAA + SAM vs. control; TAA +
SAM vs. TAA (p < 0.05*, p < 0.01 ** and p < 0.001 ***).

Table 1. Effect of TAA and TAA+SAM on (SPH)/(PE) ratio in plasma and lipoproteins fractions.

Experimental condition	Plasma	HDL	VLDL-LDL
Control	7.6	8.16	6.94
TAA	3.76	3.09	5.12
TAA + SAM	8.35	11.85	6.46

recently demonstrated in cirrhosis independent of its etiology (43, 44).
Decreased level of S-adenosyl-L-methionine in liver has been found in the
hepatocarcinogenesis induced by 2-fluorenylacetamide (45). Development of
hyperplastic nodules and neoplasia induced by diethylnitrosamine has also
been associated with a low SAM content.

264

Acknowledgements

This work was supported by grants from CAICYT, FIS and EUROPHARMA.

REFERENCES

1   Kanel, G. C., Radvan, G. and Peters, R. L. (1983) Hepatology, 3, 343-348.
2   Papadopoulos, N. M. and Charles, M. A. (1970) Proc. Soc. Exp. Biol. N. Y. 134, 797-799
3   Glickman, R. M. and Sabensin, S. M. (1988) in The Liver Biology and Pathobiology (Arias, I. M., Jakoby, W. B., Popper, H., Schachter, D.,
4   Barnard, G. F., Erickson, S. K. and Cooper, A. D. (1984) J. Clin. Invest 74, 173-184
5   Barnard, G. F., Erickson, S. K. and Cooper, A. D. (1986) Biochim, Biophys. Acta 879, 301-312
6   Alexopoulos, C. B., Blatsios, B. and Augerinos, A. (1987) Cancer, 60, 3065-3070
7   Vitols, S., Bjorkholm, M., Gahrton, G. and Peteson, C. (1985) Lancet 2, 1150-1154
8   Ananth Narayan, K. and Morris, H. P. (1972) REBS Letters, 27, 311-315
9   Zimmermann, T., Franke, H. and Dargel, R. 91986) Exp. Pathol. 36, 109-117
10 Ferreyra, E. C., Fenos, O. M. and Castro, J. A. (1980) 16, 205-214
11 Osada, J., Aylagas, H., Sanches-Vegaso, I., Gea, T., Millan, I. and Palacios-Alaiz, E. (1986) Toxicology Letters, 32 97-106
12 Chieli, E. and Valvaldi, G. (1984) Toxicology, 31, 41-52
13 Pap, A. and Varro, V. (1981) Acta Med. Acad. Aci. Hung. 34 (8), 381-384
14 Becker, F. (1983) JNCI, 3, 553-558
15 Palacious-Alaiz, E., Osada, J., Aylagas, H. and Cascales, M. (1984) Abst. 362, 16th FEBS Meeting, Moscu
16 Palacious-Alaiz, E., Osada, J., Aylagas, H. and Santos-Ruiz, A. (1982) Rev. Esp. Fisiol. 38, supl., 135-140
17 Palacious-Alaiz, E., Osada, J., Cascales, M. and Santos-Ruiz, A. (1982) Rev. Esp. Oncologia. 29, 505-510
18 Stramentinoli, G. (1987) R. B. C. S1, 67-80
19 Russell, W. G., Benderson, J. and Albers, J. J. (1983) in Selected Methods of Clinical Chemistry (Cooper, G. R. ed.) Vol 10, pp. 91-99. American Association for Clinical Chemistry, Washington D. C.
20 Alsasua, M. T. and Palacios-Alaiz, E. (1979) in Advances in the Biochemistry and Physiology of Plant Lipids (Appelquist, L. A. and Liljenber, C. eds.) pp. 251-256. Elsevier, Amsterdam
21 Rouser, G., Siakotos, A. N. and Fleischer, S. (1969) Lipids, 5, 494-496
22 Trinder, P. (1969) Ann. Clin. Biochem. 6, 24
23 Wahlefeld, A. W. and Bergmeyer, H. V. (1974) in Methods der Enymatischen Analyse, 3th, t. II. p. 1878. Verlag Chemie, Weinheim
24 Roschlau, P., Bernt, E. and Gruber, W. (1975) Abst 1. 9th Int. Congr. on Clin. Chemistry, Toronto
25 Stahler, F. (1977) Med. Lab., 30, 29
26 Ananth Narayan, K. (1971) Int. J. Cancer, 8, 61-70
27 Franke, H., Zimmermann, T. and Dargel, R. (1983) Virchows Arch (Cell Pathol), 44, 99-113
28 Franke, H., Zimmermann, T. and Dargel, R. (1985) Virchows Arch (Cell Pathol),48, 277-288
29 Bremer, J., Greenberg, D. M. (1961) Biochim, Biophys. Acta 46, 205-216
30 Yao, Z. and Vance, D. E. (1988) J. Biol. Chem. 263, 2998-3004
31 Hamilton, R. L., Regen, D. M., Gray, E. and Le-Quire, V. S. (1968) Lab. Invest. 16, 305-319
32 Higgins, J. A. and Fielsend, J. K. 91987) J. of Lipid, Res., 28, 268-279
33 Sanchez-Ramos, B., Arce, C., Mendez, M. T. and Palacios-Alaiz, E. (1985) Abst. 310, XII congr. SEB. Valencia, p. 257

34 Innerarity, T. L., Wisgraber, K. H., Stanley, C., Rall, J. R. and Mahley R. W. (1987) Acta Med. Scand. suppl. 715, 51-59

35 Palacios-Alaiz, E. (1987) in Aspectos Bioquimicos y Farmacologicos en Disfunciones Hepaticas (CSIC, ed) pp. 233-256. Madrid

36 Russell, D. W. (1987) Acta Med. Scand. 715, 39-44

37 Brown, M. S., Goldstein, J. L. (1986) Science, 232, 34-47

38 Huy, D. Y., Brecht. W. J., Hall, E. A., Friedman, G., Innerarity, T. L. and Mahley, R. W. (1986) J. biol. Chem. 261, 4256-4267

39 Barnard, G. F., Erickson, S. K. and Cooper, A. D. (1984) Hepatology, 4, 1030

40 Sabensin, S. M., Bertram, P. D. and Freeman, M. R. (1980) Adv. Intern. Med., 25, 117-146

41 Malgat, M., Maurice, A. and Baraud, J. (1986) J. of Lipid Res., 27, 251-260

42 Vance, J. E. and Vance, D. E. (1986) J. Biol. Chem., 261, 4486-4991

43 Martin Duce, A., Ortiz, P., Cabrero, C. and Mato, J. M. (1988) Hepatology, 8 (1), 65-68

44 Mato, J. M., Martin Duce, A. and Ortiz, P. (1988) Rev. Clin. Esp. 182 (1), 1 - 3

45 Tsukada, K., Abe, T., Kuwahata, T. and Mitsui, K. (1985) Life Sciences, 37, 665-672

46 Garcea, R., Pascale, R., Daino, L., Frassetto, S., Cozzolino, P., Ruggiu M. E., Vannini, M. G., Gaspa, L. and Feo, F. (1987) Carcinogenesis, 8 (5) 653-658

INDEX